Mechanisms and Machine Science

Volume 65

This book series establishes a well defined forum for monographs, edited Books, and proceedings on mechanical engineering with particular emphasis on MMS (Mechanism and Machine Science). The final goal is the publication of research that shows the development of mechanical engineering and particularly MMS in all technical aspects, even in very recent assessments. Published works share an approach by which technical details and formulation are discussed, and discuss modern formalisms with the aim to circulate research and technical achievements for use in professional, research, academic, and teaching activities. This technical approach is an essential characteristic of the series. By discussing technical details and formulations in terms of modern formalisms, the possibility is created not only to show technical developments but also to explain achievements for technical teaching and research activity today and for the future. The book series is intended to collect technical views on developments of the broad field of MMS in a unique frame that can be seen in its totality as an Encyclopaedia of MMS but with the additional purpose of archiving and teaching MMS achievements. Therefore the book series will be of use not only for researchers and teachers in Mechanical Engineering but also for professionals and students for their formation and future work. Indexed in SCOPUS, Ei Compendex, EBSCO Discovery Service, OCLC, ProQuest Summon, Google Scholar and SpringerLink.

More information about this series at http://www.springer.com/series/8779

Giuseppe Carbone · Marco Ceccarelli
Doina Pisla
Editors

New Trends in Medical and Service Robotics

Advances in Theory and Practice

Springer

Editors
Giuseppe Carbone
Laboratory of Robotics and Mechatronics
University of Cassino and Southern Lazio
Cassino, Italy

Doina Pisla
Research Center for Industrial Robot
Technical University of Cluj-Napoca
Cluj-Napoca, Romania

Marco Ceccarelli
DICeM
University of Cassino and Southern Lazio
Cassino, Italy

ISSN 2211-0984 ISSN 2211-0992 (electronic)
Mechanisms and Machine Science
ISBN 978-3-030-00328-9 ISBN 978-3-030-00329-6 (eBook)
https://doi.org/10.1007/978-3-030-00329-6

Library of Congress Control Number: 2018953824

This Springer imprint is published by the registered company Springer Nature Switzerland AG
The registered company address is: Gewerbestrasse 11, 6330 Cham, Switzerland

Preface

Medical and service robotics are getting a growing market and an increasing interest from research and applications viewpoints. The aging of population and shortage of resources are increasing the demand of novel robotic solutions and devices, as, for example, in assistance in medical treatments and rehabilitation. Human–robot interactions and performance evaluations are attracting significant interest to validate the effectiveness and acceptability of the proposed solutions.

This proceedings book systematizes the contributions in the approached field that have been accepted for presentation at MESROB 2018, the 6th International Workshop on New Trends in Medical and Service Robotics. MESROB 2018 has been held at the School of Engineering of the University of Cassino and South Latium in Cassino, Italy. This conference event continues a successful series of workshops that have been started in 2012 in Cluj-Napoca, Romania. It has been followed by MESROB 2013 at Institute "Mihailo Pupin" in Belgrade, Serbia, MESROB 2014 at EPFL in Lausanne, Switzerland, MESROB 2015 at IRCCyN in Nantes, France, and MESROB 2016 as co-organized by University of Innsbruck and Joanneum Research in Graz, Austria. This workshop series is also sponsored by IFToMM, the International Federation for the Promotion of Mechanism and Machine Science with even a collaboration through the IFToMM Technical Committees on Biomechanical Engineering, Robotics and Mechatronics, and Computational Kinematics.

The content of MESROB 2018 book covers a wide range of aspects and topics such as assistive devices, exoskeletons, surgical robotics, locomotion systems, human–robot interactions, and performance evaluations. It also includes a special session on Agewell project that is an European co-funded project. These contributions are provided as a collection of 34 papers that were selected among the 45 submitted contributions on the basis of a blind peer review process. Therefore, we wish to express our gratitude to the authors, the reviewers, and Scientific Committee for their valuable contribution to ensure the scientific quality of

MESROB 2018. We are also grateful to Nathalie Jacobs, Anneke Pot, and the staff of Springer for their editorial support.

Giuseppe Carbone
Marco Ceccarelli
Doina Pisla

Organization

Program Chairs

Giuseppe Carbone	University of Cassino and South Latium, Italy
Marco Ceccarelli	University of Cassino and South Latium, Italy

Contents

Assistive Devices

Differential System for Limb Rehabilitation

Federico Ceresoli[1], Francesco Aggogeri[1], Cinzia Amici[1], Alberto Borboni[1(✉)], Rodolfo Faglia[1], Nicola Pellegrini[1], Monica Tiboni[1], Massimo Antonini[2], Davide Fausti[2], Maurizio Mor[2], Gianluigi Petrogalli[2], and Angelo Vertuan[2]

[1] Mechanical and Industrial Engineering Department, Università degli Studi di Brescia, Via Branze 38, 25123 Brescia, Italy
alberto.borboni@unibs.it
[2] Polibrixia Srl, via Branze 45, 25123 Brescia, Italy
maurizio.mor@polibrixia.it

Abstract. This work proposes a new device for lower and upper limb rehabilitation. The device is portable and modular and can be adopted to realize different system configurations and to perform different exercises. A design description and an analysis are shown, a working prototype is under construction and tests on humans are planned.

1 Introduction

When a neurological or orthopedic injury emerges, some patients exhibits pain and/or a reduction of the functionalities of one or both lower limbs. For this reason, different rehabilitation approaches are adopted on the patient and different robotic device were realized to help the rehabilitation process. Different tests show how these devices are effective to improve the functional performances of the patient [1].

There are two main categories of devices currently on the market:

- Devices for the rehabilitation of the lower limbs – They tend to a global rehabilitation of the limbs, present a more or less complex architecture aimed to move a single limb at a time, and in this case the patient sits or lays while the physician engages the limb needing for the rehabilitation treatment with the device [2–4];
- Devices for the reeducation to the walk – They present a more complex architecture with respect to the previous ones, as they are composed of two macro-components: the cantilever system, to keep the patient in a vertical position, and the system for the motion of the limbs, allowing the gait simulation thanks to proper regulations [5–7].

It is quite clear how different and distant these two types of rehabilitation devices are, as they are conceived to perform a specific rehabilitation task.

This paper aims to describe a new modular and flexible device (LEPRE) for the rehabilitation of the lower limbs and for the walking reeducation. It can be used for the rehabilitation of the lower limbs, as well as the upper ones thanks to the connection of specific attachments to the system. Moreover, the rehabilitation treatment can be performed in both an active and passive mode, depending on whether or not the actuators

G. Carbone et al. (Eds.): *New Trends in Medical and Service Robotics*, MMS 65, pp. 3–10, 2019.
https://doi.org/10.1007/978-3-030-00329-6_1

of the device are used. It is based on a belt-based Cartesian positioning system; in the system, two electric motors, properly mounted and linked with specific belts and pulleys, allow drive the motion of a cart placed on sliding guides. The extremity of the limb to be rehabilitated is connected to the cart or to a pulley through a Velcro strip o through a bar.

2 Problem Definition

The lower limb (Fig. 1) in charge of the walk of the human being and guarantees for the vertical standing bearing the weight of the whole body and distributing it to the ground.

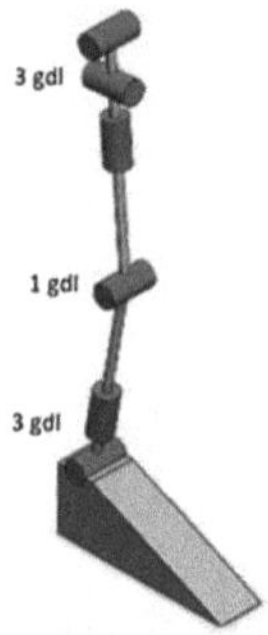

Fig. 1. Mechanical representation of the lower limb

The main articulations are:

- The hip, linking the pelvis with the femur
- The knee, connecting the thigh to the shank
- The ankle, linking the shank to the foot

The gait is a complex movement, since, in vertical position, the lower limbs bear all the body weight and the supporting area, the foot, is very small [8, 9].

These conditions force the locomotor system to involve all the parts of the body in the control process, in order to reach the desired motion without losing the equilibrium state.

During the walk, the body continuously moves around equilibrium positions, although these various positions are not stable ones; torques and forces reciprocally exchanged among the different limbs are passed to the trunk through articulations, ligaments and muscles.

Moreover, the contribute of inertial forces must be also considered; these forces may be generated by upper limbs and articulations of the spinal column and, combined with the lower limbs actions, may produce the desired gait.

During the gait, main tasks performed by the locomotor apparatus are:

- Generation of a propulsive force;
- Stability control at a higher level, although the postural conditions keep changing
- At each step, absorbing of the shock generated by the impact with the ground
- Energy conservation during the previous phases, limiting muscles effort.

When body advances, one limb sustains while the other advances until the next stand; then, the two limbs switch their role and both the feet lay in contact with the ground while transferring the weight of the body from one limb to the other.

A single sequence of these tasks is called as gait cycle.

Given its complexity, two main phases can be distinguished in the gait: the stance, when the foot is in contact with the ground, and the swing, when the foot is lifted, allowing the limb to advance.

Moreover, the stance phase can be divided in three sub-phases as well:

- A starting double stance, at the beginning of the step cycle, when both the feet lay in contact with the ground after the starting contact;
- The single stance, starting when the opposite foot leaves for the swing;
- A final double stance, beginning with the starting contact of the contra-lateral foot, and ending when the limb, initially in stance, leaves for the swing.

The typical distribution of the contact phases with the ground is 60% for the stance and 40% for the swing approximatively.

The temporal division for the stance sub-phases is 10% for each double stance, and 40% for the single stance [10].

3 Solution and Kinematic Model

The system grounds on a continuous-belt differential mechanism. As Fig. 2 depicts, motion along the X axis is assured by the action of the actuators and the motion profile assigned to them, while motion along the Y direction is allowed by an appendix properly engaged with one of the pulleys located on the same trailer. In this way, motion can be performed in all the XY plane.

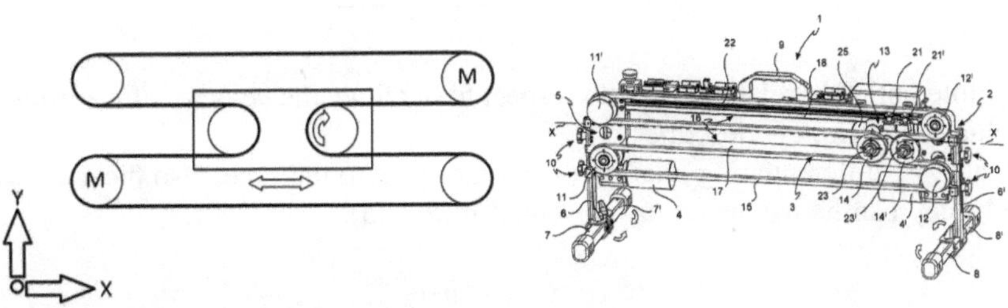

Fig. 2. Representation of the belt-based differential system

The system presents a mono-axial kinematic structure and two degrees of freedom: the translation of the trolley and the relative motion between belt and pulleys respectively (Fig. 3).

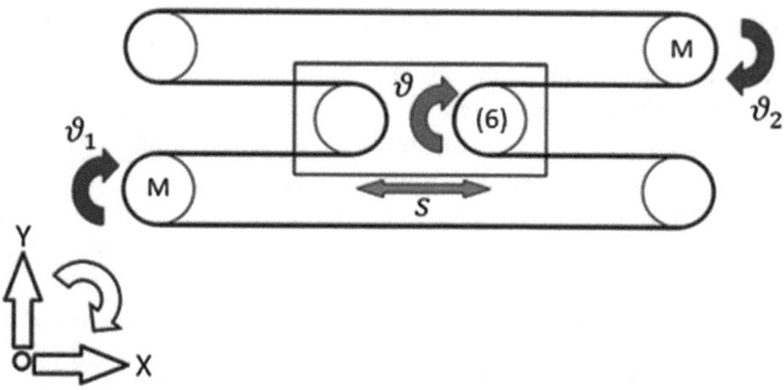

Fig. 3. Kinematic scheme of the belt-based differential system

– *Motion of the trolley*

Referring to Fig. 3, when pulley 2 is kept fixed and pulley 1 is free to move, trolley displacement can be expressed as relation (1) depicts.

In the same way, whether pulley 1 is fixed and pulley 2 rotates, the linear displacement described by the relation (2) is obtained.

$$s_1 = \frac{\vartheta_1 r_1}{2} \tag{1}$$

$$s_2 = \frac{\vartheta_2 r_2}{2} \tag{2}$$

When both the pulleys 1 and 2 rotate, the global displacement of the trolley becomes equal to the sum of the previous equations, as described by relation (3).

$$s = s_1 + s_2 = \frac{\vartheta_1 r_1 - \vartheta_2 r_2}{2} \tag{3}$$

By differentiating relation (3) with respect to the time, the velocity of the trolley can be evaluated, as Eq. (4) describes.

Finally, by differentiating relation (4) with respect to the time, also the acceleration of the trolley can be evaluated, as presented in Eq. (5).

$$v = \frac{\dot{\vartheta}_1 r_1 - \dot{\vartheta}_2 r_2}{2} \tag{4}$$

$$a = \frac{\ddot{\vartheta}_1 r_1 - \ddot{\vartheta}_2 r_2}{2} \tag{5}$$

– *Pulley rotation*

Rotation of the pulley 6 is given by the relation (6).

$$\vartheta = \frac{\vartheta_1 + \vartheta_2}{2} \tag{6}$$

Consequently, velocity and acceleration can be evaluated as presented in relations (7) and (8).

$$\omega = \frac{\dot{\vartheta}_1 + \dot{\vartheta}_2}{2} \tag{7}$$

$$\alpha = \frac{\ddot{\vartheta}_1 + \ddot{\vartheta}_2}{2} \tag{8}$$

In the followings, the dynamic analyses of the presented differential system are reported, under the hypotheses: (i) of a horizontal force directly applied to the trolley, and (ii) of a force applied to the trolley through an appendix on which the user foot is placed.

4 Dynamic Analysis

– *Dynamic analysis under the hypothesis of a horizontal force applied to the trolley*

In order to study the dynamics of the system and therefore estimate the belt tension and the actuators torques, the mechanism has been divided, so that the forces acting on the different segments of the belt could be highlighted.

In this way, eight unknowns arise, as Fig. 4 shows: six of them, generated by the belt tensions along the various segments, and two given by the opposing torques.

Therefore, from the equilibrium equations to rotation and translation of the components of the mechanism, the eight relations (9–16) can be chosen to create an eight-variables solving system.

$$T_4 R_4 - T_5 R_4 - J_4 \ddot{\vartheta}_4 = 0 \tag{9}$$

$$C_2 - J_2 \ddot{\vartheta}_2 - T_4 R_2 + T_2 R_2 = 0 \tag{10}$$

$$T_2 R_6 - T_6 R_6 + J_6 \ddot{\vartheta}_6 = 0 \tag{11}$$

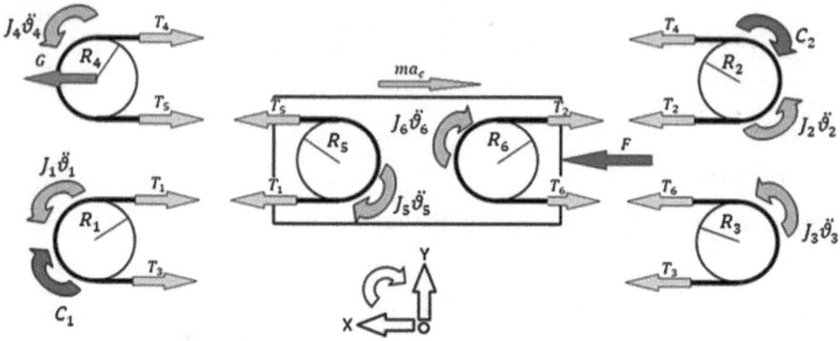

Fig. 4. Dynamic scheme of the belt-based differential system, with forces applied to the trolley

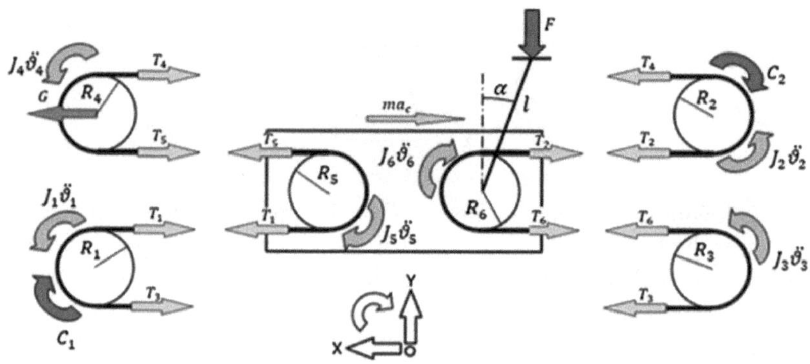

Fig. 5. Dynamic scheme of the belt-based differential system, with forces applied to the crank

$$T_3R_3 - T_6R_3 - J_3\ddot{\vartheta}_3 = 0 \tag{12}$$

$$C_1 - J_1\ddot{\vartheta}_1 - T_3R_1 + T_1R_1 = 0 \tag{13}$$

$$T_1R_5 - T_5R_5 + J_5\ddot{\vartheta}_5 = 0 \tag{14}$$

$$G - T_4 - T_5 = 0 \tag{15}$$

$$F + T_5 + T_1 - T_2 - T_6 - ma_t = 0 \tag{16}$$

- *Dynamic analysis under the hypothesis of a force applied to the crank constrained to a pulley of the trolley:*

In this case, the new forces configuration allows choosing the Eqs. (17–24) for the creation of a solving system.

$$T_4R_4 - T_5R_4 - J_4\ddot{\vartheta}_4 = 0 \tag{17}$$

$$C_2 - J_2\ddot{\vartheta}_2 - T_4R_2 + T_2R_2 = 0 \tag{18}$$

$$T_2R_6 - T_6R_6 + J_6\ddot{\vartheta}_6 + Fl\sin\alpha = 0 \tag{19}$$

$$T_3R_3 - T_6R_3 - J_3\ddot{\vartheta}_3 = 0 \tag{20}$$

$$C_1 - J_1\ddot{\vartheta}_1 - T_3R_1 + T_1R_1 = 0 \tag{21}$$

$$T_1R_5 - T_5R_5 + J_5\ddot{\vartheta}_5 = 0 \tag{22}$$

$$G - T_4 - T_5 = 0 \tag{23}$$

$$T_5 + T_1 - T_2 - T_6 - ma_t = 0 \tag{24}$$

Figure 6 shows an example of the ideal movement of an ankle of a normal human subject and the movement realized by the proposed device, where only some acceptable differences can be observed. The foot of the subject is connected to the device at the upper side of the bar 1 shown in Fig. 5.

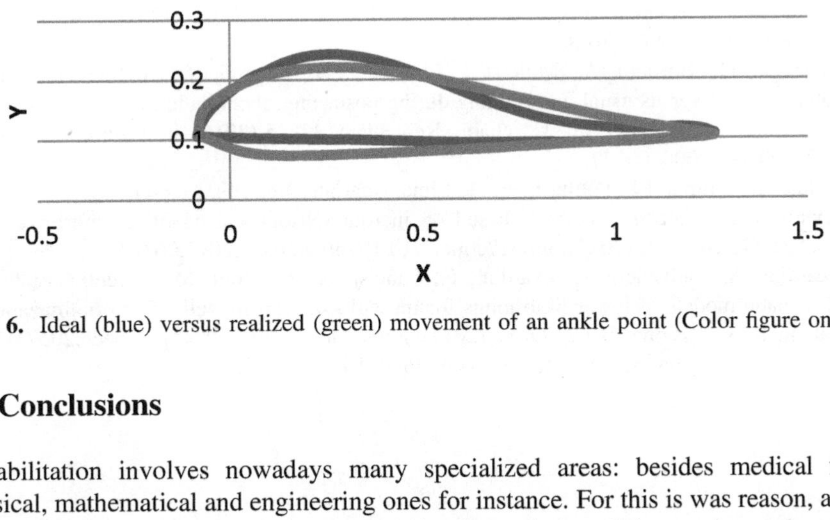

Fig. 6. Ideal (blue) versus realized (green) movement of an ankle point (Color figure online)

5 Conclusions

Rehabilitation involves nowadays many specialized areas: besides medical field, physical, mathematical and engineering ones for instance. For this is was reason, a new concept of device for the gait reeducation has been designed, innovative and optimal for this complex task. This was possible thanks to the adoption of a belt-based differential system, which allows flexibility in uses and regulation strategies. This versatility is primary due to the possibility to realize different kinds of motions, acting on the rotations of the pulleys connected to the actuators, on their dimensions, or on the control electronics.

According to what exposed in this brief dissertation, the belt-based differential system, and more in detail LEPRE, can be in conclusion identified as a valid solution for rehabilitative applications thanks to its relative simplicity and versatility.

References

1. Bruni, M.F., Melegari, C., De Cola, M.C., Bramanti, A., Bramanti, P., Calabrò, R.S.: What does best evidence tell us about robotic gait rehabilitation in stroke patients: a systematic review and meta-analysis. J. Clin. Neurosci. **48**, 11–17 (2018). https://doi.org/10.1016/j.jocn.2017.10.048
2. Goher, K.M., Fadlallah, S.O.: Design, modeling, and control of a portable leg rehabilitation system. J. Dyn. Syst. Meas. Control **139**(7), 071013-071013-071015 (2017). https://doi.org/10.1115/1.4035815
3. Rastegarpanah, A., Saadat, M., Borboni, A.: Parallel robot for lower limb rehabilitation exercises. Appl. Bionics Biomech. (2016). https://doi.org/10.1155/2016/8584735
4. Zhang, W., Zhang, S., Ceccarelli, M., Shi, D.: Design and kinematic analysis of a novel metamorphic mechanism for lower limb rehabilitation. Mech. Mach. Sci. **36** (2016). https://doi.org/10.1007/978-3-319-23327-7_47
5. Colombo, G., Joerg, M., Schreier, R., Dietz, V.: Treadmill training of paraplegic patients using a robotic orthosis. J. Rehabil. Res. Dev. **37**(6), 693–700 (2000)
6. Sacco, K., Belforte, G., Eula, G., Raparelli, T., Sirolli, S., Geda, E., Geminiani, G.C., Virgilio, R., Zettin, M.: P.I.G.R.O.: An active exoskeleton for robotic neurorehabilitation training driven by an electro-pneumatic control. Mech. Mach. Sci. **49** (2018). https://doi.org/10.1007/978-3-319-61276-8_89
7. Taveggia, G., Borboni, A., Mulé, C., Villafañe, J.H., Negrini, S.: Conflicting results of robot-assisted versus usual gait training during postacute rehabilitation of stroke patients: a randomized clinical trial. Int. J. Rehabil. Res. **39**(1), 29–35 (2016). https://doi.org/10.1097/MRR.0000000000000137
8. Caldas, R., Mundt, M., Potthast, W., de Lima, Buarque, Neto, F., Markert, B.: A systematic review of gait analysis methods based on inertial sensors and adaptive algorithms. Gait Posture **57**, 204–210 (2017). https://doi.org/10.1016/j.gaitpost.2017.06.019
9. Leardini, A., Belvedere, C., Nardini, F., Sancisi, N., Conconi, M., Parenti-Castelli, V.: Kinematic models of lower limb joints for musculo-skeletal modelling and optimization in gait analysis. J. Biomech. **62**, 77–86 (2017). https://doi.org/10.1016/j.jbiomech.2017.04.029
10. Perry, J.: Gait Analysis: Normal and Pathological Function (1992)

Rehabilitation Bed with a 3D Profile of the Support Surface

N. E. Grishchenko(✉), R. N. Polyakov, I. A. Snimchsikova, and M. A. Polyakova

Orel State Univercity named after I.S. Turgenev, 29-110, Naugorskoe shosse, Orel, Russia
overdose18@mail.ru

Abstract. The article presents a concept of creating a mechatronic isosurface with a controlled position in height and pressure for using in the rehabilitation equipment of medical institutions. The use of the synergetic principle of designing parallel drives of the surface bearing elements allows achieving a desired position of a patient in a variety of clinical events. The use of electromechanical drive allows achieving the observance of strict conditions of the operational safety of the medical equipment.

Keywords: Mechatronic actuators · Screw drive · Rehabilitation
Anatomy degree · Lateral therapy · Pressure reducing support surface
Fowler's position

The development of mechatronics as a new paradigm of designing any machines and mechanisms with integrated electronic and information-measuring components leads to the creation of new units and the improvement of the existing ones, the realization of which was impossible some time ago. For example, the serial production and operation of direct shift gearbox in civil vehicles have been made possible through the transfer of the control function to the sensor unit connected to the hydraulic actuators.

The synergetic integration of mechanical, electrical and information-measuring components leads to the significant minimization of weight-and-dimensional characteristics due to the use of ultra-dense assembly methods, the elimination of the intermediate gears leads to the increased accuracy of the output element movement. One of the main vectors for making mechatronic products is medical engineering. Devices for analysis, automatic therapy and rehabilitation, surgical robots are just some examples of the mechatronics applications. In medical practice, one of the main processes is the rehabilitation of a patient after surgery, regardless of its complexity and importance to health.

According to the data from the World Health Organization the robotic devices are widely used in the rehabilitation process of patients after stroke, spinal injuries, spinal cord injuries and other complicated clinical events. The paper [1] describes the structure of a new robotic system for multi-task rehabilitation purposes in post-stroke condition and gives a detailed analysis of the achievements in the field of kinetic physiotherapy [2].

G. Carbone et al. (Eds.): *New Trends in Medical and Service Robotics*, MMS 65, pp. 11–16, 2019.
https://doi.org/10.1007/978-3-030-00329-6_2

The paper [3] describes a rehabilitation system for orthopedic patients. It is based on the combination of robotic devices and virtual reality, which can significantly improve the effectiveness of rehabilitation procedures. The papers [4, 5] describe the developments of the Russian scientists in building a multifunctional exoskeleton for providing various procedures of the kinetic therapy: from the elementary head and torso rotation, flexion and extension of the limbs to the complicated procedures of verticalization and independent walk.

The paper [6] describes the design of an upper limb rehabilitation robot that can be configured to perform arm or wrist exercises. The robot allows to make accurate movements and to get predictable efforts to achieve greater efficiency and shorter rehabilitation period. The introduction of the mechatronic systems into the medical rehabilitation equipment can significantly improve the efficiency of the rehabilitation procedures and reduce the recovery time for patients.

There are up to 50,000 events of spinal cord injuries that occur in the Russian Federation annually [7]. The rehabilitation procedures in the post-surgery period are very vital for such patients. If after a complicated surgery a patient remains motionless for a long time, the irreversible changes occur not only in the skin (bedsores) [8], but also in the internal organs. Therefore it is very important to develop a rehabilitation bed which has the functions of rehabilitation, massage, changes in the position of a patient and supervision for his posture

In the development of modern engineering and medicine, the increase in accuracy figures and reduction of the metal consumption of the final products is of great importance. The development of technology is accompanied by a constant increase in the requirements for technical systems and their individual elements, e.g. complex indicators as reliability, energy efficiency, durability, safety, etc. The main disadvantage is the possibility of only 7 surfaces with manual control, which change the tilt angle only in the longitudinal direction of a body, which can be a fact of an incomplete contact. The necessity to develop a multifunctional rehabilitation bed with a 3D profile of the support surface is caused by the variety of diseases and rehabilitation procedures, in which a patient requires fixation in a certain spatial position.

The average medical institution can require hundreds of medical beds. At the same time, the equipment used has a small number of adjustment surfaces and manual drives, which is justified due to the low price, but at the same time has significant limitations on clinical events.

The development proposed is aimed to improve the efficiency of the rehabilitation procedures for the prevention of pathologic processes leading to the disability in various types of severe diseases (tetraplegia, paraplegia, diseases of the locomotion system and the cardiovascular system, etc.) due to the precise positioning of patients during a rehabilitation period, reducing the pressure on the parts with maximum pain by distributing the pressure on the support surface evenly due to greater anatomy of a body [9], as well as combining the functions of a rehabilitation bed and mechanotherapy treatment.

The structural and functional scheme of the rehabilitation bed with a 3D profile of the support surface is shown in Fig. 1.

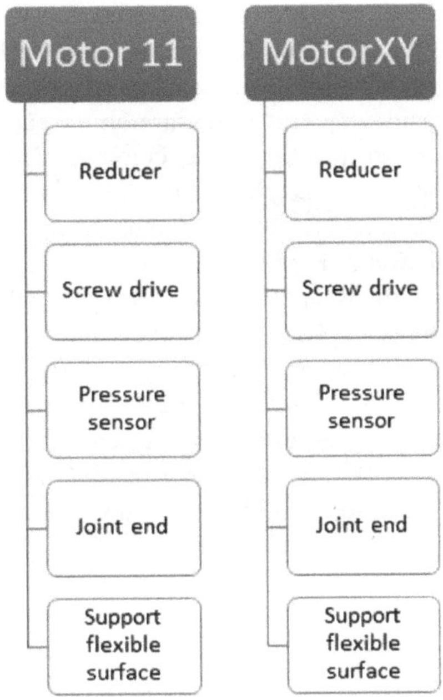

Fig. 1. Structural and functional scheme of the rehabilitation bed.

The approach gives the opportunity to combine mechanotherapy units and a rehabilitation bed, which increases its functionality and use in medical treatment and due to the more flexible programmed control of the patient's position with the use of the remote control, it is possible to achieve greater independence of severely ill patients. It is possible to achieve a more sparing initial intensive care, for example, in case of tetraplegia and paraplegia to help a patient make at least some movement with the help of actuators.

Another great advantage of the rehabilitation bed is the proper weight distribution of burn patients along the support surface to reduce pain. A more anatomy degree of the rehabilitation bed designed and the feedback in the form of built-in tensometric sensors will monitor the most comfortable position for severely ill patients.

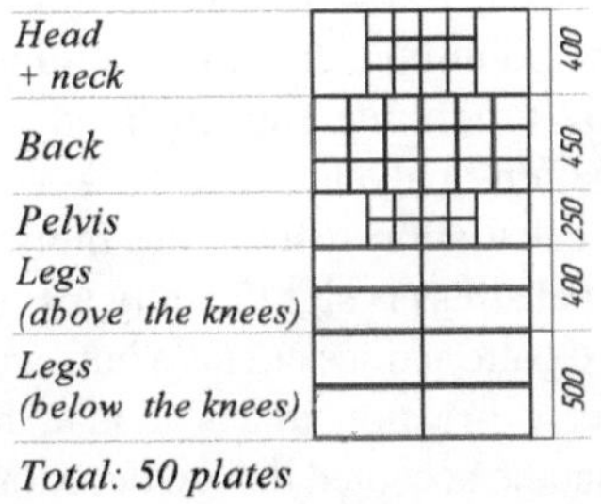

Fig. 2. Structural and functional scheme of the rehabilitation bed.

The rational separation of movable sections depending on the part of a patient's body and the required degree of mechanical action allows achieving a greater contact with a patient's body and conducting the necessary rehabilitation procedures (Fig. 2).

This approach allows achieving equal pressure on the support parts of a body both in a horizontal and any other position. The common name for such a group of equipment is derived from this principle – isosurface with adjustable height of the support points. The self-braking screw drives are used as drives. They are much cheaper than ball screw gear and do not require additional braking.

The sample calculations for wear-resistance show that when combining materials: screw – steel 41Cr4, nut – bronze CuSn11P-C even under the lack of lubrication, the required wear resistance is ensured. According to [10] for materials CuSn11P-C [*pv*] = 10 MPa·m/s, [v] = 8 m/s. Even when a patient weighs 120 kilos and a safety factor is 2, the design pressure along the turns of a screw does not exceed 2 MPa. The sliding speed does not exceed 0.3 m/s when the rehabilitation bed is changed from a fully horizontal position to the so-called Fowler's position (Fig. 3) for 40 s.

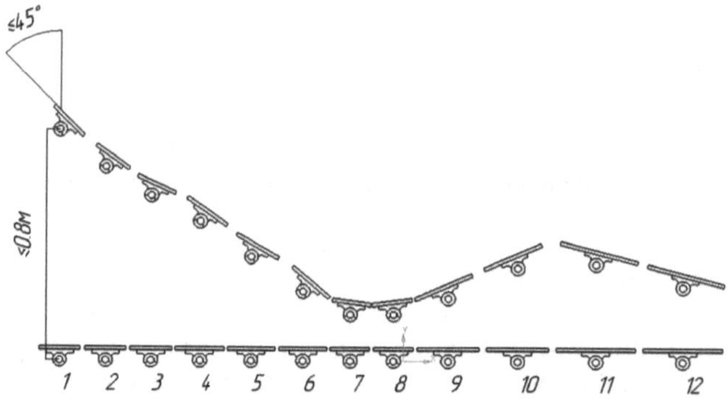

Fig. 3. The options for the profile of location of the output elements of the drives.

To adapt the bed to the individual characteristics of each person, an original design of a self-aligning joint is provided, which allows taking the optimal position depending on the patient posture. The solutions given allow combining mechanotherapy units and a rehabilitation bed, which increases its functionality and use in the treatment process. Figure 4 shows 3D images of the support elements for the Trendelenburg position (a), and the combined position when a leg is broken and for shampooing (b).

Another critical element is the screw in the maximum extended position, which limits the motion of the screw point connected to the support point and can cause the loss of stability. Figure 5 shows the finite-element model of the loaded screw and the results *of* its stress & deformation analysis.

As can be seen from the calculation results, the motion of the screw point do not exceed 0.7 mcm, and the normal stress is significantly less than those that are permitted. At the same time, the computed pattern does not take into account the bilateral constraints between sections in the transverse direction, that is, in fact, the motions given will be even smaller. Thus, it can be said that the accepted design decisions of the parallel drives of the rehabilitation bed meet the necessary criteria of strength and stiffness.

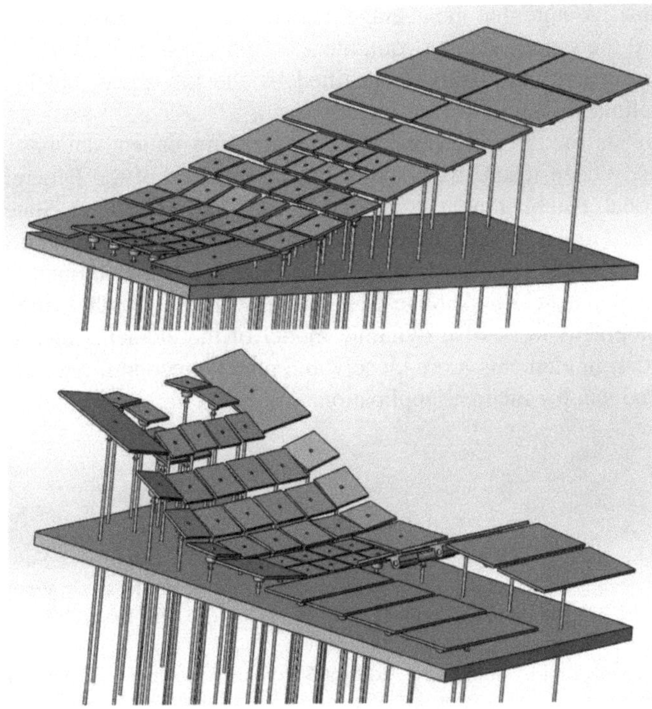

a)

b)

Fig. 4. 3D images of possible positions of elements in different clinical events.

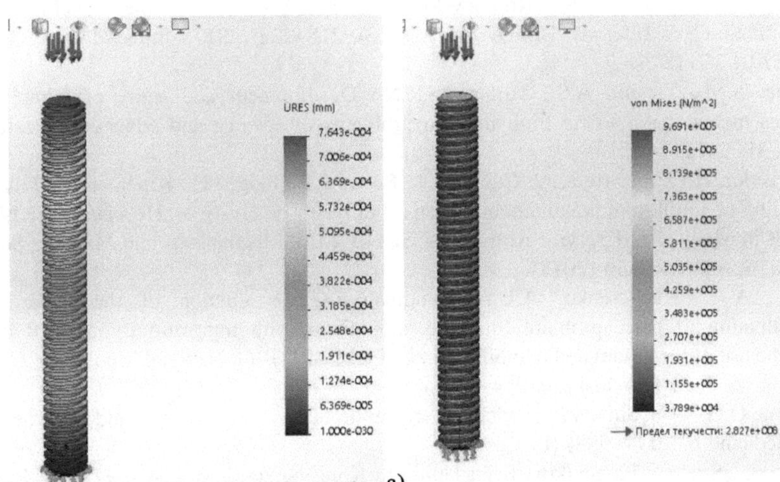

a) b)

Fig. 5. The results of normal stress (a) & deformation analysis (b).

In conclusion, we note that the average medical institution can require hundreds of medical beds. At the same time, the equipment used has a small number of adjustment surfaces and manual drives, which is justified by the low price, but has significant limitations on clinical events.

The agreements for the provision with the sites and patient database for clinical trials have already been reached with the Medical Institute of the Federal State Budgetary Educational Establishment of Higher Education "The Orel State University named after I. S. Turgenev", the Health Administration of the Orel region "Bolkhov Central District Hospital" and the Federal State Budgetary Educational Establishment of Higher Education "the Kursk State Medical University". Further, the authors will suggest a development work of a dynamic model of the elements of a rehabilitation bed, study the design elements more closely and give the economic feasibility of such mechatronic products for medical application.

References

1. Ceccarelli, M., Pîslă, D., Graur, F., Ottaviano, E., Vaida, C., Ungur, R.: Salvatore grande, monica pop, design and operation issues for parallel robotic devices in the rehabilitation of stroke patients. Int. J. Mech. Control. **11**(1), 51–58 (2010). JoMaC Special issue for RAAD09
2. Pop, L.: Textbook of Physiotherapy. Editura Medicala Universitara Iuliu Hatieganu Cluj, Ediţia (2006)
3. Dutsch, J., Latonio, J., Burdea, G., Boian, R.: Rehabilitation of musculoskeletal injuries using the rutgers ankle haptic interface: three case reports. In: Eurohaptics Conference, vol. 6, pp. 1–4. Birmingham, UK (2001)
4. Jatsun, S.F., Jatsun, A.S., Korenevskiy, N.A.: Experience in the development of rehabilitation exoskeletons. Biomed. Eng. **51**(3), 218–222 (2017). https://doi.org/10.1007/s10527-017-9718-3s
5. Yatsun, S.M., Yatsun, A.S., Turlapov, R.N.: Development and study of rehabilitation devices mehanoteraii lower limb modern problems of science and education. Basic Res. **12**(9), 1909–1911 (2014)
6. Mancisidor, A., Zubizarreta, A., Cabanes, I., Bengoa, P., Jung, J.H.: Kinematic and dynamic modeling of a multifunctional rehabilitation robot UHP. In: Husty M, Hofbaur M. (eds) New Trends in Medical and Service Robots. MESROB 2016. Mechanisms and Machine Science, vol 48. Springer, Cham (2018)
7. Vagin, A.A., Kucherenko, A.D., Tilyudina, E.G.: Evaluation of the dynamics of rehabilitation of basic problem functions in patients with traumatic spinal cord disease with the use of the automated rehabilitation bed KMF-01. Bull. National Med. Surg. Cent. N. I. Pirogova **5**(2), 114–120 (2010)
8. Kogan, O.G., Naydin, V.L.: Medical rehabilitation in neurology and neurosurgery. M. Medicine p. 304 (1988) [In Russian]
9. Bonev, L., Sychev, P., Bankov, S.T.: Guide to Kinesiotherapy. p. 354 (1978) [In Russian]
10. Anuryev, V.I.: Directory of the constructor-machine builder: V.1. - M.: Machinostroenie, p. 920 (2001) [In Russian]

Performance of a Thermal Bipedal Walker on Inclined Surfaces

Lingchao Su(✉), Takeru Nemoto, and Akio Yamamoto

Department of Precision Engineering, School of Engineering, The University of Tokyo, Tokyo, Japan
{lingchao, t-nemoto, akio}@aml.t.u-tokyo.ac.jp

Abstract. Being actuated by thermal deformation of bimetal sheets, a quasi-dynamic bipedal walker, which was introduced previously, can walk on a horizontal hot surface utilizing thermal energy. Its performance on inclined surfaces, however, has not been studied well in previous studies. In this study, we analyze the gait of a thermal walker walking up/down or along a certain contour line on an inclined surface. Experimental results show that the thermal walker can walk within such situations, and its stride changes due to the slope angle. However, the walking direction was found backward on upward slopes, which was unintended. Compared with previous studies, the results imply that it is more difficult for a walker with smaller stride to climb upward, suggesting necessity of further investigations on walker characteristics on inclined surfaces.

Keywords: Bipedal walker · Bimetal · Thermal walker · Thermal actuator Slope

1 Introduction

The development of robots has never ceased from the day they were born. In recent years, work on robots has improved the performance of robots in many aspects. In response, the improvement has changed our life to a great extent. For example, cleaning robots have become popular as soon as it came out, for its convenience and contribution to releasing people from boring floor-cleaning tasks. For more dexterous tasks, humanoid robots are expected to be brought into our daily environments in the near future.

Since most of the robots use battery as their energy source, their operating time relies on battery life, which is not very long with current technologies. To extend the operating time, locomotion efficiency needs to be improved, and, for walking robots, the mechanism of passive dynamic walkers is one of the possible solutions.

Passive dynamic walkers that only use potential energy to walk down a flat slope was introduced by McGeer in 1990 [1]. The original passive walkers have a quite obvious limitation that they can only walk down a slope [2]. When partially actuated in a certain way, however, passive dynamic walkers are changed into quasi-passive dynamic walkers, which can even walk on a horizontal surface. Many researchers have put their sights on quasi-passive dynamic walkers in recent years. Among them is a thermal bipedal walker proposed by Nemoto and Yamamoto [3]. Taking thermal

G. Carbone et al. (Eds.): *New Trends in Medical and Service Robotics*, MMS 65, pp. 17–24, 2019.
https://doi.org/10.1007/978-3-030-00329-6_3

deformation of bimetal sheets as driving force, the thermal walker has succeeded in walking on a hot horizontal surface. The prototype walker reported in [3] required considerably high surface temperature, such as 150 °C. However, by lowering the required temperature in future developments, the walker is expected to be utilized in daily environments, e.g., on hot ground surfaces in summer.

The report [3] investigated the performance of the walker only on a flat horizontal surface. However, practical situations are supposed to be much more complicated; the ground is usually inclined and uneven. Therefore, walking performance on inclined or bumpy surfaces need to be clarified. Nemoto has studied the gait of a thermal bipedal walker for walking up and down slopes in [4], which reports that the walker used in [4] walks more easily upward than downward. This research extends the work in [4], using a new prototype; the new prototype used in this work has been designed to realize more stable gait [5]. Using a high-speed camera to observe the walking gait, this work investigates walking performance on surfaces that are inclined in pitching directions. Moreover, performance along a contour line of an inclined surface was also investigated in this study. The results in this work in comparison with [4], show that walking performance on an inclined surface is influenced by the design to a great extent.

2 Structure and Prototype

The thermal bipedal walker investigated in this work utilizes self-oscillation of bimetal sheets, which is reported in [6]. When a rounded bimetal sheet is placed on a hot surface, the sheet automatically oscillates, which is the self-oscillation. As shown in Figs. 1 and 2, the walker has a rounded bimetal sheet on the bottom surfaces of two feet, which makes the whole body roll in sideways when placed on a hot surface, by the self-oscillation phenomenon. As it rolls, one leg will be lifted up and the body stands on the other leg, which is called stance leg. The lifted leg will be naturally swung forward by gravity, as it has a mass that is shifted backward. The swung leg will be landed as the body rolls back and become the new stance leg, making the body move one step forward. By repeating this sequence, the body walks continuously on a hot surface. More details of the walking principle can be found in [3].

The prototype walker used in this work, shown in Figs. 1 and 2, is made up of two polyacetal legs, which are connected to a rotational shaft via ball bearings that allows the legs to swing along the walking direction. In order to accomplish the process of self-oscillation, the bottom surfaces of the two aluminum feet, together, are shaped into a part of a cylinder with a radius of 9 cm. Bimetal sheets with the curving rate of 13.7×10^{-6} [°C^{-1}] are attached to each bottom surface of the feet, with the larger thermal expansion side being outside. The mass attached to each foot is shifted backward for 1 mm from the center line of the leg, such that the leg will naturally swing forward when lifted up. The total length of the walker is 17 cm and the weight is 310 g. Compared to the previous walker in [3, 4], the leg swing amplitude is much smaller in this prototype, with the angle of around 2°.

For motion tracking, six marks are assigned at shaft height, and two at the top of the legs, as shown in Fig. 1.

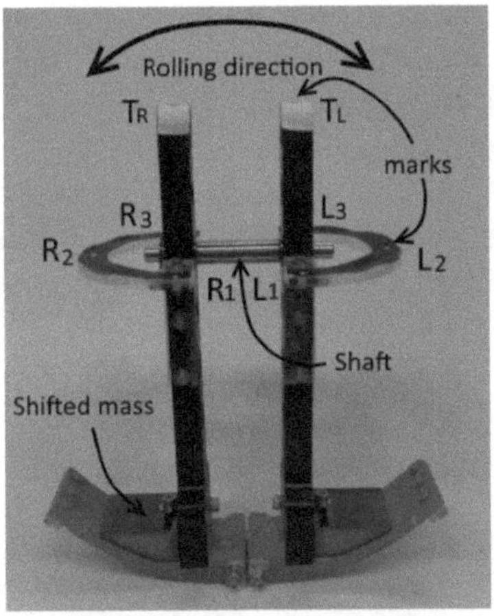

Fig. 1. Thermal walker (front)

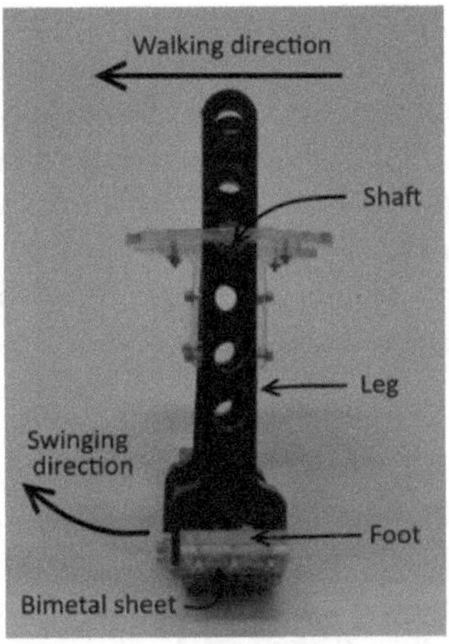

Fig. 2. Thermal walker (side)

3 Experiments

The experiment setup is shown in Fig. 3. A high-speed camera (Keyence, VW-6000) was set right above a heater, to record the gait of the walker with the frame rate of 250 fps. The high-speed camera system observed and traced the marks on the walker shown in Fig. 1. The measurement began at about 2 or 3 s after placing the walker on the heater surface. Due to the limitation of the recording time, the gait was measured for about 13 s. The heater, whose surface temperature was set to 140 °C, was arranged on a gonio stage, such that its inclination angle, φ, can be changed, see Fig. 3. The walker walked toward either negative X direction or positive Y direction, in the following experiments. Here, Fig. 3 shows the walking path, when the walker is walking toward negative X direction.

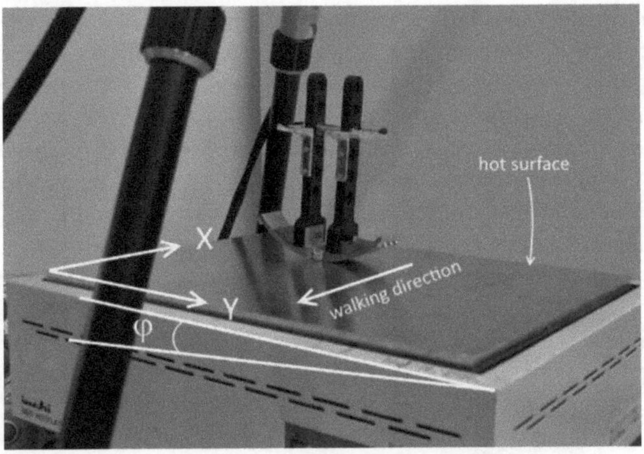

Fig. 3. Experimental setup

Two experiments were carried out. In one experiment, the walker walked toward negative X direction (along a contour line). The slope angle, φ, was changed from -5 to $4°$, with a step of $1°$, which was kept constant while the walker was walking. In the other experiment, the walker walked up or down a slope in Y direction. The angle φ was changed from -2 to $2°$ with a step of $0.5°$. In both experiments, the body was initially oriented to X- or Y-directions, but the walking direction was sometimes slightly deviated, as it walked. However, the deviations were ignored in this work and will not be discussed in the rest of this paper.

4 Results and Discussions

The marks on the walker were traced in the two experiments to analyze body displacement and leg swings. Figure 4 shows an example of the leg motions measured from R1, L1. In this example, the walker walked toward negative X direction with the slope angle of $-1.0°$. Here, positive slope angle means the walker slants to its left leg when naturally stands on the slope. From the plot, it is clear that the thermal walker rolled in Y direction by the self-oscillation of bimetal sheets, shown in Fig. 4(b), and at the same time, the whole body walks toward negative X direction as in Fig. 4(a). The stepwise motions of the leg motions observed in Fig. 4(a) came from switching of stance legs, which is regarded as stride. In order to make it clearer, 2D trajectories of both legs while walking toward negative X direction, are plotted in Fig. 4(c).

The stride for different slope angles are calculated and plotted in Fig. 5, where the plot (a) is for walking toward negative X and the plot (b) is for walking toward Y direction. The strides were measured in the stepwise motions of both legs, such as Fig. 4(a), and averaged for five steps in a steady state. The error bars represent the maximum and minimum. For Y directional walking, positive inclination angle means downward slope.

For Y direction walking in Fig. 5(b), it was found that the stride increases almost linearly with the slope angle for downward slopes. The stride linearly changed also for negative (upward) slopes; however, the walker walked backward in this case (negative stride means backward steps). The walking performance was apparently much better for downward slopes. However, this result is somewhat different from the previous work [4] using the old prototype. In the previous work, it was reported that the walker walked forward on both upward and downward slopes and the walking was found slightly more stable for upward slopes. The difference would come from different leg (and/or body) swing amplitudes. The previous walker had much wider leg swing amplitudes (in other words, much wider stride). The results obtained in these two studies are suggesting that the walking performance on upward slopes are greatly affected by the leg swing amplitude and wider amplitude would be better for climbing up slopes.

For negative X directional walking shown in Fig. 5(a), the longest stride was obtained for almost horizontal surface. Although, in principle, the walker should perform the same for positive and negative inclinations, the results shown in the plot is not symmetric. The symmetry of the walker structure and setup, which in fact was not perfect, would be the reason of this phenomenon. The reason for the decreased stride

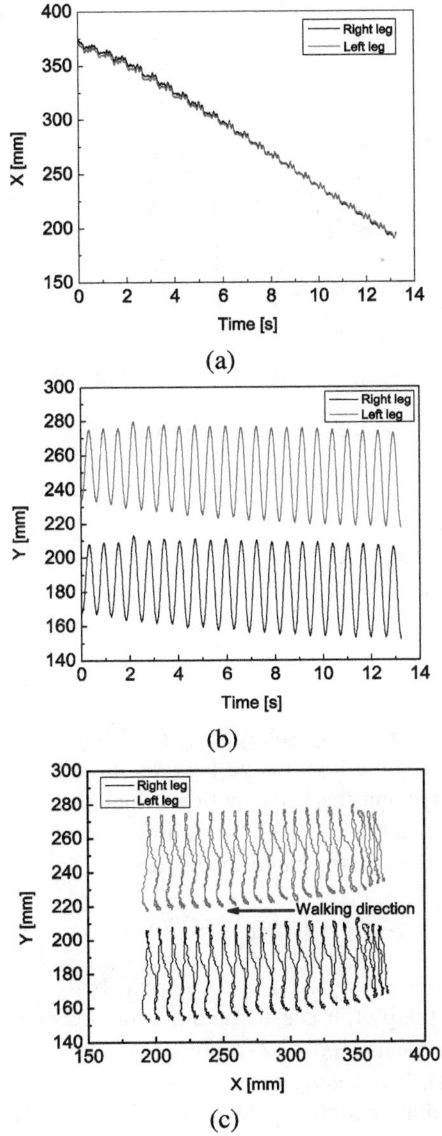

Fig. 4. Leg motions (walking toward negative X direction)

on inclined surfaces would be an inappropriate leg swing time. In the prototype used in this work, the leg is designed to have natural oscillation period that is almost same as the body rolling period, as such a condition can maximize the stride on a horizontal surface. On inclined surfaces, the leg swing period would change from that on horizontal surface and would be unbalanced.

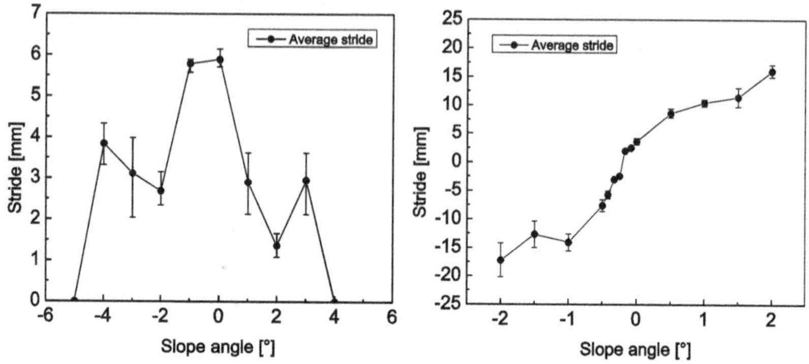

(a) Walking towards negative X direction (b) Walking towards Y direction

Fig. 5. Stride for different angles

In order to study this phenomenon, leg swings and body motion need to be analyzed. Here, in X direction, the body displacement is represented by x_b, which is calculated from X positions of the marks R2 and L2;

$$x_b = \frac{x_{R2} + x_{L2}}{2}$$

Leg swings were calculated by subtracting the body displacement from the displacements of the marks at the tops of the legs, as the motions of the top marks are combination of leg swings and the body motion. Thus, the relative motions of the top marks against body motion, are calculated as follows, representing leg motions, where x_{TL} and x_{TR} represent X-positions of two top marks.

$$x_L = x_{TL} - x_b \quad x_R = x_{TR} - x_b$$

Figure 6 shows the leg swings, x_{TL} and $x_{TR,}$ and body motion, x_b, calculated for slope angle of –1.0°. In the plot, it is estimated that the leg is swinging forward during the top mark is moving toward positive X direction (that are shown using pink and cyan background colors). It is clearly observed in the plot that the right leg (red line) has larger swing period than the left leg (blue line). This unbalance would be the cause of the decreased stride. However, depending on the leg designs, the walker may walk better on inclined surfaces. Therefore, the performance needs to be investigated with many different designs to fully reveal the walker characteristics.

Finally, it should be mentioned that, in some particular conditions of the X-direction walking experiment, walker gait drastically changed when it was on the hot plate for long time (such as 20 to 30 s); the walker eventually stopped and began walking backward. This interesting phenomenon was observed for negative X direction walking, with the slope angle between −1.0° and −3.0°. This is probably due to the structure of the walker, and overheat of the bimetal sheet can also be one of the reasons.

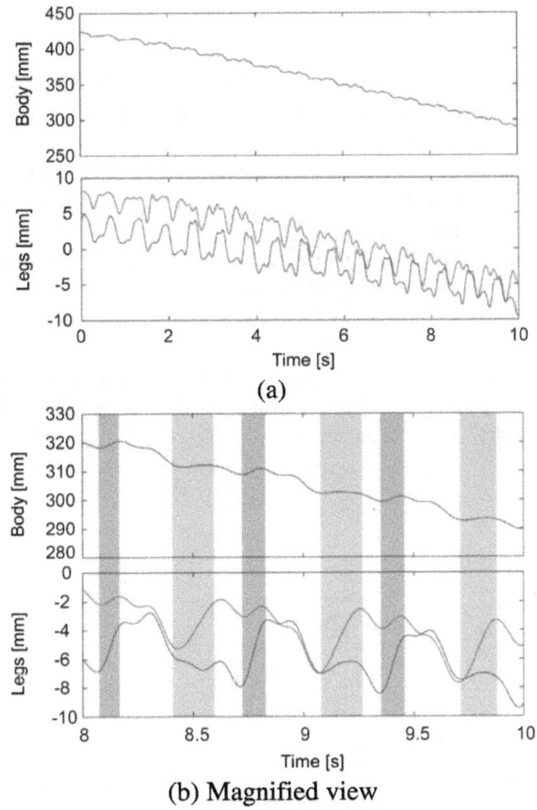

Fig. 6. Leg and body motions for negative-X walking (Slope angle = −1.0°, data filtered by LPF) Leg motions in plot (a) are supposed to have constant baselines; the decreasing trend would be due to the camera angle. (Color figure online)

5 Conclusions

In this paper, the walking performance and gait of a thermal walker on inclined surfaces were investigated in two directions; one direction is along a contour line and the other is upward/downward. The results show that the thermal walker can successfully walk along certain contour line on a flat slope; however, the stride was reduced compared to the horizontal surface. It was also confirmed that the walker can walk down a slope with increased stride. However, it was found that the walker walked backward on upward slopes, which is somewhat contradicting the results of the previous study. The difference would come from different leg designs between the studies, and therefore, further investigation is needed using many different walker designs, in future studies.

Acknowledgement. This work was supported in part by JSPS KAKENHI Grant Number JP18K03903.

References

1. McGeer, T.: Passive dynamic walking. Int. J. Robot. Res. **9**(2), 62–82 (1990)
2. Wisse, M., et al.: Passive-based walking robot. IEEE Robot. Autom. Mag. **14**(2), 52–62 (2007)
3. Nemoto, T., Yamamoto, A.: Thermobot: a bipedal walker driven by constant heating. In: Proceedings of 2015 IEEE/RSJ International Conference on Intelligent Robots and Systems (IROS), pp. 983–988 (2015)
4. Nemoto, T., Yamamoto, A.: Gait of thermobot on heated slope. In: Proceedings of the 22th Jc-IFToMM Symposium, pp. 24–28 (2016). (in Japanese)
5. Nemoto, T., Yamamoto, A.: Prototyping a thermal walker that can walk on a hot horizontal surface with a simple gait. Robomech J. **5**, 16 (2018)
6. Harada, Y., Suehara, T., Ueno, T., Higuchi, T.: Development of an actuator using thermal deformation. In: Proceedings of 2006 Japan Society for Precision Engineering Spring Meeting, pp. 803–804 (2006). (in Japanese)

Effectiveness of Physical Exercises in the Treatment of Scoliosis - Mathematical Approach

A.-M. Vutan, C. M. Gruescu(✉), V. Ciupe, D. Margineanu, and E.-C. Lovasz

Politehnica University of Timisoara, Timişoara, Romania
anamariavutan@yahoo.com, {corina.gruescu, valentin.ciupe, dan.margineanu, erwin.lovasz}@upt.ro

Abstract. The treatment of scoliosis is based on physical exercises with light and medium severity of deformation. The effectiveness of the exercises is assessed by means of Cobb angle variation. In order to compute the values of the Cobb angle, a mathematical model of the spine is generated. A set of eight sensors, type accelerometer, is attached to the vertebrae in significant points of the column. Its shape is approximated with a high degree polynomial, which allows the determination of the inflection points and Cobb angles. The study consists in assessment of three exercises frequently used in treatment protocols.

Keywords: Spinal deformation · Mathematical model of the spine Cobb angle · Physical exercises

1 Introduction

Idiopathic scoliosis is a deviation in frontal plane with unknown cause. It mainly affects children over eight years old until bone maturation is achieved. The lateral inclination of the column is accompanied by the vertebral rotation vertebrae to the convexity of the curve [1]. Generally, there is a primary curve showing the highest rotation and one or more secondary (compensatory) curves [2]. When vertebral rotation is located at the thoracic region, the rotation of the vertebral body determines the movement and deformation of the entire rib cage, causing the rib hump [1]. The child with idiopathic scoliosis is evaluated periodically by the orthopedic doctor. The protocol requires clinical and radiologic evaluation every 6 months, from the moment of diagnosis until the end of bone maturation. Thus, the Cobb angle is calculated on the X-ray, representing the intersection of the perpendiculars to the tangent line of the upper plateau of the superior vertebra of the curvature and the tangent line to the inferior plateau of the lower vertebra of the curve [2]. Depending on the degree of inclination and rotation of the spine, the orthopedist determines the treatment to follow. The protocol for a scoliosis (Stagnara protocol) requires: scoliosis specific exercises (if the Cobb angle is below 25°) or specific exercises and orthopedic treatment wearing a brace (if Cobb angle is between 25° and 50°) or surgery (if Cobb angle is above 50°) [1]. Specific exercises programs for scoliosis will be part of the patient's life immediately after

G. Carbone et al. (Eds.): *New Trends in Medical and Service Robotics*, MMS 65, pp. 25–34, 2019.
https://doi.org/10.1007/978-3-030-00329-6_4

diagnosis, until the end of the growth age, sometimes even for the entire life. For severe scoliosis, pre-surgical and post-surgical exercises are indicated in order to maintain muscle tone and good amplitude of the movements in the spine.

Effectiveness of physical exercise has been discussed by specialists in more articles over the past 30 years. Some authors, especially orthopedists, consider conservative treatment (exercise, brace, electrotherapy, massage, vertebral manipulation, insoles [3]) to be ineffective. They indicate the passive observation of the patient until the surgical intervention is required [4]. On the other hand, the physiotherapists mainly support the implementation of specific exercise programs for scoliosis immediately after diagnosis [5, 6]. There are studies showing the protocols for scoliosis in the US and the UK that include general toning exercises, posture and corrective postural training [7]. Other studies have discussed protocols used in countries in Europe (Germany, Italy, and Turkey) that include exercises specifically designed for scoliosis (especially the Schroth method [3, 8–12].

The shape of the spine is got, traditionally, as an x-Ray image. During the last decades, new approaches of investigation in a non-invasive manner for the human body developed. These new techniques are based on thermography, ultrasonic mapping, 3D scanning and 3D reconstruction using Kinect sensor [13], a wearable monitoring system based on inertial sensors [14] and others.

In a first study [15], the authors of this study presented an innovative method of measuring the Cobb angle in a scoliosis using accelerometer sensors. The same authors tried to apply the assessment method to more patients with scoliosis and the values obtained from the measurements were compared with the values calculated by radiologists on x-Ray. Because the values were similar, it was therefore attempted, in this study, to applicate the method to read the changes that the Cobb angle undergoes during some of the exercises described in the literature. Further reading of the values can help determine the effectiveness of exercises in the case of scoliosis.

2 Mathematical Approach and Equipment

During physical exercises, the spinal column changes its shape both in frontal and lateral plans. However, for pursuing the Cobb angle, which is relevant for scoliosis, it is necessary and sufficient to acknowledge the changes in shape within the frontal plan. From mathematical point of view, the task of the researcher is to study the projection of the column onto the frontal plan. Thus, the mathematical purpose is to get a plane curve and to record its changes during an exercise prescribed by the kinetotherapist, so that the variation of the Cobb angle to be monitored and assessed.

The projection of the column onto the frontal plane is traditionally got by means of radiography. The present study proposes an alternative method, which uses a set of sensors applied on the vertebrae [15]. The sensors, type accelerometer, provide 3D coordinates of the points where they are attached [15]. At a healthy, normal person, the shape of the column in the frontal plan varies from a straight line, in straight posture to a curve with one curvature, in any bended position. With scoliosis, the shape is also curved in straight posture, with one or two curvatures.

The bigger the number of precision points is the better approximation of the modelling curve is got. However, a big number of points implis more sophisticated mathematics. The authors decided to use a number of eight sensors, which is reasonably big to provide complex shapes.

The set of nine angles φ_i and eight lengths $l_{i,\ i+1}$ (i = 1... 8) - Fig. 1 - is used to determine an approximation function. Further processing, meaning finding of the inflection points, allows the computation of Cobb angles.

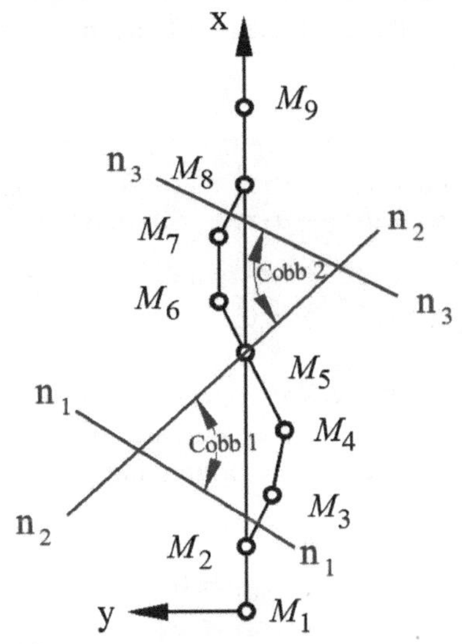

Fig. 1. Nine points on the deformed spine and Cobb angles [13]

The theory of function approximation starts from the following statement of problem: a function y = f(x), $x \in$ X and $y \in$ Y is, usually not known in its analytical expression, but is available as a set of pairs (x, y) [16]. The task is to find a function h (x) $\sim$ y for all pairs (x, y). The approximation function may be chosen among different classes of approximation functions, which are:

- polynomials
- exponentials
- logarithmics
- trigonometric functions
- linear combinations of the above types.

For the given problem, the mathematical model of the spine is chosen the class of polynomials, for the following reasons:

- the "C" or "S" shape of the column can be easily represented with a high degree polynomial
- the approximation can be as closely as desired, according to Weierstrass approximation theorem

- working with polynoms is relatively easy.

Other classes of approximation functions are unsuited to model the spine or unreasonably difficult to apply from calculus point of view.

A polynomial of n degree is an expression:

$$P_n(x) = a_0 + a_1 x + a_2 x^2 + a_3 x^3 + \ldots a_n x^n, \quad (1)$$

Where $a_i,\ i = \overline{0,n}$ are constants and x is the variable. If (n + 1) pairs of values $(x_i, y_i),\ i = \overline{0,n}$ are known, the coefficients of the polynomial result from the following system of equations:

$$\begin{cases} a_0 + a_1 x_0 + a_2 x_0^2 + \ldots a_n x_0^n = y_0 \\ a_0 + a_1 x_1 + a_2 x_1^2 + \ldots + a_n x_1^n = y_1 \\ a_0 + a_1 x_2 + a_2 x_2^2 + \ldots + a_n x_2^n = y_2 \\ \vdots \\ a_0 + a_1 x_n + a_2 x_n^2 + \ldots + a_n x_n^n = y_n \end{cases} \quad (2)$$

The function h = $P_n(x)$ is the seeked approximation function.

For the specific given problem, the coordinates of the nine points are computed with the relationships, which consider the orthogonal reference system attached to the first lower point:

$$\begin{cases} x_{M1} = 0 \\ y_{M1} = 0 \\ x_{Mi+1} = x_{Mi} + l_i \cos\varphi_i \quad i = \overline{1,8} \\ y_{Mi+1} = y_{Mi} + l_i \sin\varphi_i \quad i = \overline{1,8} \end{cases} \quad (3)$$

The approximation polynomial of eight degree is given by the expression:

$$h = a_0 + a_1 x + a_2 x^2 + a_3 x^3 + a_4 x^4 + a_5 x^5 + a_6 x^6 + a_7 x^7 + a_8 x^8, \quad (4)$$

where a_i result from the system of Eqs. (2).

Equation (4) provides the shape of column. The graphs drawn with the mathematical model were compared with the x-Ray images and, thus, validated.

Further mathematical processing of the eighth degree polynomial is quite heavy and therefore, it was approximated with a smaller degree polynomial. In order to get three inflection points, a fifth degree polynomial would be suited, but the attemp to use it provided too large approximation errors. From mathematical handling point of view the best fit was got with a fourth degree polynomial:

$$g = b_0 + b_1 x + b_2 x^2 + b_3 x^4 + b_4 x^4 \quad (5)$$

The roots of the second derivative of Eq. (5), if they result within the domain X and a third point, which was chosen the first or last value of the variable x provide the inflection points of the polynomial function. Considering the inflection points $I_1(x_1, y_1)$,

$I_2(x_2, y_2)$ and $I_3(x_3, y_3)$, the gradient of the normal to the polynomial (5) is given by the relations:

$$m_1 = -\frac{1}{g'(x_1)}, \qquad m_2 = -\frac{1}{g'(x_2)}, \qquad m_3 = -\frac{1}{g'(x_3)}, \tag{6}$$

where g' is the first derivative of the polynomial (6).

The Cobb angles (Fig. 1), as angles between two normal lines, can be computed as follows:

$$Cobb1 = \arctan\left|\frac{m_1 - m_2}{1 - m_1 m_2}\right|, \qquad Cobb2 = \arctan\left|\frac{m_2 - m_3}{1 - m_2 m_3}\right|. \tag{7}$$

The angles φ are recorded as functions of time during the exercises performed by the patient under the supervision of the kinetotherapist and thus, it is possible to get the variation of the Cobb angles.

3 Performing Measurements and Interpreting Data

Measurements were performed within a recovery clinic with the consent of patients and their parents. Initially, data acquisitions were taken from orthotic position, the same position in which the patient performed x-ray exposure. Data taken from the accelerometer sensors in the orthostatic position were compared with the results calculated by the doctors on x-ray. The subject under study has a double dorsal scoliosis with the Cobb angles calculated by the radiologist: 22° lumbar and 28° dorsal. Of the mathematical calculation using the polynomial of 8th degree, revealed the Cobb angle on the lumbar area of 20.7^{o} and the Cobb angle on the thoracic area of 25.4^{o}. The values obtained being very close to those calculated on radiography and the very similar curves could be taken into account for the desired experiment. In Fig. 2 it can be seen the deviation of the spine on x-ray and comparatively, the graph obtained by the mathematical method.

The treatment for scoliosis involves only physical exercise for scoliosis with Cobb angle below 25° (light scoliosis), and for medium scoliosis, with a Cobb angle between 25 and 50°, the wearing a brace for more than 18 h a day is recommended besides specific exercises [1]. The exercises performed by the scoliotic patient are prescribed by a physiotherapist.

There are many approaches to perform exercise for scoliosis. Depending on the physiotherapist's training, one can choose between exercises from Schroth method (most commonly used in Germany), DoboMed method, from Cotrel method, from Klapp method, from Mézièr method, stretching and toning exercises or postural correction exercises.

In this study, three of the most commonly used exercises in scoliosis recovery are analyzed using the equipment made up of accelerometer sensors to track how the Cobb angle changes during treatment.

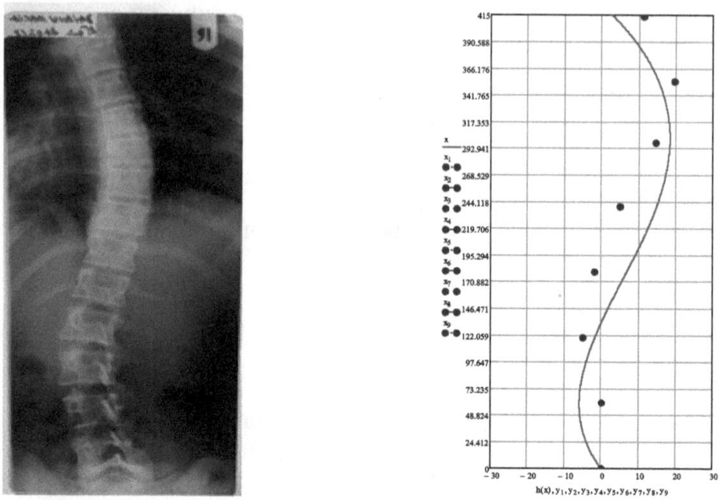

Fig. 2. On the left, the x-ray of the double scoliosis and on the right, the model of the spine – the polynomial of 8th degree

- The first exercise was chosen from Schroth method - "muscular cylinder standing". The patient is positioned in front of the mirror with the soles apart and hands on the hips, bends the trunk and kneels the knees. In this position the patient performs the Schroth tybreaths then tilts the In this position, Schroth type breathing continues.
- The second exercise was chosen from Mezier method. The patient is in a position with the lower legs apart and the arms raised parallel to the ground, he maintains lumbar lordosis activating the lumbar erector muscles, makes elongations and corrective breathing with derotation of the shoulders, rib hump and pelvis.
- The third exercise was chosen from the hanging position to the wall bars with arms apart and legs together: the pacient keep all the sagittal and frontal corrections for several seconds and performs deep breaths.

For a better understanding of the description of the exercises, these are below schematically represented in Fig. 3.

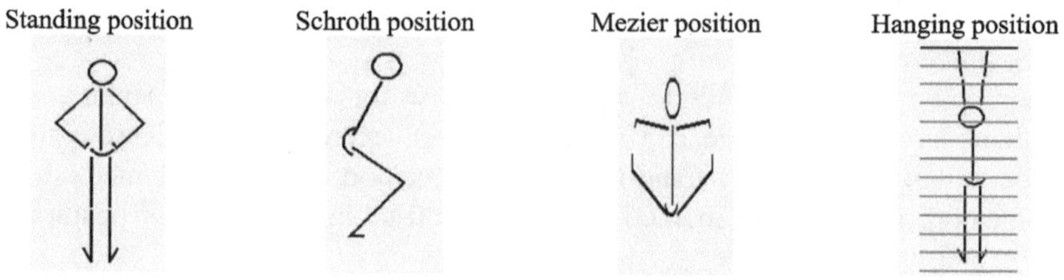

Fig. 3. Model of the exercises

For testing, on the back of the patient were positioned the accelerometer sensors at equal distances between them. The first sensor was positioned at the level of the first sacral vertebra S1 and the last one at the level of the last cervical vertebra C7, as can be seen in Fig. 4.

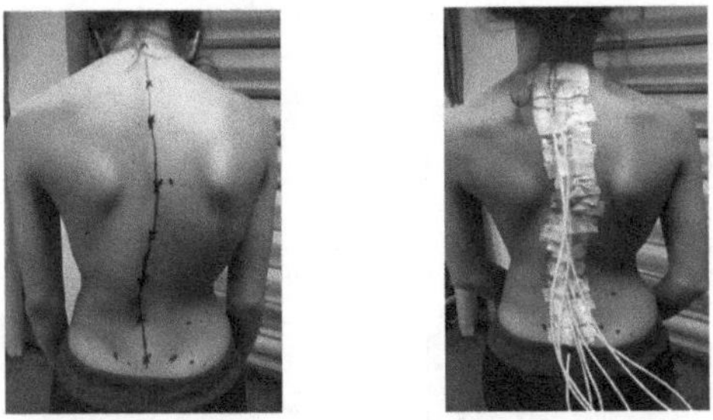

Fig. 4. The position of sensors for testing

For a first test there were used the exercises described above having as a starting point, the vertical position of the column. The distances between the sensors along the column were known from the beginning and the values of the angle φ have been read by the sensors. The values of angles φ_i and the distances l_i were introduced into Mathcad 14, where the computation and graphs were further developed. Further, the curves obtained from the computation are shown in Fig. 5 (Table 1).

The values of the Cobb angles obtained during the test performing the exercises were compared with the value of the Cobb angle obtained on the x-ray. In all 3 exercises taking into study, the Cobb angles decreased compared to the reference value.

From the table and the graphs it can be observed that the position in which both Cobb angles decrease the most is part of the Schroth method. Also, a significant decrease of the angular values can be observed in the Mézièr position, but after the curve described by the polynomial of 8th degree calculated during the exercise, an increase of the compensatory curves can be observed. In the hanging position there was only an improvement in the value of the Cobb angle in the dorsal area while in the lumbar zone the Cobb angle is increased. This last position is in the open kinetic chain and requires from the subject under test a strong control of the lower part of the body. The position in which both Cobb angles decrease the most is part of the Schroth method. Also, a significant decrease of the angular values can be observed also in the meder position, but after the curve described by the grade 8 polynomial calculated during the exercise, an increase of the compensatory curves can be observed. in the hanging position there was only an improvement in the value of the Cobb angle in the dorsal area while in the lumbar zone the Cobb angle is increased. This last position is made in an open kinetic chain, it is instable for the lower limbs and pelvis and requires from the testing subject a strong control of the lower part of the body (perhaps this is why the Cobb angle in the lumbar region showed a worsening of the value compared to the one in the standing position). Analyzing the results we can conclude that the person tested is the most appropriate to perform the Schroth based exercise in which the Cobb

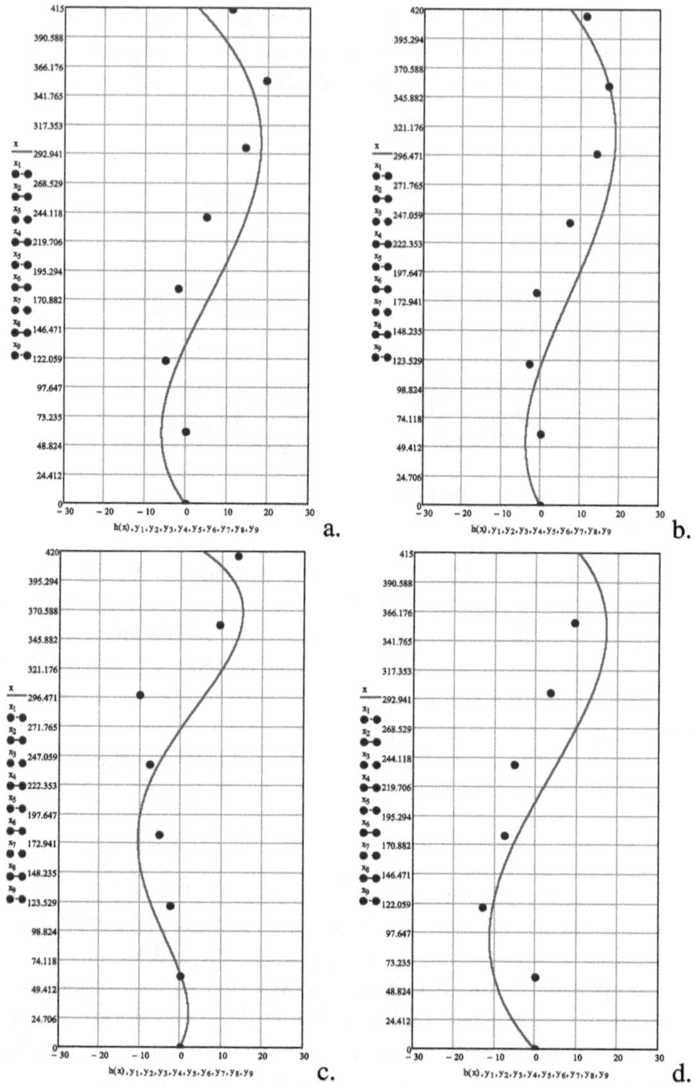

Fig. 5. Model of the spine - polynomial of 8th degree – standing position (a), Schroth position (b), Mézièr position (c) and hanging position (d)

Table 1. The values of the Cobb angles on the x-ray and those computed from the mathematical approach

Values of Cobb angles	X-Ray	Standing position	Schroth Method	Mézièr Method	Hanging on wall bars
Cobb 1	22°	20.747°	16.032°	16.558°	23.878°
Cobb 2	28°	24.536°	19.041°	19.041°	22.099°

angles showed a significant improvement while the exercise from Méziér method and the hanging on the wall bars will have to be executed with increased attention to compensatory curves and of the lumbar muscle control.

4 Conclusions

The present study came as a complement to the experimental test previously conducted by the same team of authors. This time it was attempted to track Cobb angles in three basic exercises in the recovery of scoliosis. The use of accelerometer sensors equipment has demonstrated the ability to track Cobb angle changes during exercises. This equipment enables the analysis of changes in the segments of the column involved in the exercise. In this way, after the exercise, the therapist can analyze which area of the column worked correctly and in which areas the curves worsen. Also, the equipment enables the therapist to choose the most effective exercises for each subject. For the time being, only three basic exercises have been tested in which the vertebral column is positioned vertically. We look forward to trying to analyze other exercises in which the starting positions are different from the vertical ones because the scoliosis recovery programs use different positions such as quadruped, kneeled or seated.

From mathematical point of view, the polynomial approximation was validated. Further studies will involve other classes of functions of approximation.

References

1. Jianu, M.: Pediatric scoliosis, Publishing House and Typography Pro, Bucuresti, pp 12–70 (2010)
2. Pană, I., Vlădăreanu, M., Roventa, N., Mihăiță, I.: Radiology I- Spinal column, p. 156. Didactic and Pedagogical Publishing House, Bucureşti (2000)
3. Fusco, C., Zaina, F., Atanasio, S., Romano, M., Negrini, A., Negrini, S.: Physical exercises in the treatment of idiopathic scoliosis: an updated systematic review. Physiother. Theory Pract. ISICO **27**, 80–114 (2011)
4. Hawes, M.C.: The use of exercises in the treatment of scoliosis: an evidence-based critical review of the literature. Pediat. Rehabil. **6**, 171–182 (2003)
5. Maruyama, T., Kitagawa, T., Takeshita, K., Makizuki, K., Nakamura, K.: Conservative treatment of adolescent idiopathic scoliosis: can it reduce the incidence of surgical treatment? Pediat. Rehabil **6**, 215–219 (2003)
6. Negrini, S., Fusco, C., Minozzi, S., Zaina, F., Romano, M.: Exercises reduce the progression rate of adolescent idiopathic scoliosis: results of a comprehensive systematic review of the literature. Disabil. Rehabil. **30**, 772–785 (2008)
7. Romano, F., Minozzi, S., Bettany-Saltikov, J.A., Zaina, F., Chockalingam, N., Kotwicki, T., Maier-Hennes, A., Negrini, S.: Exercises for adolescent idiopathic scoliosis. Cochrane Database Systematic Reviews, 8. Art. No. CD 007837 http://dx.doi.org/10.1002/14651858.CD007837.pub2
8. Weiss, H.R.: Influence of an in-patient exercise program on scoliotic curve. Italian Journal of Orthop. Traumatol. **18**, 395–406 (1992)

9. Weiss, H.R., Weiss, G., Petermann, F.: Incidence of curvature progression in idiopathic scoliosis patients treated with scoliosis in-patient rehabilitation (SIR): an age- and sex-matched controlled study. Pediat. Rehabil. **6**, 23–30 (2003)
10. Otman, S., Kose, N., Yakut, Y.: The efficacy of Schroth 3-dimensional exercise therapy in the treatment of adolescent idiopathic scoliosis in Tutky. Saudi Med. J. **26**, 1429–1435 (2005)
11. Lehnert-Schroth, C.: Three-Dimensional Treatment for Scoliosis. A Physiotherapeutic Method for Deformities of the Spine. The Martindale Press, Palo Alto, California (2007)
12. Rigo, M., et al.: Scoliosis intensive out-patient rehabilitation based on Schroth method. Stud. Health Technol. Inform. **135**, 208–227 (2008)
13. Ciobanu, O., Ciobanu, G., Aiello, A.: An application of Kinect sensor for scoliosis and kyphosis screening. In: Proceedings of the International Conference on Medical and Biological Engineering (CMBEBIH), Conference Paper, pp. 713–717. Springer (2017)
14. Voinea, G.-D., Butnariu, S., Mogan, G.: Measurement and geometric modelling of human spine posture for medical rehabilitation purposes using a wearable monitoring system based on inertial sensors. Sensors **17**(1), 3 (2017)
15. Vutan, A.-M., Ciupe, V., Gruescu, C.M., Lovasz, E.-C.: Experimental method for dynamic evaluation of spinal column deformation exercises, New Advances in Mechanism and Machine Science, Mechanisms and Machine Science 57, https://doi.org/10.1007/978-3-319-79111-1_13, ISBN 978-3-319-79110-4
16. Stalph, P.: Analysis and Design of Machine Learning Techniques. Springer Fachmedien Wiesbaden (2014). ISBN 978-3-658-04936-2

Simplified Design of a Device for Wrist Rehabilitation

G. García, V. Ramírez(✉), O. Ramírez, J. L. Rueda, and C. R. Torres

Instituto Politécnico Nacional, Escuela Superior de Ingeniería Mecánica y Eléctrica, Sección de Estudios de Posgrado e Investigación, Unidad Profesional Adolfo López Mateos, Edif. 5, 2do piso. Col. Lindavista, 07320, México City, Mexico
vramirezv0700@gmail.com

Abstract. It is well known that from body articulations, the wrist one possesses the highest injury index, because it is the most employed in quotidian (repetitive) activities. This research proposes a novel two-degrees of freedom device a for wrist rehabilitation, aimed to improve function and mobility of affected wrist. In order to accurately reproduce wrist articulation movements (adduction-abduction, flexion-extension and rotation), the device is endowed with just one servomotor and a robust control algorithm, which is implemented through an Arduino UNO embedded system. Novelties of this device are its easiness of building, its cheapness, its portable compact design, as well as the intuitiveness of providing therapy through it.

Keywords: Rehabilitator wrist · Carpal tunnel hand · Exercises therapy

1 Introduction

The articulation of the wrist is considered one of the most complex articulations of the human body; at the same time, it allows a big quantity of the movements and is capable to endure significant forces on it, like compression, torsion and bending without suffering a destabilization. Also, is used to perform the quotidian activities and can present a higher index of injury compared with other articulations. According to the statistics presented in 2016 by the "Secretary of Work and Social Prevention" 385,826 work accidents were recorded in Mexico, of which 104, 213 correspond of the wrist and hand kind to injuries, representing 27% of the total of the accidents, representing the greater index with respect to the injuries presented in other anatomical regions [1].

1.1 State of Art

The importance of have an apparatus that helps improve mobility and the function of said articulation is denoted. Currently, there are designs of wrist and hand rehabilitators specialized in improve tendinitis and the carpal tunnel. From 1969 to 1994, some uncontrolled mechanical designs of wrist and hand rehabilitators were documented to strengthen the muscles that compose the articulation. In 1997, devices were made for wrist therapy based on the rotation of the articulation [2–6]. By 1998, devices to treat

G. Carbone et al. (Eds.): *New Trends in Medical and Service Robotics*, MMS 65, pp. 35–42, 2019.
https://doi.org/10.1007/978-3-030-00329-6_5

the symptoms associated with carpal tunnel syndrome was development [7, 8]. In the 21st century, controlled mechanical designs of 2, 3 and/or 8° of freedom were patented, just as the first exoskeletons appear [9–13].

In Mexico, a patient with Tendinitis, tumours, syndromes and wrist dystrophies can only have access to limited designs of orthoses and splints. Therefore, this work presents a prototype of a simplified rehabilitator for injuries of this kind.

2 Antecedents

The wrist is the distal joint of the upper limb that allows the union of the forearm with the hand, is composed of the distal extremity of the ulna and radius, the eight bones of the carpus and the bases of the five metacarpals [14]. It has two degrees of freedom and with the rotation of the forearm on its longitudinal axis adds a third degree of freedom, so that, the hand can be oriented at any angle to manipulate from little objects too heavy objects [15]. In addition, the ligaments work to reinforce key areas. They are divided into intrinsic and extrinsic, the former limit mobility and stabilize the base of the hand, while the latter are important stabilizers of the movements of the wrist [16]. The wrist has nine flexor tendons that travel through a fibrous osseous tunnel in the wrist called the carpal tunnel and is formed by the carpal bones and the transverse carpal ligament. Also, by three tendons that flex the wrist [17].

The movements of the wrist are carried out around two axes: an axis AA', transversal on which the movements of flexion-extension are performed and an axis BB', anteroposterior on which the movements of adduction-abduction are performed (Fig. 1) [18].

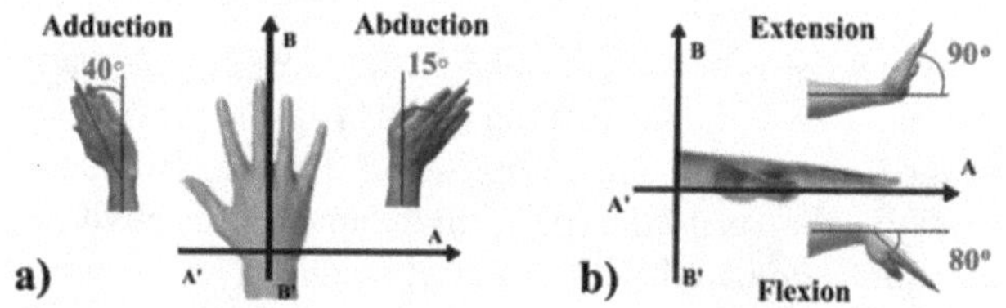

Fig. 1. Axes and amplitude of movement of the wrist.

The active flexion-extension movement with the extended fingers is rarely to exceed 90° and is smaller with the fist closed, 65° approximately. The active extension is smaller than 80°, whatever the position of the fingers. The passive extension is maximum of 95°, when exists a radial inclination, while the combination of flexion-passive ulnar inclination with the extended fingers can reach approximately 125°. The amplitude of the abduction movement (radial inclination) does not exceed 15°. The amplitude of adduction (radial-ulnar inclination) is smaller than 40° [19].

The common pathologies being: fracture of some carpal bone, ligaments sprains, tendinitis (inflammations of the tendons) and neuropathies due to trapping of a nerve (carpal tunnel syndrome). Any condition or injury that affects the stability of the carpal bones or the normal function of the tendons of the wrist, will reduce the grip strength.

3 Materials and Methods

The parameters to be covered with the design of the equipment were stablished as follows: the rehabilitator is capable to perform the movements of both Extension-Flexion and Abduction-Adduction. A relation with the standard size and weight of a hand was determinated. The number of necessary repetitions of the therapeutic exercises of the hand was defined, as well as angles, time and angular speeds.

The design of the wrist rehabilitation prototype was carried out in a CAD computer program, where each of the pieces that compose it was developed and an assembly of all of them was generated to obtain a virtual prototype. The movements are executed by a transmission system of angular speed and forces between parallel pulleys. This mechanism can transmit the movement with high efficiency according to the pulley position. Therefore, the position configurations applied in the device were: the same direction of rotation (Fig. 2a) and non-parallel axes (Fig. 2b) [20].

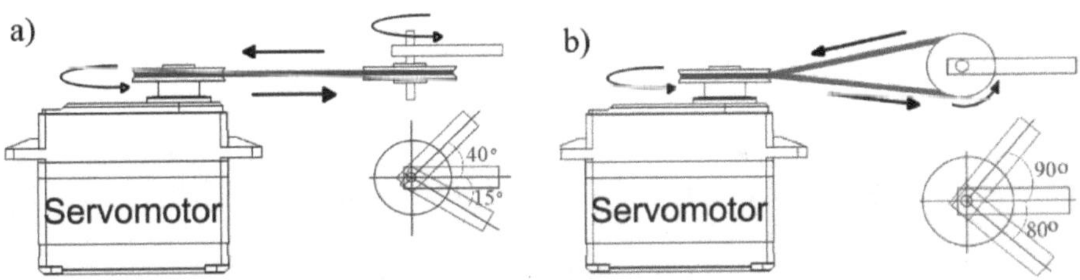

Fig. 2. Representation of the various transmission configurations per band.

First, the adequate dimensions of the band, the torque needed to move a person's hand, as well as the speed and the frequency of movement of the actuator was calculated. Considering the following dimensions of a hand: long = 0.22 m and width = 0.11 m. Taking into consideration the mass of a hand ($Mass_h$) of a 100 kg person ($Mass_T$), the force necessary to move the hand was determined and from (1) the torque that the actuator must have was obtained.

$$Mass_h = (0.7) * Mass_T = 0.7 \text{ kg}$$

$$F = (Mass_h)(aceleration) = 6.86 \text{ N}$$

$$T = rF \sin\theta = 1.50 \text{ Nm} \tag{1}$$

The assessment of the mechanical (2) and electrical (3) power necessary to perform the movements were made.

$$\theta_{max} = 90^{\circ} = 1.57\ \text{rad}$$

$$\omega = \frac{(1.57\ \text{rad})}{5\ \text{seg}} = 0.314\ \text{rad/seg}$$

$$W = T * \omega = 0.471\ \text{W} \tag{2}$$

$$W = V * I = 0.23\ \text{W} \tag{3}$$

To form an estimate the transmission of movement in the mechanism. It is obtained the tangential velocity (4), as well as the appropriate length of the band, is considered the pulley diameters (D, d), in this case the relation of transmission was 1:1 [21], so, the value of D and d are equal (5) and the contact arc (6) that is used to determine the frequency of the flexion (7).

$$n = 0.314 \frac{\text{rad}}{\text{seg}} = 2.99\ \text{rpm}$$

$$V = \frac{D * \pi * n}{60000} = 4.69 \times 10^{-3}\ \text{m/s} \tag{4}$$

$$L = 1.57(D + d) + 2A + \frac{(D + d)^2}{2A} = 247.05\ \text{mm} \tag{5}$$

$$\alpha = 180^{\circ} - \frac{60(D - d)}{A} 180^{\circ} = 180^{\circ} \tag{6}$$

$$F = \frac{V * Z * 1000}{L} = 37.91\ \text{flex} \cdot \text{seg} \tag{7}$$

The movement is controlled by an embedded system, the program contained in the prototype, determines the actions, angles and waiting times of the operations to be performed for the rehabilitation of the wrist. Figure 3 shows the operation process of the device.

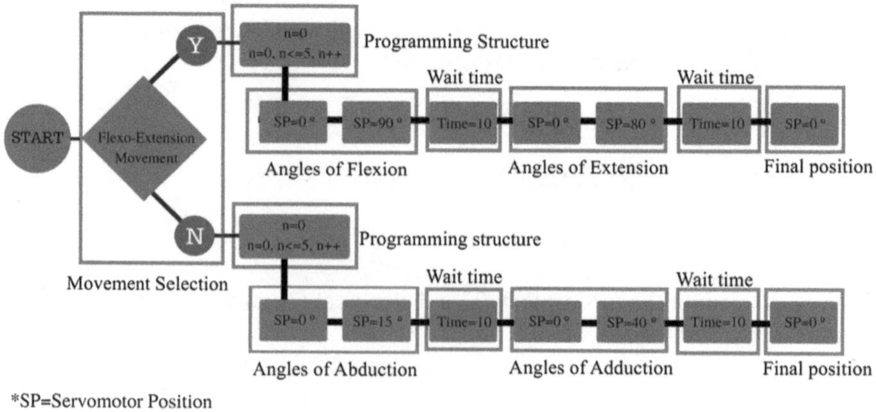

Fig. 3. Flowchart of programming

For the execution of the programming, an Arduino UNO® control system and a servo motor were used, that allows the operations of the prototype of the wrist rehabilitator.

4 Results

The design consists of mechanism of movement made up of a base piece, a set of brackets and joints, a couple of pulleys and belts that be powered by an actuator. The design of the rehabilitation device was simplified, thinking of the manufacture of each of the parts that compose it to be easily transportable.

The prototype allows two types of movements: on a vertical axis, be held flexion-extension, the movement that was achieved employing the configuration of the non-parallel axes band transmission and in a horizontal axis be performs the adduction-abduction of the wrist articulation, using the transmission configuration in the same direction of rotation (Fig. 4).

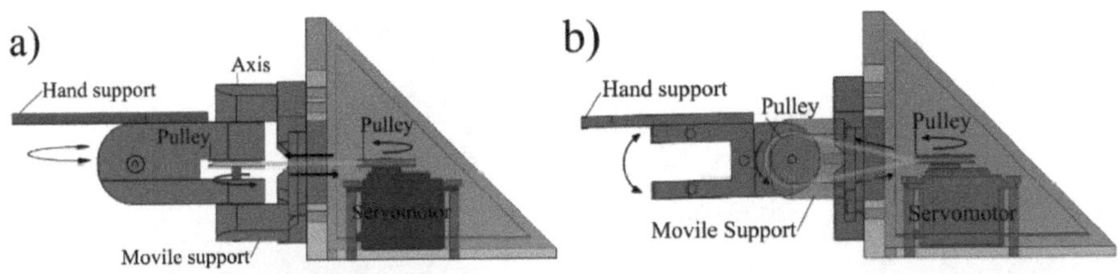

Fig. 4. Pulley Mechanism employed in the device. (a) Adduction-abduction movement. (b) Flexion-extension movement.

The periodicity of the different movements is carried out, modifying in 90° position of the axes of the pulleys, which is feasible due to the flexibility of the transmission of movement by bands. The wrist rehabilitation movements in flexion-extension are intended to recover the mobility of the hand and wrist after a period of immobilization or trauma, whatever the cause if a doctor prescribes it. In Fig. 5(a) is shown the graphic behaviour of the actuator obtained from the program corresponding to the flexion-extension movement cycle created in programming platform for Arduino UNO®.

Lateral mobility of the wrist is recovered with the open hand and fingers extended, directing hand outward and inward for periods of 5 to 10 s at each movement before relaxing the articulation. Figure 5(b) shows the graphic behaviour of the actuator obtained from the program corresponding to the Adduction-Abduction movement cycle created in programming platform for Arduino UNO®.

A preliminary prototype has been made with 3D printing technology, using the methodology described in the literature [22]. This work is focused on the evaluation of kinematic and dynamic parameters required for the proper use and function of the wrist rehabilitation device (Fig. 6).

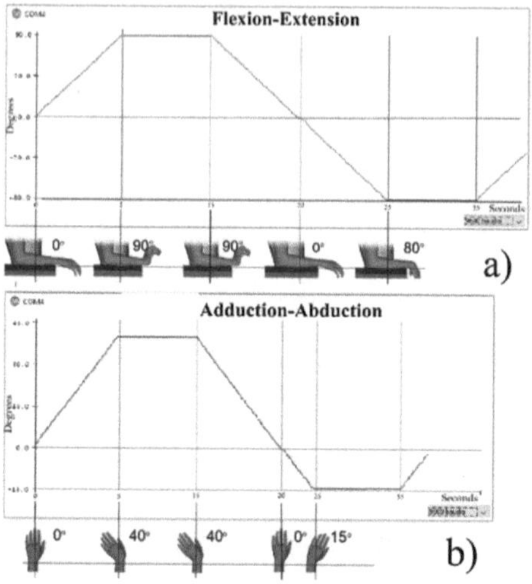

Fig. 5. (a) Flexion-extension movement cycle. (b) Adduction-abduction movement cycle

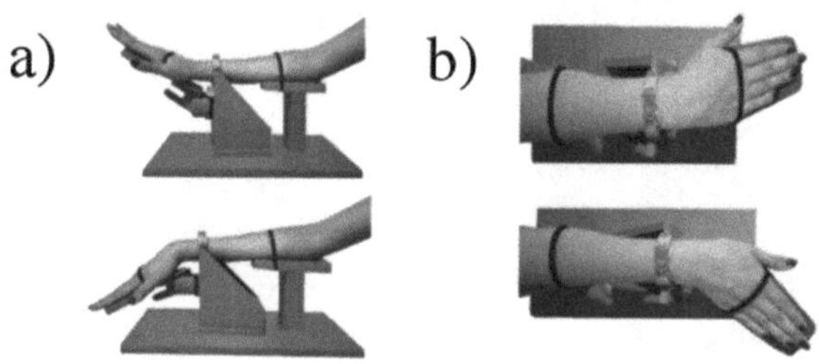

Fig. 6. Movements that can be made by the wrist rehabilitation prototype, (a) Flexion-extension and (b) adduction-abduction.

5 Discussions

A person with tendinitis of the Quervain type on the left wrist tested the operation of the rehabilitation device, the user's comments about the setup of the equipment pointed out to its easy use and transportation. In addition, the person suggests that the device allows the passive movement of the joint relieving pain, as a result of this, the user can activate active therapy of the wrist. The user indicated that the device plays correctly the flexion-extension and abduction-induction movements with a moderate speed that not caused him any type of pain. Geometric characteristics and the dimensions are suitable for people with thin arms. Therefore, the ergonomics of the arm support, as well as the general device, should be redesigned, resized and improved.

6 Conclusions

The development of this research represents a contribution to the field of the progress of wrist rehabilitation devices. As a method of transmission of movement, a servomotor was used due to the ease of manipulating the prototype. The control of the system, in general, was set up by an Arduino UNO® programming board due to its low cost and easy interphase. In addition, rehabilitation routines could be created, making a combination between movements and different speeds over a time period.

As future work, the development of other stages that help to improve the function of the present mechanism is proposed. Also, it is projected to change the actuator by an electromechanical device of discrete angular displacements. Figure 7 shows the process to be followed for the subsequent functioning of the device.

Fig. 7. Process for the subsequent operation of the rehabilitation device.

In general, this first design has expanded the vision to improve the rehabilitation device, as well as, has allowed to observe in detail the important elements for the passive manipulation of the joint and the needed to attend different types of the wrist injury. The prototype manufactured was designed for a specific arm-wrist-hand size, so to realize a test with other anthropomorphic references, it is required to escalate each element and develop an ergonomic arm support system.

References

1. Finished Work Risks registered in the IMSS.: Secretary of labour and social security STPS: Sector Statistics (2017). www.stps.gob.mx/gobmx/estadisticas/. Accessed 4 Dec 2017
2. Hillman, M.: Rehabilitation robotics from past to present - a historical perspective. In: Advances in Rehabilitation Robotics, pp. 25–44. Springer-Verlag, Heidelberg (2004)
3. Krebs, H.I., Hogan, N.: Therapeutic robotics: a technology push. Proc. IEEE Inst. Electr. Electron. Eng. **94**(9), 1727–1738 (2006)
4. Wisneski, K.J., Johnson, M.J.: Trajectory planning for functional wrist movements in an ADL-oriented, robot-assisted therapy environment. In: IEEE International conference on robotics and automation, pp. 3365–3370 (2007)
5. Dovat, L., Lambercy, O., Gassert, R., Maeder, T., Milner, T., Leong, T.C., Burdet, E.: HandCARE: a cable-actuated rehabilitation system to train hand function after stroke. IEEE Trans. Neural Syst. Rehabil. Eng. **16**(6), 582–591 (2008)
6. Dovat, L., Lambercy, O., Gassert, R., Burdet, E., Leong, T.C.: HandCARE2. a novel cable interface for hand rehabilitation. In: Proceedings IEEE Virtual Reality Conference, vol. 64 (2008)
7. Hwang, C.H., Seong, J.W., Son, D.S.: Individual finger synchronized robot assisted hand rehabilitation in subacute to chronic stroke: a prospective randomized clinical trial of efficacy. Clin. Rehabil. **26**(8), 696–704 (2012)

8. Sale, P., Lombardi, V., Franceschini, M.: Hand robotics rehabilitation: feasibility and preliminary results of a robotic treatment in patients with hemiparesis. Stroke Res. Treat. **2012**, 1–5 (2012)
9. Rodríguez, L., Cano, R., Cuesta, A., Alguacil, I., Molina, R.: Robotic therapy for the rehabilitation of the upper limb in neurological pathology. Rehabilitation **48**(2), 104–128 (2014)
10. Al-Fahaam, H., Davis, S., Nefti-Meziani, S.: Wrist rehabilitation exoskeleton robot based on pneumatic soft actuators. In: International Conference for Students on Applied Engineering (ICSAE), pp. 491–496 (2016)
11. Dauria, D., Persia, F., Siciliano, B.: Human-computer interaction in healthcare: how to support patients during their wrist rehabilitation. In: IEEE Tenth International Conference on Semantic Computing (ICSC), pp. 325–328 (2016)
12. Takaiwa, M.: Wrist rehabilitation training simulator for P.T. using pneumatic parallel manipulator. In: IEEE International Conference on Advanced Intelligent Mechatronics (AIM), pp. 276–281 (2016)
13. Lu, Z.J., Wang, L.C.B., Duan, L.H., Lui, Q.Q., Sun, H.Q., Chen, Z.I.: Development of a robot MKW-II for hand and wrist rehabilitation training. In: The Annual IEEE International Conference on Cyber Technology in Automation, Control and Intelligent Systems, pp. 302–307 (2016)
14. Ariznavarreta, C., Cachofeiro, V., Cardinali, D., Escrich, E., Gil, P., Lahera, V., Mora, F., Romano, M., Tamargo, J.: Human Physiology, 3rd edn. McGraw-Hill, New York (2005)
15. Chabas, J.F., Legré, R.: Sprains and dislocations of the carpus. Locomotor Apparatus EMC **14**(46), 1–15 (2011). Elsevier
16. Medina, C., Benet, M., Marco, F.: The joint complex of the wrist: anatomical and biomechanical aspects, characteristics, classification and treatment of the distal radius fracture. Medisur **14**(4), 430–446 (2016)
17. Esplugas, M., Garcia, E.M., Lluch, A., Llusá, P.M.: Role of muscles in the stabilization of ligament-deficient wrists. J. Hand Ther. **29**(2), 166–174 (2016)
18. Waldt, S., Eiber, M., Wörtler, K.: Methods of measurement and classification systems in musculoskeletal radiology. Editorial Panamericana Medical, pp. 107–112 (2013)
19. Voegeli, V.: Basic Biomechanics Lessons of the Locomotor System. 2nd edn., pp. 162–169. Springer (2004)
20. Brotzman, S.B.: Extensor Tendon Injuries. Injuries of Wrist and Hand, pp. 1–51. Elsevier (2012)
21. Budynas, R.: Design in mechanical engineering of shigley. 9th edn., pp. 847–851. McGrawHill, New York (2008)
22. Cafolla, D., Ceccarelli, M., Wang, M.F., Carbone, G.: How to use 3D printing for feasibility check of mechanism design. Int. J. Mech. Control. **17**(1), 3–12 (2016)

Surgical Robotics

Design of a Teleoperation Scheme with a Wearable Master for Minimally Invasive Surgery

Gizem Ateş, Ronny Majani, and Mehmet İsmet Can Dede(✉)

Izmir Institute of Technology, Izmir, Turkey
{gizemates,candede}@iyte.edu.tr, ronnymajani@gmail.com

Abstract. Minimally invasive surgery is increasingly being preferred over conventional surgery, however many problems still persist in longer surgeries such as pituitary surgeries, where surgeons are still required to hold an endoscope in their hand for prolonged periods of time. Many modern approaches have recently been proposed in literature to reduce the surgeon's effort. In this paper we extended upon these previous attempts and presented a promising solution; a real time teleoperation scheme with 3 different modes of operation, composed of a wearable ring system that captures and transmits voluntary hand motions over a wireless connection to a slave system. Accordingly, this slave system processes the received data to generate velocity demands for the robot endoscope controller. Finally, the feasibility of the proposed modes of operation are demonstrated and compared by measuring their learning curve and effort by running a set of training simulations on human subjects.

Keywords: Teleoperation · Assistive surgical robotics
Human interface device · Minimally invasive surgery

1 Introduction

Minimally invasive surgery (MIS) is a methodology that was developed to reduce the side effects of traditional open surgeries. However, it requires higher dexterity of the surgeons. To compensate for this increased requirement, the use of robot systems has become more preferable [5]. There are some commercial systems which are already in use today. We can group these surgical systems as autonomous and teleoperated robot systems. A review of autonomous medical robotic systems was conducted by Moustris et al. [3]. Furthermore, one of the earliest examples of teleoperated robotic surgical systems in endoscopic area was AESOP voice controlled endoscope manipulator [4]. We can further categorize the teleoperated surgical systems as surgeon off-site systems like the Da Vinci surgical robotic system [1] and on-site assistive robotic systems like Viky [6]. Da Vinci is a master/slave, system that consists of a surgeon's computer console to control the robotic tower that supports three or four robotic arms as the slave

G. Carbone et al. (Eds.): *New Trends in Medical and Service Robotics*, MMS 65, pp. 45–53, 2019.
https://doi.org/10.1007/978-3-030-00329-6_6

component. On the other hand, Viky offers a voice controlled "third hand" that handles the endoscope for the surgeon during a laparoscopic procedure.

This paper introduces a new on-site assistive surgical robot teleoperation system for minimally invasive pituitary surgeries. The next section explains the teleoperation system which is used for navigation of the endoscope by our novel master device. The operation modes of the teleoperation system is later explained which is followed by the description of the algorithm developed for these modes. Finally, the simulator constructed for user evaluations of the operation modes is introduced and the results are discussed by comparing the operation modes in terms of usability, learning curve and effort in use.

2 Teleoperation System Overview

The teleoperation system presented in this section allows the surgeons to control an endoscope remotely using their hand within a constrained environment (i.e. holding a tool that is partially inside a person's nose). This system is composed of three parts as; master, slave and communication channels which are depicted in Fig. 1. The master consists of 2 parts:

- A ring that embeds a 6 DoF (Degree of Freedom) IMU (Inertial Measurement Unit) and Bluetooth module and is worn on the surgeon's index finger of the hand that will hold the non critical tools during the surgery. It transmits the angular velocity and acceleration of the surgeon's hand motions at a rate of 100 Hz.
- A foot pedal that triggers the teleoperation scheme and sets the operation mode.

The slave consists of 3 parts:

- A microcontroller that implements the teleoperation scheme's algorithm described in Sect. 4, in order to generate a set of velocity demands for the endoscope robot controller.
- Endoscope Robot Controller: a microcontroller that processes the generated endoscope velocity demands and utilizes motor drivers and signal processing to control the endoscope robot accordingly.
- Endoscope Robot: a manipulator with its joint actuators and sensors that handles an endoscope. This robot has 2 rotational DoF (pitch and yaw) and 1 translational DoF (dolly) around a remote center of motion [7].

There are two communication channels:

- Wireless: transmits the angular velocity and acceleration vectors measured by the ring to the microcontroller.
- Wired: transmits the pedal state to the microcontroller.

The teleoperation scheme renders hand motions by integrating the angular velocity that the ring transmits. It calculates the angular position difference

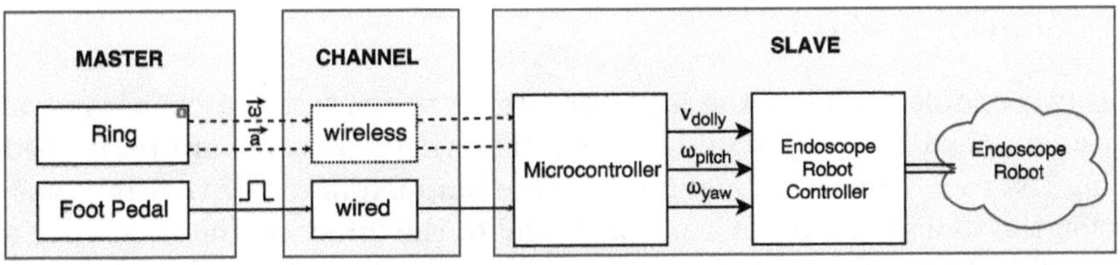

Fig. 1. Overview of the teleoperation system

since the foot pedal was pressed, and then maps these angles to the proportional velocity demands for the endoscope according to the current system state. This scheme is also known as the displacement to velocity teleoperation scheme as described in [2].

There are 3 system states: *OFF*, *ROTATION*, and *DOLLY*. In the *OFF* state the pedal is released, the teleoperation scheme is inactive, and the angle accumulator is reset. In the *ROTATION* and *DOLLY* states, the pedal is pressed, and the teleoperation scheme is active. The active system state is assigned according to the way the surgeon chooses to press the foot pedal which is described in Sect. 4.

A visualization example of the angle accumulation of the ring's pitch motion and its mapping to a velocity demand for the endoscope's pitch motion is given in Fig. 2.

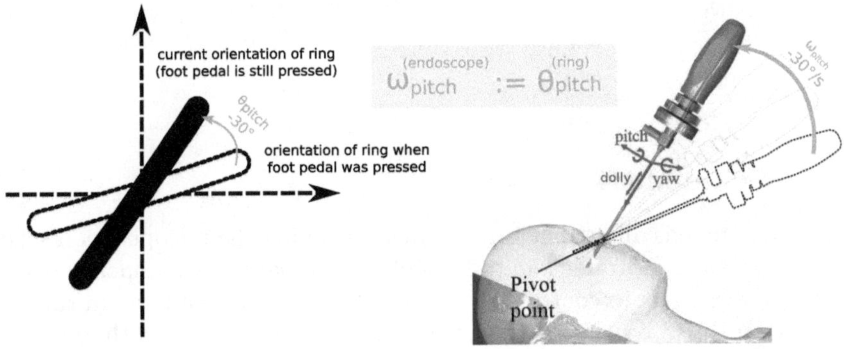

Fig. 2. Visual example of angle accumulation and its mapping to a velocity demand

3 Operation Modes

We propose a teleoperation scheme with 3 operation modes, two modes where the fingers are coupled to a tool *(Coupled Modes)*, and one mode where the index finger is uncoupled *(Uncoupled Mode)*.

3.1 Coupled Modes

In the two coupled modes, the surgeon is able to control the endoscope while maintaining their full grip around a tool. The ring's motions are performed by rotating the tool. The rotational motions are mapped in a way that it seems the direction the tool's tip motion is proportional to the direction the endoscope's tip motion. To allow for such motions, axis alignment is performed to compensate for the ring frame's misalignment with respect to the Earth's frame while the surgeon is holding the tool. Axis alignment will reorient the angular velocities measured by the ring so they are rendered in the Earth's frame.

The two variations of coupled mode differ in the mapping of the endoscope's dolly motion. The mapping of each mode's motions is described in Tables 1 and 2:

- coupled 3D mode: when the system is in the DOLLY state, the ring's roll motion is mapped to the endoscope's dolly motion.
- coupled 2D mode: when the system is in the DOLLY state, the ring's pitch motion is mapped to the endoscope's dolly motion, meaning we controlling a 3 DoF Robot using only 2 DoF motion.

Table 1. Mapping of axis aligned ring motions to endoscope velocity demands for coupled mode 3D

Mode	Aligned ring motion	Endoscope velocity demand
ROTATION	θ_{pitch}	ω_{pitch}
	θ_{roll}	ω_{yaw}
DOLLY	θ_{yaw}	v_{dolly}

Table 2. Mapping of axis aligned ring motions to endoscope velocity demands for coupled mode 2D

Mode	Aligned ring motion	Endoscope velocity demand
ROTATION	θ_{pitch}	ω_{pitch}
	θ_{roll}	ω_{yaw}
DOLLY	θ_{pitch}	v_{dolly}

3.2 Uncoupled Mode

In this mode, no axis alignment is performed, meaning the teleoperation scheme is invariant to the ring frame's initial orientation. The surgeon depends on their index finger's motions to control the endoscope. They are rendered in the ring's own frame. Intuitively, the surgeon would be holding a tool with their hand except for the index finger which would be free to move when the surgeon wishes to control the endoscope. Also due to the physical difficulty of performing roll motion while the surgeon's hand is gripping a tool, the pitch motion of the ring is mapped to the dolly motion of the endoscope when the system is in the DOLLY state. The mapping of the ring's motions to the endoscope's motions are described in Table 3.

Table 3. Mapping of ring motions to endoscope velocity demand for uncoupled mode

Mode	Ring motion	Endoscope velocity demand
ROTATION	θ_{pitch}	ω_{pitch}
	θ_{roll}	ω_{yaw}
DOLLY	θ_{pitch}	v_{dolly}

4 Teleoperation Scheme's Algorithm

The teleoperation scheme's algorithm is visualized as a linked chain of nodes in Fig. 3, each with its own functionality. In this system, there are two sources of input: the ring worn by the surgeon, and a foot pedal. The ring transmits 6 DoF information: 3 DoF angular velocity vector $\vec{\omega}$, and 3 DoF acceleration vector $\vec{a}$. The foot pedal works as a binary trigger with two states: *pressed*, and *released*. Axis alignment is only done in the coupled modes. In uncoupled mode, the acceleration vector $\vec{a}$ is not used, and the angular velocity $\vec{\omega}$ is fed directly to the angle accumulator. At the end of the chain, the output of the teleoperation scheme is passed to the endoscope robot controller.

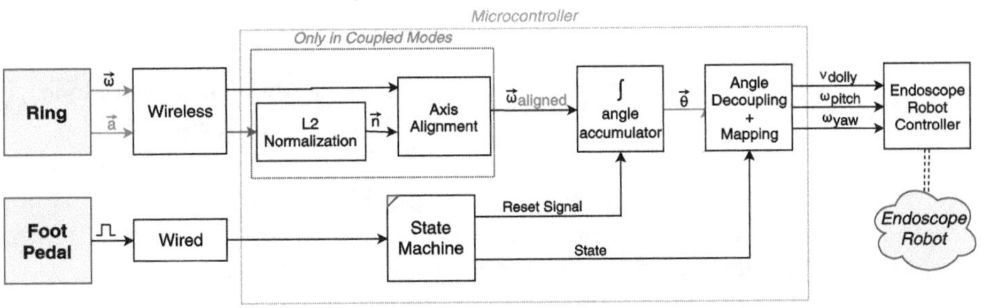

Fig. 3. An overview of the teleoperation scheme.

Axis Alignment. We compensate for the misalignment of the ring's vertical axis and the Earth frame's vertical axis (determined by the unit vector $\vec{n}$ of the ring's accelerometer data) by using a transformation matrix $\widehat{C}^{(e,r)}$ composed of two Euler rotations to transform the angular velocities measured in the ring frame $\mathscr{F}_r$ to their equivalent angular velocities in the Earth's frame $\mathscr{F}_e$.

Given the raw accelerometer acceleration vector $\vec{a}$ as input, we first calculate the gravity direction in the ring frame as a unit vector $\vec{n}$:

$$\vec{n} = \frac{\vec{a}}{\|\vec{a}\|} = n_x \vec{u_x}^{(r)} + n_y \vec{u_y}^{(r)} + n_z \vec{u_z}^{(r)} \tag{1}$$

This unit vector $\overrightarrow{n}$ is used in order to determine how much the ring is oriented with respect to the Earth's frame. Since the gravity vector is defined as $\overrightarrow{g} = g.\overrightarrow{u_z}^{(e)}$ in Earth's frame, we cannot determine the orientation about $\overrightarrow{u_z}^{(e)}$. Assuming the ring is oriented with an Euler Y-X sequence by α and β amount of rotations respectively, the transformation matrix $\widehat{C}^{(e,r)}$ from ring's frame to Earth's frame is evaluated as follows:

$$\widehat{C}^{(e,r)} = e^{\tilde{u}_y \alpha} e^{\tilde{u}_x \beta} \tag{2}$$

where, its α and β values can be determined by solving the following equation:

$$\overline{n}^{(r)} = \widehat{C}^{(r,e)} \overline{u}_z \tag{3}$$

Then we calculate the aligned angular velocity $\overrightarrow{\omega}_{aligned}$ which is the input to the accumulator node in the Coupled Modes, with the following formula:

$$\left\{\overrightarrow{\omega}_{aligned}\right\}\Big|_{\mathscr{F}_e} \triangleq \overline{\omega}^{(e)} = \hat{C}^{(e,r)} \cdot \overline{\omega}^{(r)} \tag{4}$$

Angle Accumulator. The angle accumulator converts the raw IMU gyroscope data (angular velocity $\overrightarrow{\omega}$) to the angle displacement which will be processed to be a velocity demand. It is active and accumulating when the teleoperation scheme is active, and is reset and stopped when the teleoperation scheme is deactivated. The angle accumulator is mathematically modeled in (5), with K_i being the scaling constants, and σ being the activation value and is set to 1 when the teleoperation scheme is active, and 0 otherwise.

$$\theta_i = sat\left[K_i \cdot \sigma \int \omega_i^{(j)}\, dt\right]; \quad i = pitch, yaw, roll; \quad j = \begin{cases} r & \text{in Uncoupled Mode} \\ e & \text{in Coupled Modes} \end{cases} \tag{5}$$

State Machine. The state machine is responsible for managing the system state and its transitions based on the foot pedal press states. The transition of the system's states are shown in Fig. 4 and described as follows:

- *OFF*: set when the pedal is released. The angle accumulator is reset and stopped, and the teleoperation scheme is inactive.
- *ROTATION*: set when the foot pedal is pressed and held down longer than a predefined time called the *double click tolerance* τ_{dolly}. Only motions of the ring mapped to the rotational motions of the endoscope are processed.
- *DOLLY*: set when the foot pedal is pressed 2 consecutive times, where the time elapsed between the two pedal presses is no longer than τ_{dolly}. Only the motion of the ring mapped to the dolly motion of the endoscope is processed.

Decoupling and Mapping. The final step of the scheme is converting and mapping the accumulated angle displacement to appropriate velocity demands for the endoscope robot controller according to the current system state. The mapping is fully dependent on the mode of operation and is described in their relevant Subsects. 3.1 and 3.2.

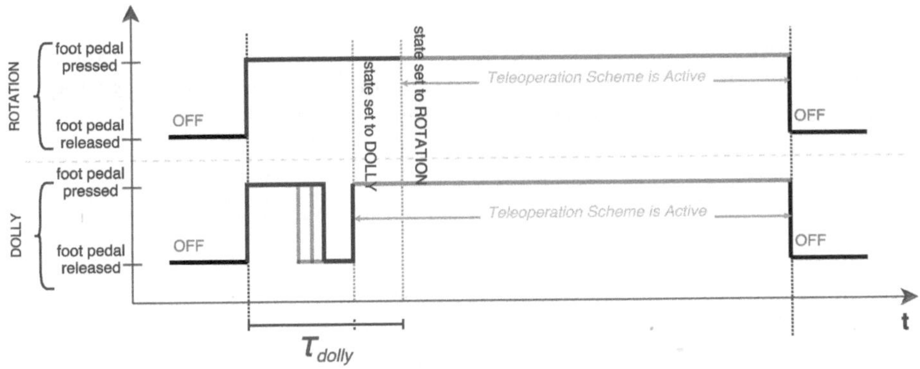

Fig. 4. Visualization of system state transitions.

5 Feasibility Study

To measure, assess, and compare the feasibility of the proposed operation modes, a training simulator is developed. The simulation environment is composed of:

- A simulation software running on a computer, that consists of a 2D virtual environment that will mock the endoscope controller, endoscope robot and its environment as shown in Fig. 5.
- In this environment, there are pre-specified circles at various coordinates with varying radii, and a special circle that the human subject controls, which we call the *"tool"*. In this environment, endoscope robot's pitch and yaw demands are processed as vertical and horizontal translations of the *tool* respectively, while the dolly demand is processed as the increase/decrease of *tool*'s radius,
- A 3D printed human skull model, with a soft rubber nose, and an aspirator tool that the test subjects hold with its tip positioned inside the nose,
- *Ring*, *Foot Pedal*, and *Microcontroller* work as previously described.

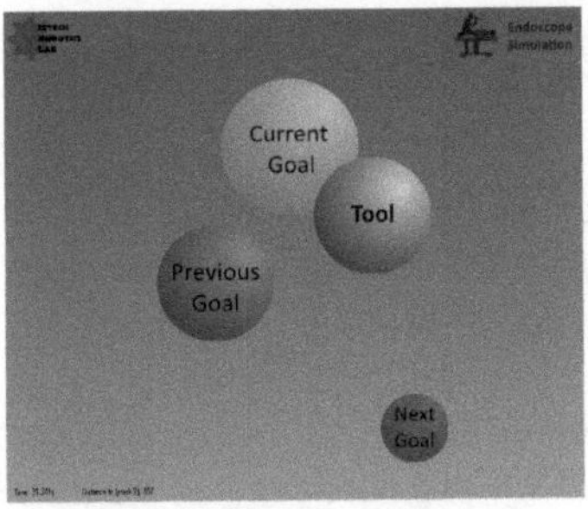

Fig. 5. Screenshot of the training simulator.

The test subjects' goal in each simulation trial is to manipulate the *tool* to overlap the pre-specified circles one at a time. The metrics used to measure the overall performance of each proposed operation are training time (the time it took to complete a whole run of the simulation), and the number of foot pedal presses per simulation trial (to measure the user's "effort").

For each different mode, we had 10 test subjects perform 10 trials with five unique layouts composed of the same amount of pre-defined circles with different radii that are color-coded to direct the user. The pre-defined circle layout is designed to have the same amount of traveling distance for each run. All users ran the same layouts in the same order allowing us to plot the average improvement in performance over time giving us the learning curve of our system in different modes (Fig. 6). We also recoded the number of pedal presses to assess the effort in completing each trial and its improvement after a training period (Fig. 7).

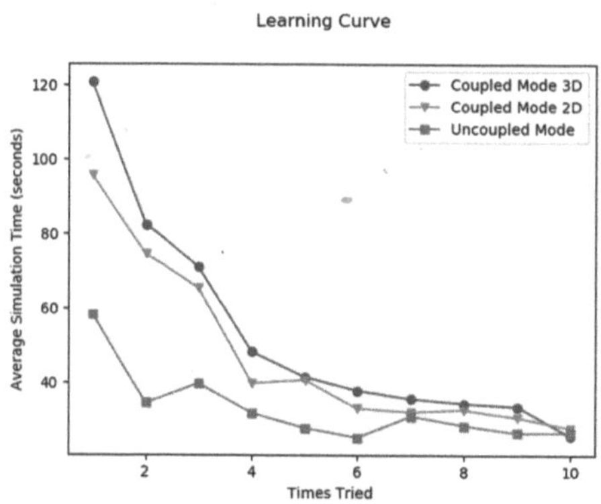

Fig. 6. Plot of average time versus num. of trials.

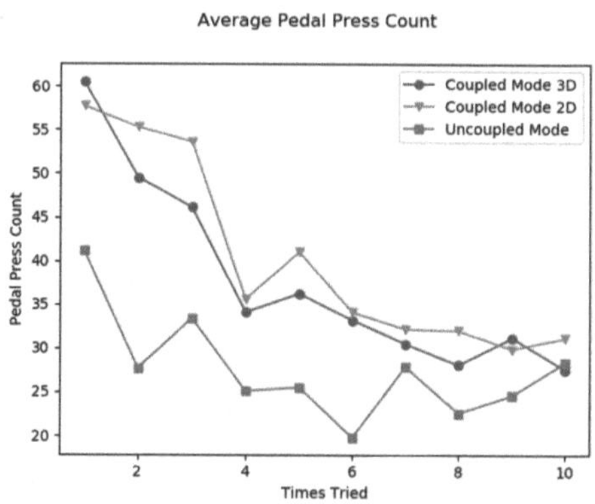

Fig. 7. Plot of average pedal presses per trial.

It is observed in Fig. 6 that in the first 5 trials, coupled mode 2D resulted in slightly shorter completion durations ($\sim$ 63 s average) over 3D ($\sim$ 73 s average) which are longer than the uncoupled mode's ($\sim$ 39 s average). Moreover, two coupled modes have almost the same longer learning curves compared to the uncoupled mode: their completion durations drop to 40 s limit in the 5th trial whereas the uncoupled mode results drop to 40 s limit in the 2nd trial. Similar to the decrease in completion durations for each consecutive trial, the average pedal press count shows a decreasing trend in all modes. In the 10th trials, all modes' completion duration converge to 25 $\sim$ 30 s and pedal press counts converge to 25 $\sim$ 30 counts.

6 Conclusion

In this study, we proposed a teleoperation scheme for an endoscope robot whose 3-DoF is fully actuated. Moreover, we proposed three different operation modes. We evaluated these modes in user studies to assess the usability and user effort. The coupled modes of operation are perceived as more difficult to adapt and requiring more user effort in comparison to the uncoupled mode of operation since they had longer learning curves. Eventually, all modes converge to similar performance levels after a learning process. In the future, an actual endoscope robot will be integrated with this teleoperation scheme.

Acknowledgement. This work is supported by The Scientific and Technological Research Council of Turkey via grant number 115E726.

References

1. Ballantyne, G.H., Moll, F.: The da Vinci telerobotic surgical system: the virtual operative field and telepresence surgery. Surg. Clin. N. Am. **83**(6), 1293–1304 (2003)
2. Dede, M.I.C., Maaroof, O.W., Ates, G., Berker, M., Isikay, I., Hanalioglu, S.: Unilateral teleoperation design for a robotic endoscopic pituitary surgery system. In: Husty, M., Hofbaur, M. (eds.) New Trends in Medical and Service Robots, pp. 101–115. Springer International Publishing, Cham (2018)
3. Moustris, G.P., Hiridis, S.C., Deliparaschos, K., Konstantinidis, K.: Evolution of autnomous and semi-autnomous robotic surgical systems: a review of the literature. Int. J. Med. Robot. + Comput. Assist. Surg. MRCAS **7**, 375–392 (2011)
4. Nathan, C.A.O., Chakradeo, V., Malhotra, K., D'Agostino, H., Patwardhan, R.: The voice-controlled robotic assist scope holder AESOP for the endoscopic approach to the sella. Skull Base **16**(3), 123–132 (2006)
5. Peters, B.S., Armijo, P.R., Krause, C., Choudhury, S.A., Oleynikov, D.: Review of emerging surgical robotic technology. Surg. Endosc. **32**(4), 1636–1655 (2018)
6. Voros, S., Haber, G.P., Menudet, J.F., Long, J.A., Cinquin, P.: ViKY robotic scope holder: Initial clinical experience and preliminary results using instrument tracking. IEEE/ASME Trans. Mechatron. **15**(6), 879–886 (2010)
7. Yaşır, A., Kiper, G.: Structural synthesis of 2R1T type mechanisms for minimally invasive surgery applications. In: Dede, M.I.C., İtik, M., Lovasz, E.-C., Kiper, G. (eds.) Mechanisms, Transmissions and Applications, pp. 31–38. Springer International Publishing, Cham (2018)

Automatic Device for Skin Biopsy

J. Meneses(✉), J. C. García-Prada, C. Castejón, and H. Rubio

Mechanical Engineering Department, Universidad Carlos III de Madrid, Avda. De la Universidad, nº30, 28911 Leganés, Spain
meneses@ing.uc3m.es

Abstract. In this work, first the "friction cam-follower" kinematic pair is presented with its corresponding equations of motion; then a biomedical device based on this kinematic pair is shown: the automatic device for skin biopsy. This article focuses on its kinematic and dynamic models, as well as the corresponding equations of motion. These equations have been implemented in a MatLab code and some results are shown. A more detailed description of this device can be found in the patent text [1].

1 Introduction

In skin biopsies currently performed [2], once local anaesthesia is practiced, the specialist manually provides a helical movement to a cylindrical blade, until it has penetrated the desired depth into the skin. The sample is then withdrawn, sometimes with the aid of a pair of clamps, and the subsequent suturing is performed. Thus, qualified personnel as well as an operating room are required, all of which makes the skin biopsy the bottleneck in the diagnosis of skin diseases. With the device proposed here, it would be possible to automatically perform the cut in few tenths of a second. This would avoid anaesthesia; an operating room would not be necessary, nor would the intervention of qualified personnel. As a consequence, operating time and resources would be significantly reduced. In addition, there are ongoing medical research [3–6] that require massive practices of skin biopsies that would greatly benefit from the use of this device.

2 The "Friction Cam-Follower" Kinematic Pair

In order to provide the blade with the required helical movement, it is coupled to the follower of the kinematic pair described in this section. Figure 1 shows the kinematics (left) and dynamics (right) schemes for the "friction cam-follower" kinematic pair (front and top views). For the shake of simplicity, an eccentric circular cam profile and a flat follower are considered. The angular position of the cam determines the follower longitudinal position:

$$z_2 = e(\cos\varphi_1 - 1); \dot{z}_2 = -e\dot{\varphi}_1 \sin\varphi_1; \ddot{z}_2 = -e\ddot{\varphi}_1 \sin\varphi_1 - e\dot{\varphi}_1^2 \cos\varphi_1 \qquad \text{(1a, b, c)}$$

G. Carbone et al. (Eds.): *New Trends in Medical and Service Robotics*, MMS 65, pp. 54–61, 2019.
https://doi.org/10.1007/978-3-030-00329-6_7

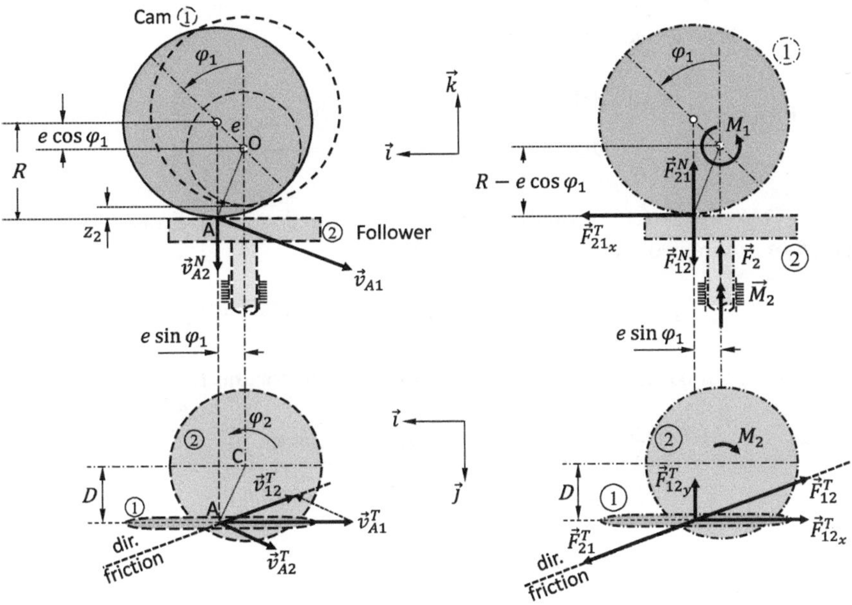

Fig. 1. Front and top views of the "friction cam (1) - follower (2)" kinematic pair. Left: kinematics, sliding velocity. Right: dynamics, forces and torques.

but not the follower angular position, since no pure rolling occur, but sliding. Actually, the sliding velocity results (see Fig. 1):

$$\vec{v}^T_{12} = \vec{v}^T_{A1} - \vec{v}^T_{A2} = [D\dot{\varphi}_2 - (R - e\cos\varphi_1)\dot{\varphi}_1]\vec{\imath} - e\dot{\varphi}_2\sin\varphi_1\vec{\jmath} \tag{2}$$

The friction force between the cam and the follower has the same direction as this sliding velocity and it is proportional to the normal force, namely:

$$\vec{F}^T_{12} = \mu\left|\vec{F}^N_{12}\right|\frac{\vec{v}^T_{12}}{\left|\vec{v}^T_{12}\right|}, \tag{3}$$

where μ is the cam-follower coefficient of friction. According to (2), the friction force in (3) is decomposed into the following x and y components:

$$F^T_{12x} = \frac{\mu\left|\vec{F}^N_{12}\right|[D\dot{\varphi}_2 - (R - e\cos\varphi_1)\dot{\varphi}_1]}{\sqrt{[D\dot{\varphi}_2 - (R - e\cos\varphi_1)\dot{\varphi}_1]^2 + e^2\dot{\varphi}_2^2\sin^2\varphi_1}} \tag{4}$$

$$\mathrm{F}^T_{12y} = \frac{\mu\left|\overrightarrow{\mathrm{F}}^N_{12}\right|e\dot{\varphi}_2\sin\varphi_1}{\sqrt{[\mathrm{D}\dot{\varphi}_2 - (\mathrm{R} - e\cos\varphi_1)\dot{\varphi}_1]^2 + e^2\dot{\varphi}_2^2\sin^2\varphi_1}} \tag{5}$$

From Fig. 1 (top-right) the rotational y-component equation of motion for the cam is (after using Eqs. 2 and 3):

$$M_1 - F_{21}^N e \sin\varphi_1 + \frac{\mu F_{21}^N (R - e\cos\varphi_1)[D\dot{\varphi}_2 - (R - e\cos\varphi_1)\dot{\varphi}_1]}{\sqrt{[D\dot{\varphi}_2 - (R - e\cos\varphi_1)\dot{\varphi}_1]^2 + e^2\dot{\varphi}_2^2 \sin^2\varphi_1}} - I_1\ddot{\varphi}_1 = 0 \quad (6)$$

Using (1c), the z-translational equation of motion for the follower is (weight is neglected or subtracted from F_2):

$$F_2 - F_{21}^N + m_2 e\ddot{\varphi}_1 \sin\varphi_1 + m_2 e\dot{\varphi}_1^2 \cos\varphi_1 = 0, \quad (7)$$

And the rotational z-component equation of motion for the follower results:

$$M_2 + \frac{\mu F_{21}^N [D(R - e\cos\varphi_1)\dot{\varphi}_1 - (D^2 + e^2\sin^2\varphi_1)\dot{\varphi}_2]}{\sqrt{[D\dot{\varphi}_2 - (R - e\cos\varphi_1)\dot{\varphi}_1]^2 + e^2\dot{\varphi}_2^2 \sin^2\varphi_1}} - I_2\ddot{\varphi}_2 = 0 \quad (8)$$

Where M_2 is the resistive external torque on the follower.

In summary, the equations of motion for the proposed pair, is a system of three second-order nonlinear differential Eqs. (6), (7) and (8). A more detailed explanation of the kinematics and dynamics of the friction cam-follower kinematic pair is given in [7].

3 The Automatic Device for Skin Biopsy

The designed mechanism, depicted in Fig. 2, has two friction cams rolling on one follower. Both are linked to the upper housing by the same revolute axis.

Each cam incorporates a pinion, and both rotate simultaneously in opposite directions by the action of a piece named "load pushbutton", provided with two rack-type gears. In turn, the "load pushbutton" slides along the housing, being actuated by the so-called "trigger spring". The contact between the follower and the cams is guaranteed by the action of the "contact spring".

The upper and lower housings are screwed together so that the depth of cut can be finely adjusted (screwing or unscrewing them) before punching. The process would be performed in two steps: load and punch. First, the load pushbutton is pressed halfway, where the triggers retain the mechanism. In that position, the cams have rotated half turn, and the follower is down (see Fig. 3a). Then the blade (that would be disposable) is attached to the end of the follower. The load step is completed by pressing the load pushbutton until its end-of-stroke, where the triggers retain the mechanism again. Now the trigger spring is compressed and the device is ready to perform the punch by activating the triggers (see Fig. 3b). After punching, the device reaches its "discharged" state; with the trigger spring at its minimum compression and the follower is up (see Fig. 3c).

The problem of ensuring the basal detachment of the skin sample, and its corresponding adhesion thereof to the blade, can be attempted with suitable blade designs.

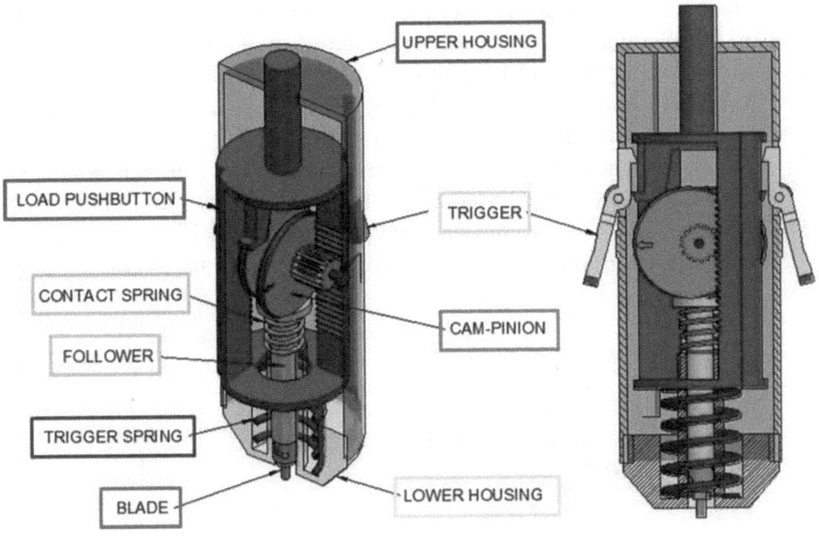

Fig. 2. Views of the automatic device for skin biopsy

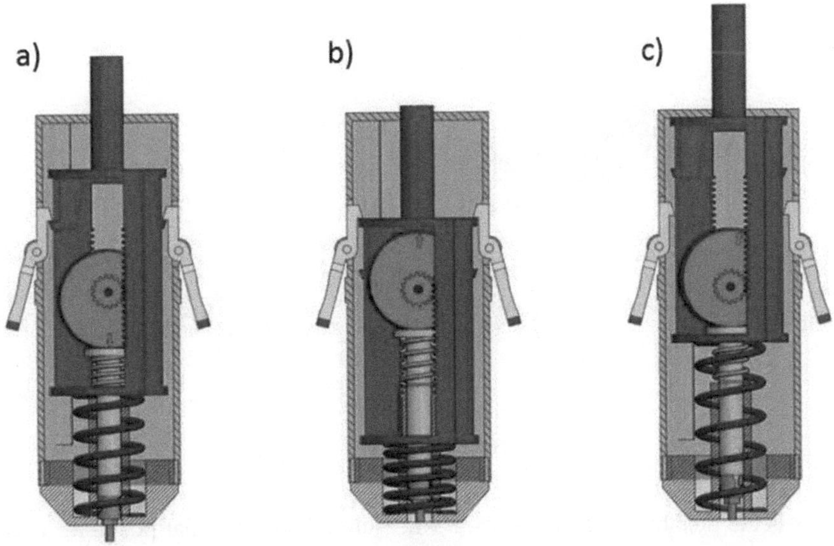

Fig. 3. Operation of the Automatic Device for Skin Biopsy: (a) position to place or remove the blade; (b) loaded position, ready to punch; (c) final position, after punch.

Skin-blade adhesion could be improved, for example, by means of a helical thread on the inner surface of the blade. Other shapes/roughness of the inner surface of the blade might be considered.

The expansion of the equations of motion presented in the previous section is carried out by introducing into the model, a second cam and the mentioned actuation sub-mechanism. Knowing the mechanical properties of the skin -which determine the torque and force necessary for cutting- this model will allow us to optimize: (i) the elastic constants of the trigger and contact springs, as well as their corresponding preloads; (ii) the significant dimensions of the cams and the follower *R* and *D*; and (iii) the value of the cam-follower coefficient of friction.

During the punching, the pushbutton slides along the housing (z-direction), actuated by the trigger spring, under the reaction forces by both cam-pinions, $-2F_{13}$ (see Fig. 4); so the equation of its motion is:

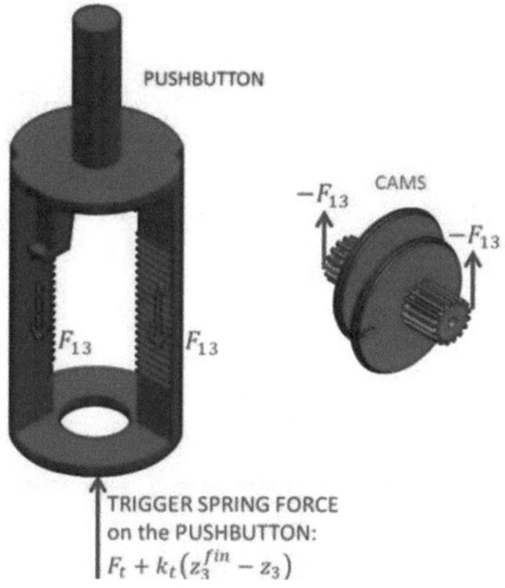

Fig. 4. Forces acting on the pushbutton and between the pushbutton and the cams

$$F_t + k_t\left(z_3^{fin} - z_3\right) - 2F_{13} = m_3\ddot{z}_3, \tag{9}$$

where z_3 is the pushbutton axial position (measured from its initial position, when the trigger spring is at its maximum compression); F_t is the trigger spring preload, that is, the load at its minimum compression, when the pushbutton reaches its end-of-stroke i.e.: $z_3 = z_3^{fin}$; k_t is the trigger spring constant, and m_3 is the mass of the pushbutton.

The gear-to-rack pair between the pushbutton and the cams determines the relation between the axial position of the former and the angular position of the latter, by means of the pitch radius of the pinions, r: $z_3 = r\varphi_1$; besides, the designed device is such that the pushbutton end-of-stroke is reached after one revolution of the cam, so: $z_3^{fin} = 2\pi r$; and Eq. (9) becomes:

$$F_{13} = \frac{F_t}{2} + \pi k_t r - \frac{k_t r}{2}\varphi_1 - \frac{m_3 r}{2}\ddot{\varphi}_1 \tag{10}$$

By multiplying F_{13} by r, the torque on each cam is obtained. Then Eq. (6) can be reused to describe the motion for each cam in the device as follows:

$$\begin{aligned} &\frac{F_t r}{2} + \pi k_t r^2 - \frac{k_t r^2}{2}\varphi_1 - F_{21}^N e \sin\varphi_1 + \frac{\mu F_{21}^N (R - e\cos\varphi_1)[D\dot{\varphi}_2 - (R - e\cos\varphi_1)\dot{\varphi}_1]}{\sqrt{[D\dot{\varphi}_2 - (R - e\cos\varphi_1)\dot{\varphi}_1]^2 + e^2\dot{\varphi}_2^2 \sin^2\varphi_1}} \\ &- \left(\frac{m_3 r^2}{2} + I_1\right)\ddot{\varphi}_1 \\ &= 0 \end{aligned} \tag{11}$$

As regards to the equations of motion for the follower in the device, Eqs. (6) to (8) must be modified taking into account that now (i) both cams exert normal and friction forces on it, so they must appear in duplicate; and (ii) the follower-cams contact is guaranteed by the spring contact (having elastic constant k and preloaded by F_0). Rearranging terms, the equations of motion for the cams-follower in the device result:

$$F_0 + ke(1 - \cos\varphi_1) - 2F_{21}^N + m_2 e\ddot{\varphi}_1 \sin\varphi_1 + m_2 e\dot{\varphi}_1^2 \cos\varphi_1 = 0, \tag{12}$$

$$M_2 - 2\mu F_{21}^N \frac{(D^2 + e^2\sin^2\varphi_1)\dot{\varphi}_2 - D(R - e\cos\varphi_1)\dot{\varphi}_1}{\sqrt{[D\dot{\varphi}_2 - (R - e\cos\varphi_1)\dot{\varphi}_2]^2 + e^2\dot{\varphi}_2^2\sin^2\varphi_1}} - I_2\ddot{\varphi}_2 = 0, \tag{13}$$

Eliminating F_{21}^N from Eqs. (11) to (13) we obtain:

$$\begin{cases} \ddot{\varphi}_1 = \dfrac{F_t r + 2\pi k_t r^2 - k_t r^2\varphi_1 + [F_0 + ke(1 - \cos\varphi_1) + m_2 e\dot{\varphi}_1^2\cos\varphi_1]\left\{\frac{\mu[D\dot{\varphi}_2 - (R - e\cos\varphi_1)\dot{\varphi}_1](R - e\cos\varphi_1)}{\sqrt{[D\dot{\varphi}_2 - (R - e\cos\varphi_1)\dot{\varphi}_2]^2 + e^2\dot{\varphi}_2^2\sin^2\varphi_1}} - e\sin\varphi_1\right\}}{m_3 + 2I_1 - \frac{\mu[D\dot{\varphi}_2 - (R - e\cos\varphi_1)\dot{\varphi}_1](R - e\cos\varphi_1)m_2 e\sin\varphi_1}{\sqrt{[D\dot{\varphi}_2 - (R - e\cos\varphi_1)\dot{\varphi}_2]^2 + e^2\dot{\varphi}_2^2\sin^2\varphi_1}} + m_2 e^2\sin^2\varphi_1} \\ \ddot{\varphi}_2 = \dfrac{M_2}{I_2} - \dfrac{\mu[F_0 + ke(1 - \cos\varphi_1) + m_2 e\ddot{\varphi}_1\sin\varphi_1 + m_2 e\dot{\varphi}_1^2\cos\varphi_1][(D^2 + e^2\sin^2\varphi_1)\dot{\varphi}_2 - D(R - e\cos\varphi_1)\dot{\varphi}_1]}{I_2\sqrt{[D\dot{\varphi}_2 - (R - e\cos\varphi_1)\dot{\varphi}_2]^2 + e^2\dot{\varphi}_2^2\sin^2\varphi_1}} \end{cases} \tag{14}$$

This system of 2nd-order differential equations in φ_1 and φ_2, describing the device dynamics, can be reduced to an equivalent system of four 1st-order differential equations that can be numerically solved using "ode45" from MATLAB ®, for a given set of parameters.

4 Numerical Results

As an example of application, in Fig. 6 some results are shown for the device parameters (distances in mm, forces in N): $e = 2, R = 15, r = 5, D = 10, F_C = F_T = 1, k_C = k_T = 0.1, \mu = 0.7$; and for different values of the external torque applied on the follower, namely the cutting torque (Fig. 5).

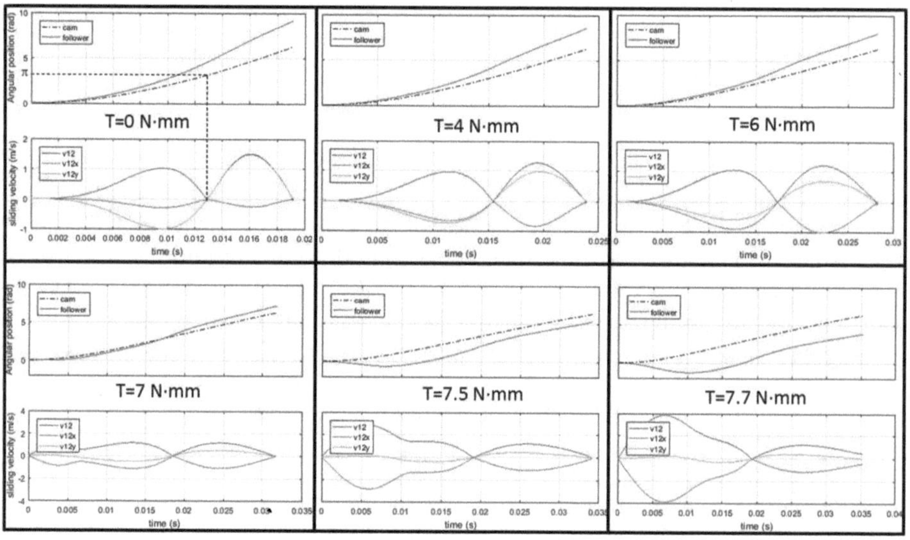

Fig. 5. Cam and follower angular position (graphics on the top) and sliding velocity (graphics on the bottom) for different cutting torque values

As can be seen, as the value of the cutting torque increases, the rotation of the follower decreases, even rotating in the opposite direction (this is an unrealistic behaviour, since the cutting torque is friction type, therefore passive). In fact, for each set of parameters, there is a cutting torque threshold value, above which the apparatus does not perform the cut. In the example, this value was 8 N·mm.

The sliding velocity increases in average as the cutting torque increases, but in all cases it vanishes at highest and lowest positions of the follower.

5 Conclusions and Future Work

A novel device to automatically perform skin biopsies, based on the "friction cam-follower" kinematic pair, is presented in this article. A prototype of the device has been obtained by additive manufacturing to demonstrate its mechanical operation (see Fig. 6), but its parts and springs have not yet been dimensioned to perform skin biopsies in humans.

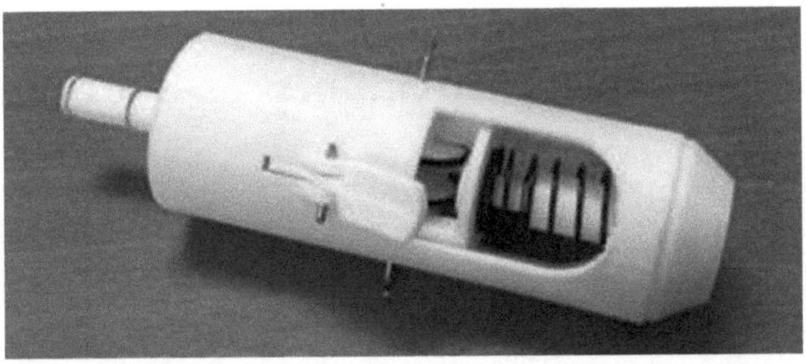

Fig. 6. A prototype of the automatic device for skin biopsy obtained by additive manufacturing

A numerical model of the device has been proposed and a code has been developed to simulate its behavior according to the geometric and dynamic parameters that characterize it. This tool will be of great help in the final design of the device, once the mechanical properties of the skin and the corresponding cutting process are studied in the future. In fact, the results shown correspond to constant values of the external torque on the follower. But an adequate modeling of the cutting process will allow us to calculate more realistic values of the blade velocity and the reaction force between the skin and the device during the punch.

Acknowledgment. The authors acknowledge the Spanish Found through the project "MAQ-STATUS", DPI 2015-69325-C2-1-R, for financial support.

References

1. Grillo, E., Vañó, S., Jaén, P., Castejón, C., Meneses, J., García-Prada, J.C., Rubio, H.: Dispositivo Automático para Biopsias Cutáneas. Oficina Española de Patentes y Marcas, No. ES2537831 B1 (12 June 2015). International Class.: A61B10/02, (2006.01) (2015)
2. Zuber, T.J.: Punch biopsy of the skin. Am. Fam. Phys. **65**(6), 1155–1158 (2002)
3. Korfitis, C., Gregoriou, S., Antoniou, C., Katsambas, A.D., Rigopoulos, D.: Skin biopsy in the context of dermatological diagnosis: a retrospective cohort study. Dermatol. Res. Pract. **2014**, 734906 (2014). https://doi.org/10.1155/2014/734906
4. Gibbons, C.H., Griffin, J.W., Polydefkis, M., Bonyhay, I., Brown, A., Hauer, P.E., McArthur, J.C.: The utility of skin biopsy for prediction of progression in suspected small fiber neuropathy. Neurology **66**(2), 256–258 (2006). https://doi.org/10.1212/01.wnl.0000194314.86486.a2
5. Gu, Z., Li, Y., Li, H.: Use of condensed nanofat combined with fat grafts to treat atrophic scars. Jama Facial Plast. Surg. **20**(2), 128–135 (2018)
6. Cheng, A.M.Y., Cheng, H.S., Smith, B.J., Stewart, D.A.: Neutrophilic dermatosis of the hands: a review of 17 cases. J. Hand Surg. **43**(2), 185-e1 (2018). American Volume
7. Meneses, J., Corral, E., Castejón, C., Rubio, H., García-Prada. J.C.: The rotary into helical transmission by the "Friction Cam-Helical Follower" kinematic pair. In: Sent to the European Conference of Mechanism Science Eucomes 2018, Aagen, Germany, 4–6 July 2018

Design of a Methodology for the Determination of the Mechanical Rib Stiffness as Injury Index

L. A. Aguilar[1(✉)], Ch. R. Torres-San-Miguel[2], M. Ceccarelli[3], I. Sánchez-Cruz[2], J. A. Leal-Naranjo[2], and C. G. Urriolagoitia[2]

[1] International University of Catalunya, Campus Barcelona, Immaculada, 22, Terré, 11-19, 08017 Barcelona, Spain
laguilarp@uic.es

[2] Polytechnic National Institute, Av. Luis Enrique Erro S/N, Delegación Gustavo A. Madero, 07738 Ciudad de México, Mexico
israelsanchez0410@gmail.com

[3] University of Cassino and South Latium, Via Gaetano di Biasio, 43, 03043 Cassino, FR, Italy
ceccarelli@unicas.it

Abstract. The analysis of the thorax's bones has become a critical issue in the evaluation of injuries on the human being. For this reason, this paper shows a methodology to evaluate the stiffness of the ribs. To compute the stiffness of the ribs, this parameter was related to the strain energy caused by a payload applied over their free-end. Then, the strain energy is computed through energetical models. In this way, it was developed a custom software to evaluate important biological conditions on the human being like the shape of the rib, the size of the cross-section, or the density of the bone. Finally, it was found that the strain energy can be displayed through 12 parameters of significance. The main parameter that modifies greatly the critical zone of breakability on the rib is the angle of The payload applied.

Keywords: Biomechanics · Numerical simulation · Human ribs

1 Introduction

The aim of this study is to formulate an algorithm to estimate the rate of damage caused on the human thorax. The damage in the human thorax is measured through the rate of injury on individual parts [1]. For example, a deathly damage in the thorax can be related either by multiple fractures on the bones (more than five ribs) or by a bilateral fail of the chest (caused by a pneumothorax, hemothorax or even other internal injuries). In earlier works, like Reference [2] the damage of the ribcage was commonly computed through the analysis of experimental tests made with dummies. In general, all the literature works to figure out how to determine the biomechanics of the injury mentioned in medical reports and reproduce it through artificial methods.

The trending on the research of the injuries on the thorax has focused only on the study of the biomechanics of the human body as a related response of the mechanical properties of the biological components. That assumption is made due sometimes the

G. Carbone et al. (Eds.): *New Trends in Medical and Service Robotics*, MMS 65, pp. 62–69, 2019.
https://doi.org/10.1007/978-3-030-00329-6_8

opposition of the entire element to be broken (the injury behavior) is close related to the behavior of an element made off with similar mechanical properties (pattern of stress-strain). But the real behavior of the strength of a material is the intrinsic capacity related with a well-defined strain rate pattern. In other words, the strength of a material is the stress-strain relation that defines a material like to be brittle or resilient. For this reason, a more accurate parameter that defines the injury behavior on biological tissues is the mechanical stiffness.

The mechanical property of stiffness of one element is defined as the opposition of the entire element to be bend it or in the worst case to be broken. In order to study that behavior, it is grouped the modifiers of the stiffness in conditions like how is it held and loaded the element, the shape and size of the element and how it is made off the element itself [3]. On the work presented by Reference [4] mentioned a fine observation that weight this hypothesis. On her work, she mentions that during the running of the impact tests, the axis of the rib was manually aligned in order to prevent a torsional effect caused by a natural twist over the rib. Thus, this paper proposes to analyze the modifiers of the rib stiffness as a parameter of relevance in the study of the thorax injury behavior. In order to work out a propose for an analytical solution to this problem formulation, an idealized model of the rib is first presented. Then, is presented a methodology to evaluate the physiological parameters described under different anthropometrical conditions in order to evaluate their biological influence on the physical response of the model.

2 The Proposed Model

The morphology of the rib was reconstructed by a mathematical model widely described in [5]. The sagittal view of the rib is shown in the simplified model shown in Fig. 1. The payload F applied on the thorax is a vector whose direction is described by the angles κ_1 for azimuthal direction and κ_2 for the elevation direction. The rectangular components of this vector are named as F_N, F_Q and F_{op}. The shape of the rib is geometrically described as a second-grade curvature. The widest side of the rib is named as B, and the largest side of the rib is named as A.

The internal components of the load are N_i for the axial component, Q_i for the shear component and M_i for the bending moment on the i-th point. These components are tangential to that point given the φ angle. The radius R of the second-grade curvature for the i-th point is given by the θ_i angle. The angle of the vertebra-rib connection is defined by φ_v. The angle β represents the named by medical literature as rib's edge curvature [6]. Finally, the i-th point over the rib is attached to the frame A2, the center of the rib is attached to the frame A1, and the general frame is A0.

The second-grade curve that represent the curvature of the rib is defined by Eq. 1

$$Ax^2 + 2Bxy + Cy^2 + 2Dx + 2Fy + G = 0 \tag{1}$$

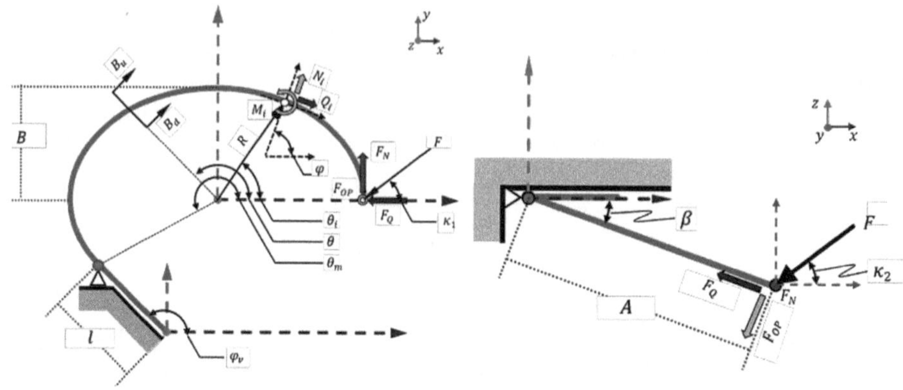

Fig. 1. Shape and parameters of the model used to describe a rib. (a) View on horizontal plane, (b) View in sagittal plane.

where A, B, C, D, F, G are constants computed by Eqs. 2 and 3

$$E = \begin{bmatrix} B & C & D & F & G \end{bmatrix} = \frac{M}{-x^2} \tag{2}$$

$$M = \begin{bmatrix} 2xy & y^2 & 2x & 2y & 1 \end{bmatrix} \tag{3}$$

On this way x_0 and y_0 are computed by Eqs. 4 and 5

$$x_0 = \frac{CD - BF}{B^2 - AC} \tag{4}$$

$$y_0 = \frac{AF - BD}{B^2 - AC} \tag{5}$$

The radius of the second-grade curve is defined by:

$$r = \frac{b}{\sqrt{1 - \left(1 - \left(\frac{b}{a}\right)^2\right)\cos\left(\theta^2\right)}} \tag{6}$$

where

$$a = \sqrt{\frac{2(AF^2 + CD^2 + GB^2 - 2BDF - ACG)}{(B^2 - AC)\left((C - A)\left(\sqrt{1 + \frac{(4B^2)}{(A-C)^2}}\right) - (C + A)\right)}}$$

$$b = \sqrt{\frac{2(AF^2 + CD^2 + GB^2 - 2BDF - ACG)}{(B^2 - AC)\left((A - C)\left(\sqrt{1 + \frac{(4B^2)}{(A-C)^2}}\right) - (C + A)\right)}}$$

By defining the eccentricity of the ellipse as e, that value can be computed by

$$e = \sqrt{1 - \left(\frac{b}{a}\right)^2} \tag{7}$$

The payload F applied on the thorax is a vector defined by Eq. 8

$$F = \begin{bmatrix} F\sin(\kappa_1)\cos(\kappa_2) \\ F\sin(\kappa_1)\sin(\kappa_2) \\ F\cos(\kappa_1) \end{bmatrix} = \begin{bmatrix} F_Q \\ F_N \\ F_{Op} \end{bmatrix} \tag{8}$$

On the other side, the cross-section plane is defined by the $B_u - B_d$ points for the angle θ. The cross section along the rib can be considered constant with the shape shown in Fig. 2. The variable a_1 represents the major axis of the internal ellipse and the a represents the major axis of the external ellipse. The variable b_1 represents the minor axis of the internal ellipse and the b represents the minor axis of the external ellipse. The parameters of the cross-section are named as A_{rea} for the Area of the cross-section and the value that relates that area to the radius of bending in curved beams is named as A_m. Equations 9 and 10 computes these values respectively

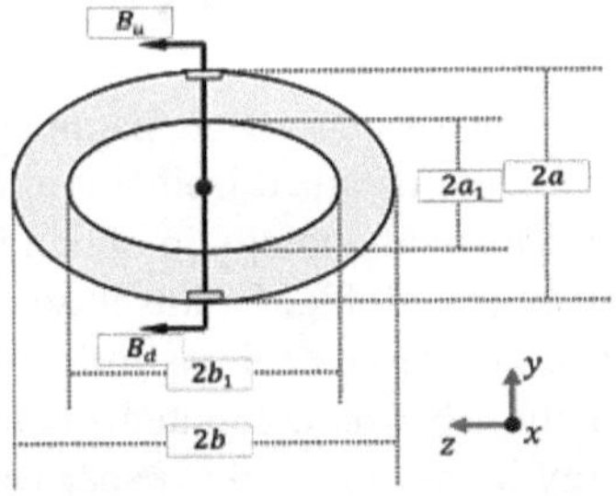

Fig. 2. Shape and parameters of the model used to describe the cross-section of the rib

$$A_{rea} = \pi(ab - a_1 b_1) \tag{9}$$

$$A_m = 2\pi\left(\frac{a}{b}R - \frac{a_1}{b_1}R - \frac{a}{b}\sqrt{R^2 - b^2} + \frac{a_1}{b_1}\sqrt{R^2 - b_1^2}\right) \tag{10}$$

The moment of inertia along the z axis is computed by

$$I_x = \frac{\pi}{4}\left(a^3 b - a_1^3 b_1\right) \tag{11}$$

On the other side, the axial A_N shear S_Q components of the loads look through the horizontal plane of the rib can be determined as function of the $\varphi(\theta)$ by Eqs. 12 and 13

$$A_N = F_Q \sin(\varphi(\theta)) + F_N \cos(\varphi(\theta)) \quad (12)$$

$$S_Q = F_Q \cos(\varphi(\theta)) - F_N \sin(\varphi(\theta)) \quad (13)$$

where the geometry model gives

$$\varphi(\theta) = \operatorname{atan}\left(-m(\theta)^{-1}\right) \quad (14)$$

$$m(\theta) = \frac{R * \left(1 - \left(1 - \left(\frac{b}{a}\right)^2\right)\cos\left(\theta^2\right)\right)^{\frac{3}{2}} - e^2 B \sin^2\theta}{-R * \left(1 - \left(1 - \left(\frac{b}{a}\right)^2\right)\cos\left(\theta^2\right)\right)^{\frac{3}{2}} - e^2 B \cos^2\theta} \quad (15)$$

Finally, the bending moment M_x can be described by

$$M_x = M_{f1} + F_Q R \sin(\theta) - F_N R(1 - \cos(\theta)) \quad (16)$$

$$M_{f1} = F_{Op} \sin\theta \quad (17)$$

3 Parameters of Significance on the Evaluation the Rib

In order to solve Eq. 1, the numerical values of the parameters A to G were computed. The numerical values to solve it were obtained from a tomography of one older people with no diseases on the thorax. The methodology to obtain those points were discussed on [5]. Using the methodology proposed, Fig. 3 plots in solid line the elliptical shape of seventh rib. Those values were computed by solving the proposed Eqs. 1 to 6. By using the Eq. 7, the eccentricity of the rib was computed obtaining a value of 0.67. The computation of the strain energy in the rib, on this paper was evaluated under 5 values of eccentricity ranged from 0 to 1. When the eccentricity value is 0, it means that the second-grade curve represents a circle, other values ranged from 0.1 to 1, means that the second-grade curve represents an ellipse. The results of this computation are plotted in Fig. 3.

Another parameter analyzed is the cross section of the rib. This parameter is described as a ratio of the cortical versus cancellous bone-tissue. This value is greatly modified by the interpretation of the data used to reconstruct a geometrical model. The distinction between both materials it is based on the porosity of the tissue. In order to evaluate the importance of this parameter, Fig. 4 shows the measured values for the idealization of the cross section presented in Fig. 2.

Figure 4a represents the measured values for the "a" variable of cross section of the rib. Figure 4b represents the measured values of the "b" variable of the cross section of the rib. In both figures the dots represents the values measured on the i-th point at θ_i degrees on clockwise sense. The solid line represents an interpolation of those values done to obtain a constant curve with no great variation of the values. The methodology proposed to evaluate each condition is summarized on the flowchart shown in Fig. 5.

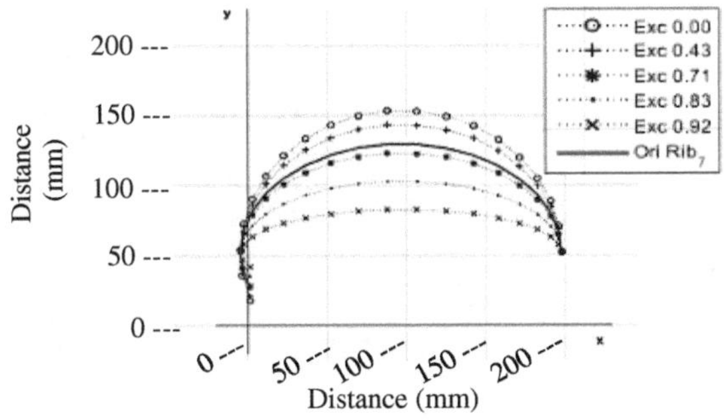

Fig. 3. Graphic representation of the eccentricity variation value on a rib

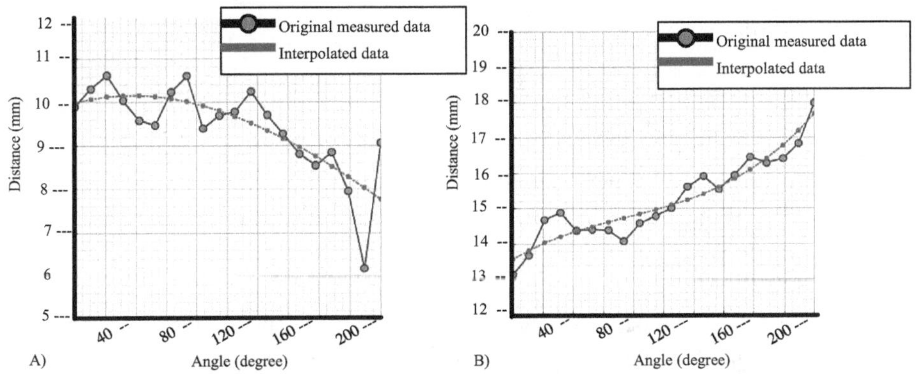

Fig. 4. Parameters of the cross section presented in Fig. 2. (A) Values for "a" value. (B) Values for "b" value

The flowchart is subdivided on 4 individual tasks in order to compute the numerical values of Eqs. 1 to 17.

The flow-chart shown in Fig. 5 presents the core of own developed software to compute the strain energy on the rib. It initiates with the introduction of the NDATOS variable. This variable defines the quantity of the samples used in Eqs. 1 and 2 to solve the second-grade curve of the rib. After that, the Eqs. 4 to 6 are computed by using a for cycle that represents the number of eccentricities that will be computed. Once is computed the shape of the ellipse and the main parameters of the second-grade curve, the next step is to introduce the numerical values of the payload applied over the rib. This value is restricted under the condition of $0 < LOAD < 400N$. After that, the angle of azimuthal and elevation of the payload vector is defined. Thus, the numerical value of the reactions is computed for all along the rib. That condition is evaluated for the 6 values of eccentricity described previously. Finally, the geometrical properties of the rib and the strain energy of the rib are computed in order to evaluate the 12 conditions stablished on this paper.

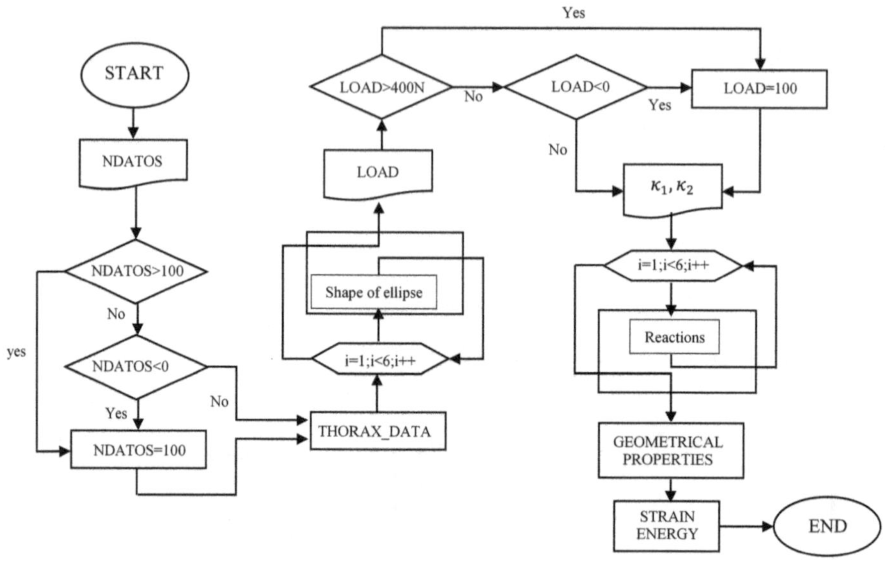

Fig. 5. The proposed procedure to evaluate the 12 conditions of the behavior of the rib

4 Numerical Results

When it is evaluated the payload angle condition or PAC, the behavior of the axial, shear and momentum components on the rib has shown no great variation on the numerical values. The zones of interest are reached at the middle of the rib around the 80 to 100° and near to the vertebra joint. Those values are dependent of the angle of the payload applied. The payload applied was restricted to only being applied on the end-tip of the rib, but modifying the angle κ_1 and κ_2 between 0 to 180°. The material properties considered to evaluate the strain energy rate on the rib were 3. It was considered the mechanical properties of the steel, and the square cross section of the rib shape as is shown in the manufacture of the Hybrid IIII dummy. It was considered either the mechanical properties of the cortical bone and the PLA mechanical properties. On the other side, the computed values of the geometrical properties of the cross-section of the rib shown a slight variation of the computed values between hollow elliptical of 30% hollow elliptical 50% compared with the solid elliptical shape. The computed results for the ribcage stiffness is increased by 13% if the cross-section is interpreted as solid ellipsoidal, but this value is increased to 92.85% if the cross section is interpreted as solid rectangular. The computed results points that PLA plastic has a computed displacement similar to cortical bone, but their yielding strength $\sigma_y = 80$ MPa is almost the half of the cortical bone $\sigma_y = 180$ MPA. The steel material has no a similar computed displacement to cortical bone, but their yielding strength $\sigma_y = 185$ MPa is quite similar to cortical bone. For this reason, the yielding strength is not a parameter of design to define the construction of artificial dummies because the biomechanical response is not the same.

5 Conclusions

The numerical results obtained with the proposed formulation gives a clear way to understand how the anthropometric values modify the ribcage stiffness. The response of this parameter was characterized as a function of 12 conditions. It was found that the most important condition that modifies the entire behavior is the cross-transversal section and the election of the material used in the construction of artificial dummies. Another important parameter that was evaluated was the eccentricity of the rib used on the simulation of numerical dummies. This parameter allows to define two regions of interest, one of them always will be at the middle of the rib. The other will change depending on the angle κ_1 and κ_2 of the payload application. Finally, the strain energy computed shows that the region that dissipate more energy is near to 200° on the rib. This area is near to the vertebras.

Acknowledgments. The firsts authors wish to gratefully acknowledge Consejo Nacional de Ciencia y Tecnología (CONACyT) and Instituto Politécnico Nacional through scholarship "Beca mixta para estancias de investigación en el extranjero" for permitting his period of study at LARM of Cassino University in the A.Y. 2016–17 within a double Ph.D. degree program.

References

1. Alén, C.C., et al.: Experimental tests in human-robot collision evaluation and characterization of a new safety index for robot operation. Mech. Mach. Theor. **80**, 184–199 (2014)
2. Lobdell, T.E., et al.: Impact response of the human thorax. In: Proceedings of human impact response, measurement and simulation, General Motors Research Laboratories, Warren, Michigan, pp 201–246 (1972)
3. Howell, L.L., Magleby, S.P., Olsen, B.M.: Handbook of Compliant Mechanisms. Wiley (2013). ISBN 978-1-119-95345-6
4. Agnew, A.M., et al.: Age-related changes in stiffness in human ribs. In: Proceedings of IRCOBI Conference, Gothenburg, pp 257–269 (2013)
5. Aguilar, P.L.A., et al.: Análisis de las propiedades geométricas de las costillas mediante tomografía axial computarizada. In: Proceedings of VII Congreso Latinoamericano de ingeniería mecánica, Cuenca, p. 6 (2014)
6. Gray, H.: Anatomy of the Human Body, Bartleby, Philadelphia (1918). ISBN 978-0-443-06952-9

Enabling Minimal Invasive Palpation in Flexible Robotic Endoscopes

Ivan Sušić[1(✉)], Azhar Zam[2(✉)], Philippe C. Cattin[3(✉)], and Georg Rauter[1(✉)]

[1] Bio-Inspired RObots for MEDicine-Lab (BIROMED-Lab), Department of Biomedical Engineering, University of Basel, Geverbestrasse 14 (Room 12.03.008), 4123 Allschwil, Switzerland
{ivan.susic, georg.rauter}@unibas.ch

[2] Biomedical Laser and Optics Group (BLOG), Department of Biomedical Engineering, University of Basel, Geverbestrasse 14 (Room 12.03.005), 4123 Allschwil, Switzerland
azhar.zam@unibas.ch

[3] Center for Medical Image Analysis & Navigation (CIAN), Department of Biomedical Engineering, University of Basel, Geverbestrasse 14, 4123 Allschwil, Switzerland
philippe.cattin@unibas.ch

Abstract. In open surgery, palpation allows surgeons to haptically examine the patient's tissue with their fingers. Hereby, palpation relates the pose of the surgeon's finger inside the wound and the finger's indentation into the tissue to the experienced forces in all directions. In this way, the surgeon is able to experience differences in tissue stiffness. Despite being a well-recognized standard procedure in open surgery, palpation has so far been unavailable for Minimally Invasive Surgery (MIS). Thus, surgeons can only rely on endoscope vision. To overcome this drawback, we aim to restore haptic feedback and thus palpation in MIS.

In this paper, we present an Encapsulated Force-sensing Device (EFD) that can be mounted on the tip of a flexible articulated robotic endoscope. The EFD houses a miniature tri-axial force sensor (TFS) together with an integrated camera mounted onto the EFD's tip. In order to enable palpation, the forces at the endoscope's tip have to be acquired together with the exact shape of the flexible articulated endoscope. Importantly, articulated endoscopes enable us to apply conventional kinematics and determine the endoscope's shape, which is not possible in endoscopes based on continuum mechanics. In a test set-up, we investigated the accuracy of the force measurement provided by our EFD. For this investigation, the robotic structure was fixed in one predefined pose and the forces from our EFD were compared to forces measured by a commercial force/torque sensor. With this test set-up, we could verify the concept of palpation (force and position measurement), even for MIS.

G. Carbone et al. (Eds.): *New Trends in Medical and Service Robotics*, MMS 65, pp. 70–77, 2019.
https://doi.org/10.1007/978-3-030-00329-6_9

1 Introduction

Endoscopes are used both as diagnostic and surgical tools to visually inspect, treat, or resect tissue inside the human body. The most common applications of endoscopes are conventional colonoscopy [1], biopsy [2], natural translumenal endoscopic surgery (NOTES) [3] and Minimally Invasive Surgery (MIS) [4]. MIS is a surgical technique, which enables surgeons to gain access to surgical sites that are difficult to reach through $1 \div 2$ cm small incisions in the skin. In MIS, surgeons have to manually manipulate the endoscope from the outside to guide it to the desired point inside the body. However, the manipulation of endoscopes inside the human body is a difficult task: the surgeon has to work within a constrained area, has to rely only on constrained visual feedback from the endoscope camera, has to cope with motion inversion effects, and the resulting loss of dexterity. Therefore, manually performed MIS requires intensive training [5]. Furthermore, the possibility to palpate the tissue is partially (stiff laparoscopes) or completely lost (flexible endoscopes based on continuum mechanics). Only by combining precise force measurement and endoscope shape measurement, allows determining the stiffness of the touched tissue, i.e. palpation. However, until today, endoscopic palpation is stuck at the level of force measurement with small force sensors on 3D stages under lab conditions [6].

Enabling endoscopes not only to provide visualization of the currently visible inside of a patient, but also to detect structures below the tissue surface would be of high interest. Particularly in tumor resection, finding the border of the tumor below the surface is of high importance [7, 8]. Thus, adding a force sensor at the tip of an endoscope seems to be a viable solution for providing force information in visual [6] or haptic ways [8]. Although the force information of a sensor at the endoscope tip can complement the visual information, the effective force direction and tissue stiffness information (including surface rendering scheme [8]) can only be determined if the pose and movement of the endoscope are known [9]. To determine the end-effector pose (i.e. position and orientation of the tip with the integrated force sensor) and movement, the kinematics of the endoscope needs to be known. This is particularly necessary for flexible endoscopes that can attain complex shapes. Determining the endoscope movement and shape at high precision is challenging, but necessary for endoscopic palpation. Also, to establish a relation between the forces at the shaft and the tip of the endoscope and vice versa (kinematic transformation), knowledge of the endoscope shape and kinematics is necessary. Even if robotic and flexible endoscopes become increasingly common, until today, the effect of haptic information on the performance of surgeons has been studied only on stiff, straight endoscopes [8, 10]. For technical reasons (combination of small force sensors and precise pose estimation), palpation measurements on flexible endoscopes have never been performed so far to our knowledge.

To overcome current limitations in MIS, the current paper aims at enabling palpation during MIS with flexible endoscopes and allowing force mapping from the endoscope tip to the shaft and vice versa. To do so, we have developed an Encapsulated Force-sensing Device (EFD) that can be mounted on the tip of a flexible articulated robotic endoscope. The EFD houses a miniature tri-axial force sensor (TFS) together with an integrated camera mounted onto the EFD's tip (details can be found in [11]). In this paper, we present the kinematics for an articulated endoscope combined with force measurements at the endoscope shaft and tip. We use the entire setup to investigate force mapping from tip to shaft and vice versa by knowing endoscope kinematics and according coordinate system transformations. Additionally, the influence of disturbances along the endoscope on the force mapping is investigated. In future, providing feedback on forces is envisioned (visually and/or haptically).

2 Materials and Methods

To show the feasibility of palpation in MIS, we used a 1:1 endoscope phantom, which is the replica of a 4-DoF articulated endoscope prototype with two consecutive Cardan joints (previously developed at the BIROMED-Lab, DBE, Uni Basel). The robotic structure of the phantom was fixed in one pose for calibration and in a second, different pose for the experiments. The size of the links l and joint distances d were taken from the original endoscope design. In this paper we show feasibility of palpation (i.e. combination of force and position measurement), only in one experiment (one pose) of the phantom. Therefore, the EFD was installed on the endoscope tip. Then, the phantom was installed on top of a commercial force/torque sensor Nano 17 (ATI Industrial Automation, USA). In a first step (calibration), the TFS in the EFD was calibrated in a straight configuration of the endoscope phantom. In a second step (experiment 1), to verify the correlation between force measurement of the EFD, a u-shape configuration of the endoscope phantom was used. Consecutively, forces were applied to the EFD. In a third step (experiment 2), we also applied disturbance forces at the shaft of the endoscope while applying forces to the EFD, to simulate interaction forces from surrounding tissue in addition. Data from the EFD and the Nano 17 were obtained in a synchronized way through an automation setup (Beckhoff Automation GmbH & Co, Germany) that was using TwinCAT3 as a platform and EtherCAT as a real-time data acquisition protocol (100 Hz). Data was processed offline in Matlab 2016b (The MathWorks, Natick, Massachusetts, USA).

2.1 Basic Concept and System Overview

To compare the forces measured at the base and the tip of the endoscope, the EFD was calibrated in a straight pose of the endoscope (see Sect. 2.2). After calibration, the endoscope shape was changed for the experiments. In the new pose, the kinematics was determined in order to compare forces between base and EFD. The inverse kinematics was calculated by Denavit-Hartenberg Parameters (DH parameters) [12]. Forward kinematics was obtained by simple matrix inversion of the inverse kinematics formulation.

2.2 Calibration Procedure

In a first step, the self-developed TFS implemented in the EFD at the tip of the endoscope had to be calibrated. Therefore, the EFD was mounted on top of the straight endoscope phantom. The phantom itself was mounted on top of a commercial Nano 17 force sensor. Forces were applied by one finger in all three directions on the EFD in the following order: $\mathbf{F}_{z^-}$, $\mathbf{F}_{x^-}$, $\mathbf{F}_{x^+}$, $\mathbf{F}_{y^-}$, $\mathbf{F}_{y^+}$. Force data was obtained from the TFS in the EFD ($\mathbf{F}_{EFD}$) and the commercial Nano 17 sensor ($\mathbf{F}_{nano}$) in a synchronized way at a data rate of 100 Hz. The data of the Nano 17 was acquired by the automation setup through an EtherCAT F/T Interface (ATI, USA) and the TFS data was digitalized by differential analog-digital converters (EL3602 slaves, Beckhoff, Germany). For the calibration process, the signals of the three unidirectional single sensors (USSs) of type HSFPAR003A (ALPS ELECTRIC CO, Japan) that form the TFS inside the EFD, were compared to the three force signals of the Nano 17. Based on the following formula, a calibration matrix (CM) was obtained by using the least squares algorithm implemented for pseudoinverse calculations in Matlab [13]:

$$\mathbf{CM}' = \mathbf{F}^{+}_{EFD}\mathbf{F}_{nano} \tag{1}$$

After applying the **CM** on the $\mathbf{F}_{EFD}$ data, the signals of our EFD should match the data of Nano 17. For our prototype, complete data matching is not yet given. However, tendencies are acceptable to show the principle of endoscopic palpation (Fig. 1).

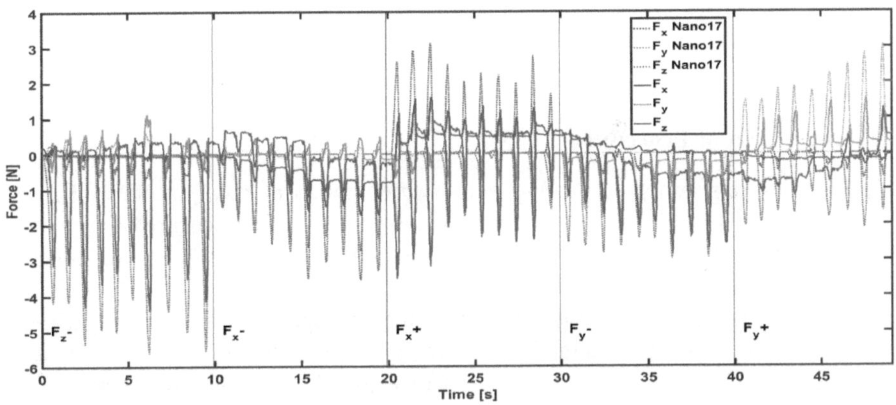

Fig. 1. The signals from the Nano 17 and our EFD after the calibration process.

3 Experiments

In total, two experiments were performed. In both experiments, the endoscope phantom was used in a u-shape configuration. The EFD was mounted at the tip of the endoscope phantom that was mounted on a Nano 17 force/torque sensor itself (see Fig. 2). The first experiment was intended to show the validity of the coordinate transformation based on DH parameters and consequently the forward and backward transformation of forces from the endoscope tip to the base and vice versa. At the same time, the experiment should show that the surgeon would be able to feel the interaction forces from the endoscope tip at the shaft. However, the forces would appear to point in a different direction if the endoscope is not in a straight configuration. The second experiment aimed at demonstrating that it is possible to perform palpation while determining the presence of disturbance forces along the endoscope shaft. Therefore, while the EFD provides the surgeon with force data for palpation during interventions, the comparison of the transformed EFD force data with the force data of the endoscope shaft allows determining the presence of disturbance forces that act along the endoscope structure.

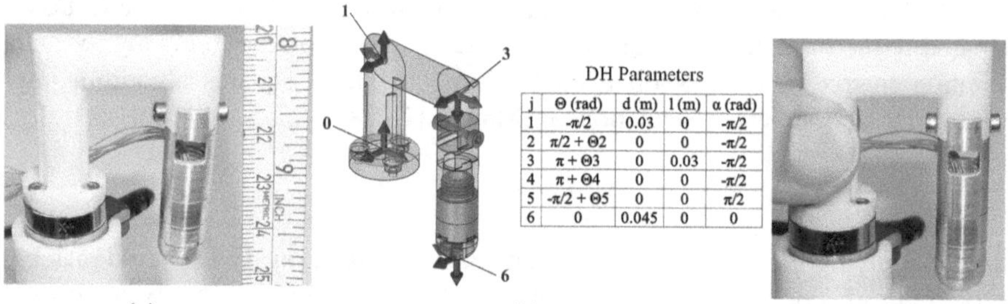

j	Θ (rad)	d (m)	l (m)	α (rad)
1	-π/2	0.03	0	-π/2
2	π/2 + Θ2	0	0	-π/2
3	π + Θ3	0	0.03	-π/2
4	π + Θ4	0	0	-π/2
5	-π/2 + Θ5	0	0	π/2
6	0	0.045	0	0

(a) (b) (c)

Fig. 2. (a) The u-shaped endoscope phantom is mounted on a Nano 17 force/torque sensor and is equipped with a self-developed Encapsulated Force-sensing Device (EFD) that measures forces in three directions. (b) Shows the position and orientation of the coordinate systems for the corresponding DH parameters. (c) Experiment 2: Disturbance on the shaft.

3.1 Experimental Procedure

During the first experiment, pulsating forces were applied manually at the EFD in the following order: $\mathbf{F}_{z^-}$, $\mathbf{F}_{x^-}$, $\mathbf{F}_{x^+}$, $\mathbf{F}_{y^+}$, $\mathbf{F}_{y^-}$. Consequently, the measured forces were compared with each other in the base coordinate frame 0 and the EFD coordinate frame 6 (see Fig. 2b).

During the second experiment, the shaft of the endoscope was manually grasped in a stable fashion while pulsating forces were applied manually at the EFD in the following order: $\mathbf{F}_{z^-}$, $\mathbf{F}_{x^-}$, $\mathbf{F}_{x^+}$, $\mathbf{F}_{y^+}$, $\mathbf{F}_{y^-}$. Consequently, the measured forces were compared with each other in the base coordinate frame 0 and the EFD coordinate frame 6 (see Fig. 2b).

3.2 Results of the Experiments

After recording the data from experiment 1 (pulsating force application at the EFD), the data from the EFD was multiplied with the calibration matrix (**CM**). For the data of the Nano 17, the forward kinematics based on DH parameters was applied. The resulting force comparison in the EFD coordinate frame (see coordinate frame 6 in Fig. 2b) can be seen in the following image.

The same data from Experiment 1 can be also compared in the base coordinate frame (see coordinate frame 0 in Fig. 2b).

After recording the data from Experiment 2 (first constant disturbance at the endoscope shaft, then additional pulsating force at the EFD), the data from the EFD

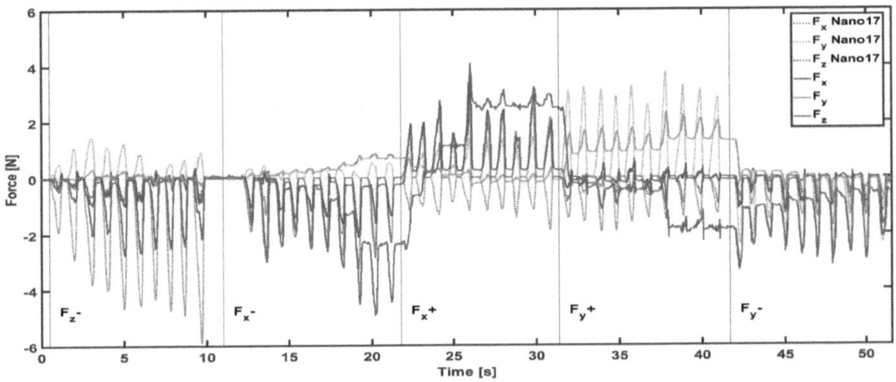

Fig. 3. Experiment 1: Force comparison in the coordinate system of the Encapsulated Force-sensing Device (EFD) located in the tip of the endoscope (frame 6).

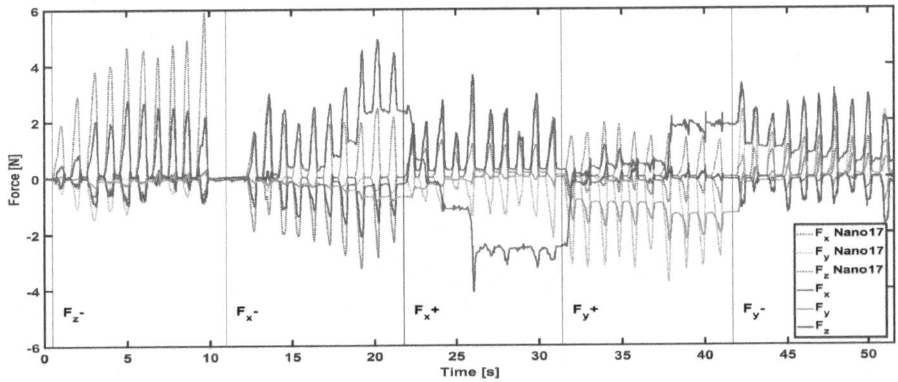

Fig. 4. Experiment 1: Force comparison in the base coordinate system (frame 0).

was multiplied with the calibration matrix (**CM**). For the data of the Nano 17, the forward kinematics based on DH parameters was applied. The resulting force comparison in the EFD coordinate frame (see coordinate frame 6 in Fig. 2b) can be seen in the Fig. 5.

4 Discussion

In this paper, we employed a self developed Encapsulated Force-sensing Device (EFD) that can be integrated in the tip of an endoscope for palpation [11]. The comparison of force data obtained with our EFD and force data obtained from a commercial Nano 17 force/torque sensor still yields important differences in performance. These differences can be clearly attributed to remaining design issues. Nevertheless, this first version of the EFD has serious advantages over the Nano 17 particularly in terms of

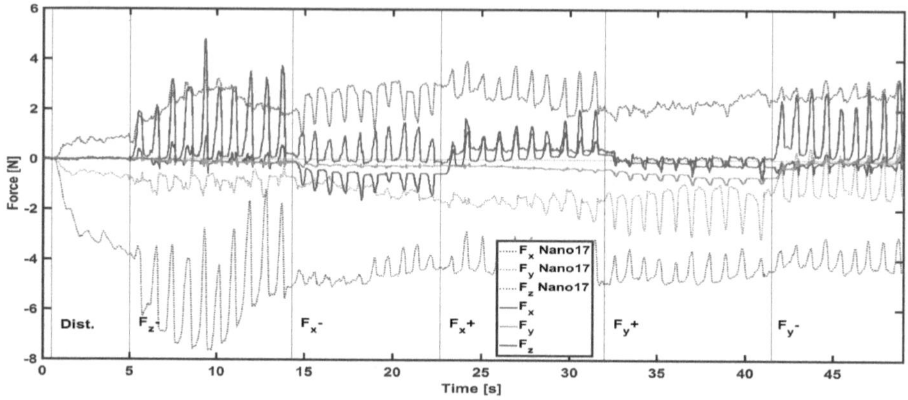

Fig. 5. Experiment 2: Force comparison in the coordinate system of the EFD.

size. Also, the tendencies of the force signals are correct, which is promising for redesigned versions of the EFD. So in principle palpation measurements with the EFD at the tip of an endoscope is possible (Fig. 6).

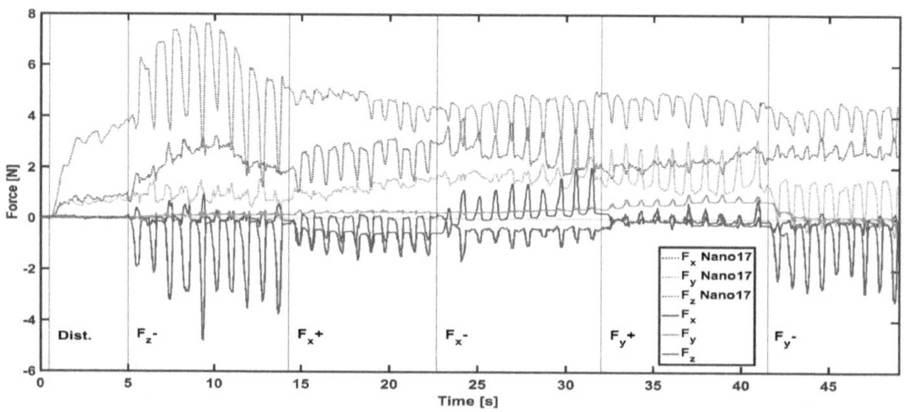

Fig. 6. Force measurement at the tip of the phantom for the second stage of the experiment and presented in the base coordinate system (frame 0).

The results of Experiment 1 show that force transformations based on the Denavit-Hartenberg convention from one force sensor to the other were correct. While this seems to be a trivial result, the application of force transformations through flexed endoscopes in palpation applications will become a difficult job for surgeons. When comparing Figs. 3 and 4, it can be easily seen that the force directions of the y-axis change (as expected). However, for a surgeon this has a big impact: with changing endoscope shape, the forces that can be felt at the shaft change, even if the force at the tip is constant. Additionally, this indicates that the combination of force measurement and endoscope shape measurement needs to be good to enable palpation with endoscopes for finally estimating tissue stiffnesses.

The results of Experiment 2 reveal the advantage of a combination in two force sensors in endoscopic palpation. While the sensor at the tip measures the palpation forces, the sensor at the shaft measures the handling and palpation forces. Accordingly, a configuration of 2 sensors like in our experiment allows detecting high (undesired) interaction forces on the endoscope while performing palpation. So, the surgeon can be warned if predefined interaction thresholds are exceeded to avoid undesired damage to the tissue along the endoscope shaft. To our knowledge, similar ideas have never been presented so far to avoid undesired harm due to endoscope manipulation.

Acknowledgements. We gratefully acknowledge funding by the Werner Siemens Foundation through the MIRACLE project.

References

1. Sanchez, W., et al.: Evaluation of polyp detection in relation to procedure time of screening or surveillance colonoscopy. Am. J. Gastroenterol. **99**(10), 1941 (2004)
2. De Vos, A., et al.: Aspects of biopsy procedures prior to preimplantation genetic diagnosis. Prenat. Diagn. **21**(9), 767–780 (2001)
3. Clark, M.P., et al.: Natural orifice translumenal endoscopic surgery in humans: a review. Minim. Invasive Surg. **2012**, 189296 (2012)
4. Rosen, M.J., et al.: Minimally invasive surgery, 2004-2005. Endoscopy **38**(02), 137–143 (2006)
5. Westebring-van der Putten, E.P., et al.: Haptics in minimally invasive surgery–a review. Minim. Invasive Ther. Allied Technol. **17**(1), 3–16 (2008)
6. Nakadegawa, T., et al.: Three-axis scanning force sensor with liquid metal electrodes. Sens. Actuators, A **264**, 260–267 (2017)
7. Karkanis, S.A., et al.: Computer-aided tumor detection in endoscopic video using color wavelet features. IEEE Trans. Inf. Technol. Biomed. **7**(3), 141–152 (2003)
8. Yoshihiro, T., et al.: Tactile sensing system including bidirectionality and enhancement of haptic perception by tactile feedback to distant part. In: World Haptics Conference (WHC) (2013)
9. Puangmali, P., et al.: State-of-the-art in force and tactile sensing for minimally invasive surgery. IEEE Sens. J. **8**(4), 371–381 (2008)
10. Wagner, C., et al.: Force feedback in surgery: analysis of blunt dissection. In: The 10th Symposium on Haptic interfaces for virtual environment and teleoperator systems (2002)
11. Susic, I., et al.: Versatile, force range-adjustable, tri-axial force sensor with integrated micro camera for the tip of endoscopic devices. In: The Hamlyn Symposium (2018, accepted)
12. Corke, P.I.: A simple and systematic approach to assigning Denavit–Hartenberg parameters. IEEE Trans. Rob. **23**(3), 590–594 (2007)
13. Keekyoung, K., et al.: Calibration of multi-axis mems force sensors using the shape-from-motion method. IEEE Sens. J. **7**(3), 344–351 (2007)

ROSA – RObotic System for Angioplasty

Pasquale F. Greco[1], Gabriuele Larocca[1], Michele Perrelli[1], Guido Danieli[1,2(✉)], Salvatore De Rosa[3], and Ciro Indolfi[3]

[1] Calabrian High Tech – CHT S.r.l., Rende, CS, Italy
pasqualegreco@me.com, gabriele.laroccall@gmail.com, {michele.perrelli, danieli}@unical.it
[2] Mechanics of Calabria University, Rende, CS, Italy
[3] Cardiology of Magna Graecia University, Catanzaro, CZ, Italy
{saderosa, indolfi}@unicz.it

Abstract. This paper describes an innovative robotic system designed to perform transcatheter angioplasty, that can be used in a number of different applications. It is based on a very simple system, including both a measuring unit, called Master, that is the one on which the physician directs the operations, and an actuation device, called Slave, located near to the patient, that actively reproduces the movements commanded by the operator at the passive Master unit. Both systems require a small additional element that takes direct contact with guides and catheters and needs therefore to be sterile on the Slave unit. The system has been designed adding also the possibility of using an alternative Master station equipped with two joysticks and a series of knobs and can drive up to three different Slaves, one dedicated to the guiding catheter, the second and third to balloon catheters and relative guide wires, that may be used also in couple. Two patents have already been presented on this topic.

Keywords: Coronary angioplasty · Avoiding ionized radiations
Least invasive surgery · Robotic angioplasty · Robotic surgery

1 Introduction

Percutaneous transcatheter coronary angioplasty represents the standard treatment for coronary artery disease [1]. The better safety profile, together with the rapid technical, procedural and scientific progress in the field of cardiovascular interventions has brought about a progressive expansion of the field to a growing number of diseases, conditions and procedures [1–5].

The exposure associated to a percutaneous coronary intervention (PCI) is comparable to around 750 chest X-rays (range: 350–2350) [6]. Only in Europe, over one million angiographies or transcatheter interventions were performed in 2001 [7]. This implies a significant professional exposure for interventional cardiologists, that need to operate beside the patient and very close to X rays source in the catheterization

This work was supported in part under PON02_PON03PE_00009_4 Optima Cardiopaths.

G. Carbone et al. (Eds.): *New Trends in Medical and Service Robotics*, MMS 65, pp. 78–90, 2019.
https://doi.org/10.1007/978-3-030-00329-6_10

laboratory. It has been calculated that the average single exposure dose per procedure is about one thousand μSV of the total patient exposure [8]. This chronic exposure to ionizing radiations is a major health hazard for interventional physicians, despite the adoption of the specific anti-radiation protection devices and has been associated to several different diseases [9]. For these reason, Corindus [10–12], just to name a few, created a robot that enables to guide the catheters during coronary procedures allowing both the catheter progress and the rotation by moving mouse and joystick. However, the method of handling joystick requires additional specific training to a doctor who instead has been trained to perform tasks manually operating directly on guides and catheters, resulting in use of additional time and financial resources. Moreover the system needs a major disposable component containing also engines, and is not as flexible as our system.

The initial idea of ROSA was instead to develop a new system whose basic mechanism had to be able to function both as a measurement of penetration and rotation of the catheter, in the control section and to actuate the same movements on the active section. Thus the Master, would reproduce a traditional operatory field for the surgeon, while the Slave in the section of implementation, acts as an actuator of the same movements giving also the possibility of independent movement of guide and catheter. The introduction of a small disposable component, while all the rest of the apparatus could also be used thousands of times, allows the system to be used in a sterile ambient. All this is now possible thanks to the innovative working scheme described below.

Thus the separation of the doctor from the patient is granted, allowing the doctor to keep the patient under continuous radiographic control, without being affected by ionizing radiation, while operating in a way in he/she has been already trained. This is also foreseen in another patent [13] then acquired by Corindus US, which developed the only competitor presently on the market.

In addition, our system [14, 15] will be able to use multiple Slaves as well as to guide the first catheter by coupling two slaves into a series, and also to allow control of two guide-catheter pairs by placing two Slaves in parallel and possibly one in series for the first catheter. Obviously then the system also allows PC control via keyboard, mouse, touch screen, and even with a new ad-hoc interface. Finally, the system will also be able to allow doctors' training to traditional manual use, while allowing them to practice on previously recorded cases through an ambient in VR representing various scenarios and relative decision making options.

In practice, the system allows all the usual and unusual operations, being virtually configurable as preferred by the doctor for the particular case, all this using small relatively low cost disposables, one for Slave used. Obviously all the advancements are independently controllable, and measured, so the system, for example, knows how much a catheter has penetrated, so it even knows, when retreating the catheter, how much it needs to call in order to be always controlled by the system. Among other things, it is possible to adjust the speeds of both feed and feedback. It should also be noted that the separation operation between the guide and the catheter is in this case particularly simple.

Of course remote controls both for the injection of contrast medium, even if not yet built.

2 Description of the Main Mechanism

The basic idea is to measure rotations and advances of the catheters carried out by the doctor and reproduce them on the patient side using two rotating systems practically identical from a mechanical point of view. The heart of the system consists of three cylindrical toothed wheels, of which the first one (in red) is used for measure the rotations imposed by the doctor to catheter and guide wire, the second, hollow, (in blue) to measure independently the catheter advancements, while the third, hollow, (in green) drives the guide-wire.

Figure 1 illustrates our basic system in a preliminary version that registers or controls guide-wires and catheter advances separately, while rotations are common. We note that the system presents a considerable number of toothed wheels of different sizes, in the middle of which there are three larger wheels, of which the front one registers the guide-wires and catheter rotations, while the subsequent ones and are dedicated to separate registration of guide-wires and catheter advances. Note the presence of three electronic components equipped with pins and that are connected, through an idle wheel, to incremental encoders. Also note that in front of the first wheel is also positioned a further element, that we will call nose, containing four small friction wheels, two of which are integral with other two small bevel gears, and which is the only disposable element of all this apparatus, besides obviously guide-wires, catheters, etc.

Fig. 1. The internal mechanisms of Master and Slave

Specially designed thrust bearing, and relative separation cages between steel balls have been positioned between the three central gears and initial and final holding frames.

The bevel gears on the nose mesh with other bevel gears that are hinged on the first central gear and are coaxially fixed to spur gears internal to the group of the three big central gears, each meshing with the internal toothing of the second or third hollow central gear. Having clarified this, note that in Fig. 1 the three central gears are colored in different colors, red the first one, allowing to follow the motion transmission of the nose rotations to its encoder, in blue the second (hollow) gear, allowing the measure of the rotations produced by the sum of the nose rotation and of the translations measured by the friction wheel fixed to the blue bevel gear, and similarly for the green path. Naturally, since the system is electronically controlled, all data on advancement and rotations are also visible on a PC monitor.

The entire system has been built using PA2200 plastic powder on a Formiga 1100 by EOS, property of Calabria University. The profiles of all spur gears have been designed using involute equations on Excel, and extruded using a CAD modeler. All bevel gears have been designed using Tredgold method obtaining 3D profile points again with Excel through the involute equations, that were joined via CAD to the center of the cones of the bevel gears, then cut to the desired size.

Figure 2 shows the mechanism as seen from the rear during the assembly of a prototype, without the final cover of closure of the internal gears and without the encoders.

Fig. 2. The internal mechanisms of Master and Slave as seen from the rear

Table 1 presents the characteristics of the various gears.

Table 1. Slave gear characteristics

	#	m	z
First Central Gear	1		
External toothing		1	70
Hollow Central Gears	2		
External toothing		1	70
Internal toothing		1	53
Internal and Transmission Gears	10	1	21
Other Transmission Gears	3	1	29
Nose Bevel Gear	2		14
Mating Transmission Bevel Gear	2		13

In order to compute the transmission ratio, we have to take into account the actual sequence of teeth numbers, from the first spur gear on the motor to the bevel on the nose, that results: (21/29) (29/70) (53/21) (14/13).

Simplifying and considering that the motor has 200 steps, for each step, taking into account the radius of the friction gear on the nose, we obtain an advancement per step of 0,096 mm. If we consider that we can control, 1/16th of step, we understand the control capability of this system. Also notice that all gears have number of teeth that are prime numbers among themselves, so that each tooth of a gear has to mesh with all the others before meshing again with the same tooth of the other gear. And this is fundamental since all gears are plastic and generated with a 3D printer, in order to obtain a smooth brake-in. One more observation is needed about the choice of these teeth numbers, with reference to the bevel gears. In fact this choice of 13 and 14 derives from the need to have taller teeth in order to ensure transmission even if the gears are not exactly positioned since this contact is not fixed. This has been modified in the following phase, using a screw to fix the nose to the first central gear.

Figure 3 shows the traditional Master to read catheter and guide motion as performed by the doctor in its present version. This will probably be also the base of the new tutorial system for instructing new doctors to this technique in conjunction with a V.R. representation of the vascular system.

Fig. 3. The traditional Master with an old nose mounted

3 Description of the Disposable "Nose"

The following Figs. 4 and 5 shows the lateral view of the CAD model of the nose, with transparent shell, in a condition of transmission of the motion the first, and open for the introduction of the catheters the second. It should be noted that the upper friction wheel, integral with the bevel gear is compressed by a lower wheel, loaded by a double-acting cam, of which the return arm, fixed to the cam, can slide inside a groove on the small frame holding the lower wheel.

Fig. 4. The disposable "nose" of the instrument in transmission.

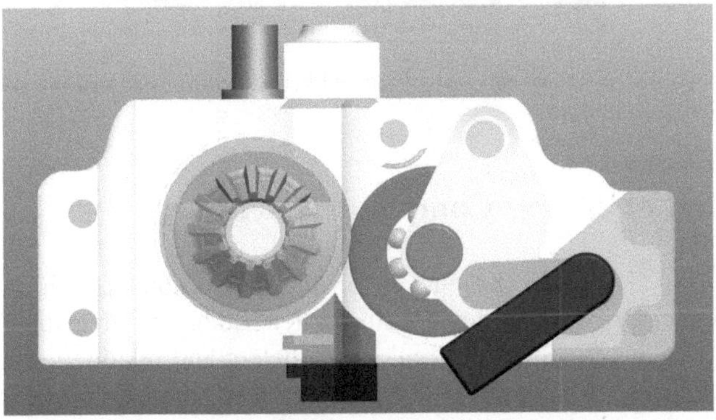

Fig. 5. The disposable "nose" open for catheter insertion

In the back portion of the nose, is present a tube (cut in the following images to enhance te mechanism, since it's long, but visible in Fig. 6) used to isolate catheters and guides from the internal mechanisms, ensuring sterility.

Thus acting on the blue levers that protrude laterally, it is possible to lower the thrusting wheels, so as to freely move the catheter or guide from one side or the other. It should be noted, in fact, that this is necessary because initially catheter and guide advance together, and only from a certain point on it is useful to separate the control, as it may be necessary later on to reunite them. Both the upper and lower wheels contain micro ball bearing in order to reduce friction. In addition, the upper wheels are covered by O-rings to increase the friction between the wheels and the catheter.

Finally, still on the subject of sterility, a special disk has been designed that covers the entire central wheel so as to allow the whole apparatus to be covered with a sterile cloth, but allowing the rotation of the nose itself without allowing the whole cloth to be dragged. This is shown in Fig. 6. The components of the nose are cleaned after production with sandblaster, then cleaned ultrasound, and finally assembled in a laminar flow hood and packed in envelop for sterilization. The tube portion will soon be substituted with a AISI 316L stainless steel tube, while sterilization will be performed with gamma radiation.

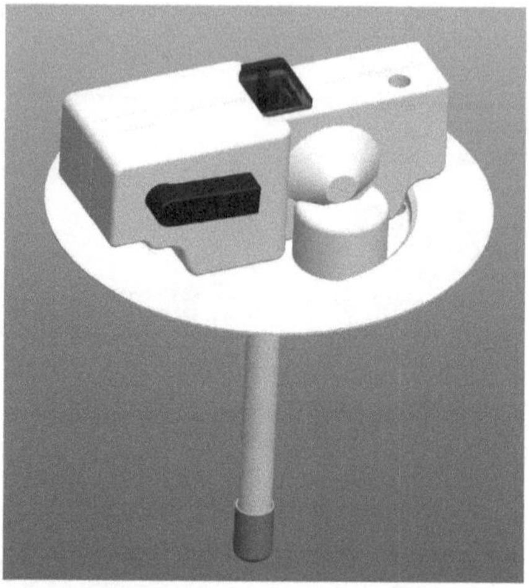

Fig. 6. The disposable "nose" of the instrument with its special disk and the tube to shelter the catheters from the mechanisms

4 Description of a First Complete System

The following Fig. 7 shows the Slave for the first catheter, which only actuate advancement and rotation of a single catheter (two degrees of actuation), while the following Fig. 8 shows the Slave for balloon catheters and relative guide wires, which presents three degrees of actuation. As can be noticed, the Slave of Fig. 7 is considerably shorter then the one in the following figure, and this is due to the fact that the first contains only two motors, positioned vertically, having only two degrees of freedom, while the second having to control advancements of two medical devices (catheter and guide-wire) and the rotation of both, needs the third motor in the center, which increases its height. Obviously the system was completed by the Master presented in Fig. 3.

Fig. 7. The initial Catheter Slave

Fig. 8. The Catheter balloon Slave

5 Description of the New Joystick Master

While the Master using the same internal mechanism can be used pushing and pulling catheters or guides in a non sterile nose, when working only with a single balloon catheter and its guide-wire, if a more complex configuration is desired, using multiple Slaves, a new joystick based Master, shown in Fig. 9, becomes a necessity.

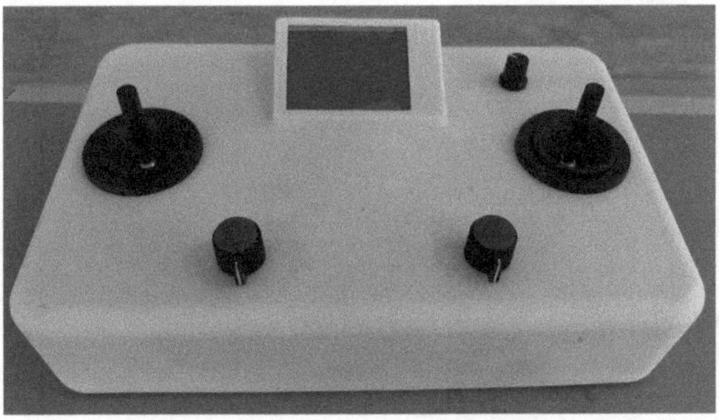

Fig. 9. The joystick Master

This system has been designed to allow the use of multiple Slaves, for instance to control the advancement of the introductory catheter, for which is necessary to control advancement and rotation of both catheters and guide-wire. Thus one Slave closer to the patient will control motion of the catheter, while a second Slave, behind it, will control the guide-wire. In this case the selector is to be positioned in option 1, and the right joystick will control advancement of the guide wire, when moved longitudinally, and its rotation when moved laterally, sending the commands to the Slave closer to the patient, while the left joystick will allow the same control on the catheter, that is moved by the Slave behind. In option 2, vice-versa, the two joystick will control catheter advancement moving longitudinally, and nose rotation moving laterally, while le left joystick will only control catheter advancement. Similarly in option 3 the control will pas to a third Slave used for a second pair of balloon catheters. Finally the two knobs present below the joysticks allow varying the ratio between joysticks displacement and catheter and guide advancement.

6 The Different Versions of the Slave

Since however our medical partners told us of the usefulness of being able to use sometimes two balloon catheters, the idea came to make a new slave that could be asymmetric, in order to allow positioning two Slaves side by side in order to minimize the distance between the catheters. This brought us to redesign completely the internal component disposition of the Slave, without changing the working principle. Meanwhile we also changed motors, passing from DC motors equipped with reducer, to small but powerful step motors. This allowed to eliminate the encoders in the new version of the Slave, since the encoders were 200 steps/complete turn, and so were the motors. Using then the possibility of controlling the 1/16, it is also possible to easily vary the ratio between motion imposed by the doctor and advancement speed.

The next two figures show the new asymmetric Slave as seen in front (Fig. 10) and from the rear (Fig. 11), which will be used also for first catheter introduction. Notice the presence of the two slots that allow hanging the Slave in two directions, also to allow positioning it on both sides of the bed.

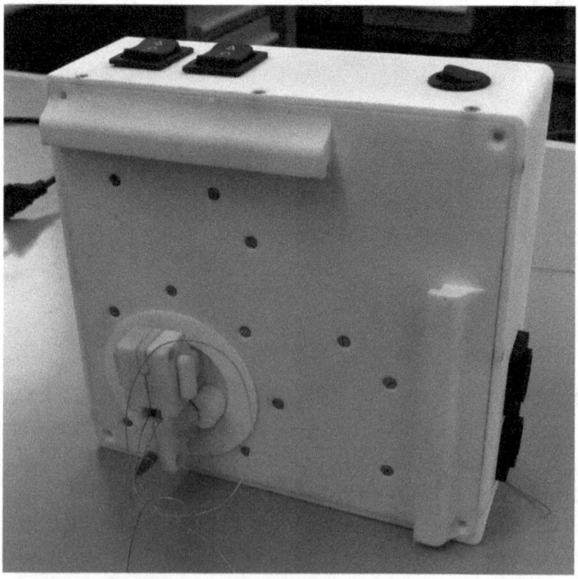

Fig. 10. The last and almost final version of the Slave, as seen from the front

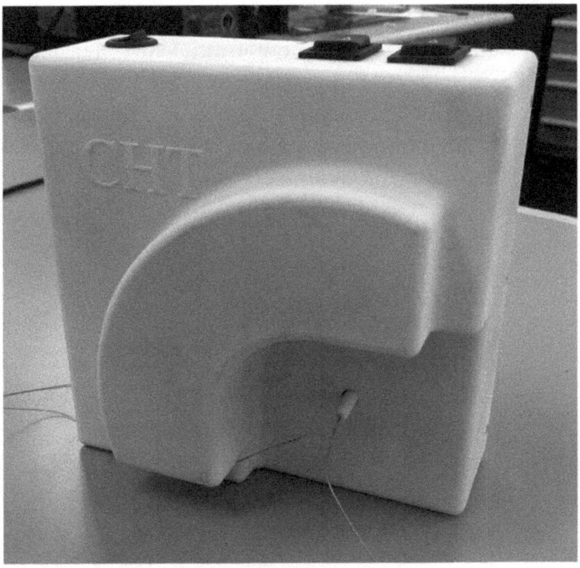

Fig. 11. The same as seen from the back

7 First in Vitro Test

The first test that was performed was to check the correspondence between motion imposed on the Master and subsequent motion obtained in the Slave. To do this, we placed Master (on the right) and Slave (on the left) side by side, we wrapped around the cable coming out of the Master near the nose a black scotch (Fig. 12), we placed the finger near the nose of the Master, then we pulled the cable in the Master a few centimeters, and we took a second picture (Fig. 13) in which are visible the positions of both the scotch tape and the finger. Obviously this might seem an artifact, but to ensure that this is for real, we also posted the movie of this on our web site www.chtsrl.com, projects, ROSA, test 1.

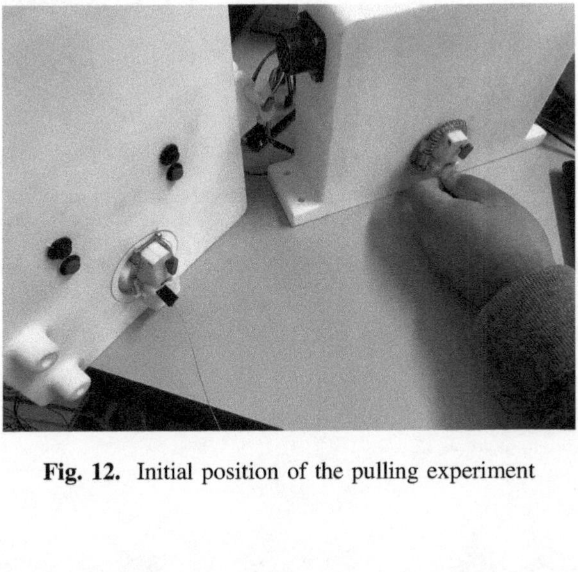

Fig. 12. Initial position of the pulling experiment

Clearly this is just a simple test, but shows that the system works. During the month of January we already decided to test our new joysticks Master and asymmetric Slave on a silicon model of the human circulatory system placed in water in order to obtain a situation similar to real, to pass immediately after to the first animal experiment.

Finally Fig. 14 shows the first trial in the operatory room on a plastic model of the arterial system filled with water and examined under X-Rays by one of our medical partners. Notice that the Master is covered by other operatory equipment, while the Slave is visible and indicated by an arrow.

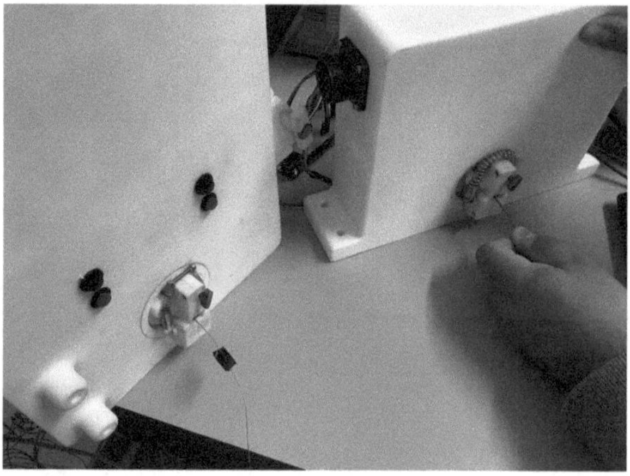

Fig. 13. Final position of the pulling experiment

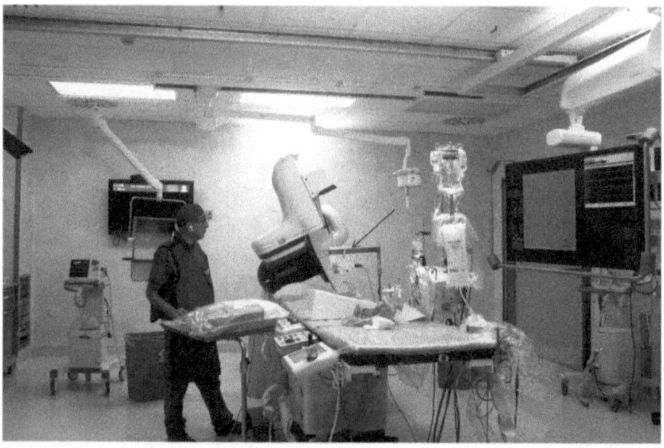

Fig. 14. Practical trial on the control capability of the system

8 Conclusions

As can be noted, we are not far from starting our final experimentation and beginning of pre-clinical trials. We believe that this system will indeed be suitable not only for angioplasty, but for a number of other procedures thank to the fact of being able to drive the initial catheter in situations much more complex than reaching the coronaries.

The system is potentially not very expensive, even if the investment to develop it was substantial. And its best characteristic, is the compactness of our nose, the only disposable part, coupled with its flexibility.

Acknowledgments. The authors wish to acknowledge Basilio Sinopoli, Attilio Caseti and Gianluca La Greca for their precious technical support, and Domenico Battaglia and Giovanni De Gaetano for their participation to the first phases of this research as design engineers.

References

1. De Rosa, S., Polimeni, A., Sabatino, J., Indolfi, C.: Long-term outcomes of coronary artery bypass grafting versus stent-PCI for unprotected left main disease: a meta-analysis. BMC Cardiovasc. Disord. **17**(1), 240 (2017)
2. Indolfi, C., Spaccarotella, C., Yasuda, M., De Rosa, S.: Climbing the hill of left main coronary artery revascularization: PCI or CABG? J Thorac Dis (2018, in Press)
3. Spaccarotella, C., Mongiardo, A., De Rosa, S., Indolfi, C.: Transcatheter aortic valve implantation in patients at intermediate surgical risk. Int. J. Cardiol. **15**(243), 161–168 (2017)
4. Indolfi, C., Sabatino, J., De Rosa, S., Mongiardo, A., Ricci, P., Spaccarotella, C.: Description and validation of TAVIApp: a novel mobile application for support of physicians in the management of aortic stenosis-management of aortic stenosis with TAVIApp. Biomed. Res. Int. (2017). https://doi.org/10.1155/2017/9027597
5. Yasuda, M., Spaccarotella, C., Mongiardo, A., De Rosa, S., Torella, D., Indolfi, C.: Migration of a stent from left main and its retrieval from femoral artery. A case report. Medicine (2017, in press). https://doi.org/10.1097/md.0000000000009281
6. Gerber, T.C., Carr, J.J., Arai, A.E., et al.: Ionizing radiation in cardiac imaging. A science advisory from the American heart association committee on cardiac imaging of the council on clinical cardiology and committee on cardiovascular imaging and intervention of the council on cardiovascular radiology and intervention. Circulation **119**, 1056–1065 (2009)
7. Togni, M., Balmer, F., Pfiffner, D., et al.: Percutaneous coronary intervention in Europe 1992–2001. Eur. Heart J. **25**, 1208–1213 (2004)
8. Picano, E., Vano, E.: Radiation exposure as an occupational hazard. EuroIntervention **8**, 649–653 (2012)
9. Picano, E., Andreassi, M.G., Piccaluga, E., et al.: Occupational risks of chronic low dose radiation exposure in cardiac catheterisation laboratory: the Italian Healthy Cath Lab study. EMJ Int. Cardiol. **1**, 50–58 (2013)
10. Beyar, R.: Transmission for a remote catheterization system. European Patent EP1755727B1, granted 11th January 2012
11. Guerrera, S.: Catheter System Cassette. US Patent USD671640, granted 26th November 2012

12. Beyar, R.: User interface for remote control catheterization, US Patent US8257302, granted 4th September 2012
13. Meglan, D.A.: User interface for remote control catheterization, US Patent US 6096004, granted 1st August 2000
14. Danieli, G., Indolfi, C., De Rosa, S., Battaglia, D., De Gaetano, G., Greco, P.F., Larocca, G., Perrelli, M.: Robotic system for angiography and angioplasty, Italian patent application # 102017000114767_0 presented 12th October 2017
15. Greco, P.F., Larocca, G., Perrelli, M., Danieli, G., De Rosa, S., Indolfi, C.: New joystick master for ROSA, Italian patent application # 102017000151259_0 presented 29th December 2017

Exoskeleton

Design and Simulation of a Leg Exoskeleton Linkage for Human Motion Assistance

I. Geonea[1(✉)], D. Tarnita[1], G. Carbone[2], and M. Ceccarelli[2]

[1] University of Craiova, Craiova, Romania
igeonea@yahoo.com, tarnita.daniela@gmail.com
[2] University of Cassino and South Latium, Cassino, Italy
{carbone,ceccarelli}@unitcas.it

Abstract. In this study, a new exoskeleton leg design for the human motion assistance and rehabilitation use is proposed. The structural solution of the leg mechanism is analysed and a mechanical design solution is developed in Solid Works. The 3D model is used for design optimization and simulation purposes. The study of the leg mechanism is complete with a kinematic analysis, solved with a computational algorithm in Mathlab software, in order to obtain numerical results with plots. The rehabilitation exoskeleton is based on a five links planar mechanism, and design to fulfil human main locomotion tasks. A proper dynamic model of the proposed rehabilitation exoskeleton system is build in ADAMS multi body dynamics software and used to obtain numerical results for exoskeleton motion parameters with plots. These results consist in motion parameters of the exoskeleton solution and exoskeleton gait trajectories, being very useful to appreciate the exoskeleton feasibility and usefulness for human motion assistance and rehabilitation purposes.

Keywords: Wearable sensors · Human gait · Exoskeleton design
Kinematics and dynamic analysis

1 Introduction

The problem of rehabilitation exoskeletons is a study subject in numerous research papers and articles [1]. The locomotion capability is need to be restored for patients with stroke, cerebral paralyze or spinal cord injury [2]. The development of human motion assistance exoskeletons started from 1960 in two main directions: for military purposes, to enhance the motion performance of soldiers, and for motion assistance of patients with rehabilitation needs [3].

The rehabilitation systems for human lower limb are grouped into five groups: (a) devices for motion assistance on a treadmill; (b) systems with mobile plate under the foot; (c) mobile systems for motion assistance; (d) stationary rehabilitation systems; (e) ankle rehabilitation systems. The main achievements in this field and challenges are presented in review articles [4, 5] where are highlighted the design specification and performances for the most relevant rehabilitation systems. The conclusion of these studies is that there is available on medical rehabilitation market a large number of rehabilitation exoskeletons: Lokomat, LOPES, ALEX, ReWalk, Atlas, Ekso, HAL,

G. Carbone et al. (Eds.): *New Trends in Medical and Service Robotics*, MMS 65, pp. 93–100, 2019.
https://doi.org/10.1007/978-3-030-00329-6_11

with different design solution [6, 7]. Most of them are open kinematics chains with motors placed on the joints (for hip, knee and ankle joints) and closed kinematics linkages with only one motor, that represent other solutions. Solutions with motors placed at joints, reproduce with accuracy the human gait, but they are expensive because of the high price for motors and control architecture. On the other side, the easy operation and low cost features of the solutions with closed kinematic linkages, makes them subject of research and development, started by collaboration with LARM Laboratory, from University of Casino and South Latium-Italy [8–10]. Aspects concerning numerical simulation, kinematic and dynamic analysis are studied in many research papers [11–16].

In this paper, the kinematic and design solution for a single DOF leg mechanism is presented. The robotic system is intended for patients who have suffered strokes. The exoskeleton prototype is designed to develop a proper ADAMS dynamic simulation model. Simulation results and human gait experimental results are compared, in order to validate the proposed linkage design.

2 Human Gait Analysis with Wearable Sensors

Actually, are available various research equipments, for human gait experimental analysis (Fig. 1).

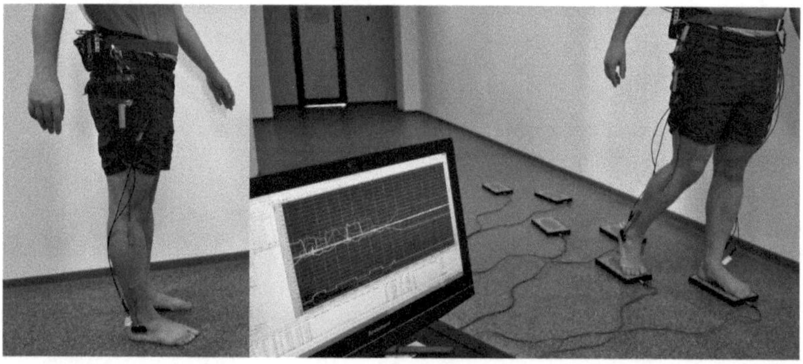

Fig. 1. Biometrics equipment with goniometers sensors, from University of Craiova.

For experimental gait measurement, the Biometrics equipment comprising data acquisition software and wearable goniometers sensors, connected to a portable DataLog hardware system is used [12–14]. By means of Bluetooth interface the measurements data are sent to the computer. The angular variation of each leg joint is measured, processed in Statistica and the gait cycles are plotted (Figs. 2, 3 and 4).

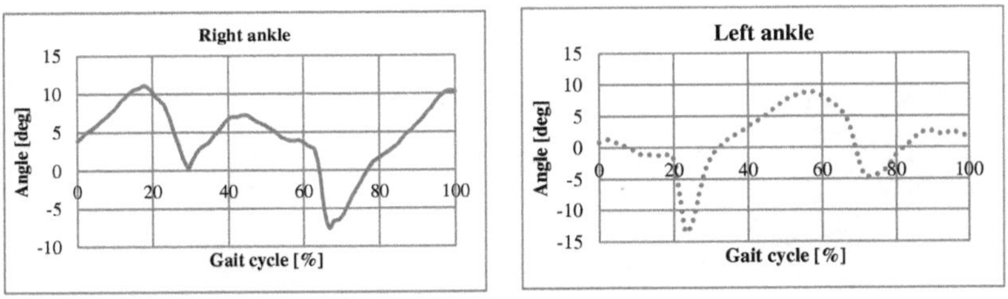

Fig. 2. Ankle joints angle variation in time.

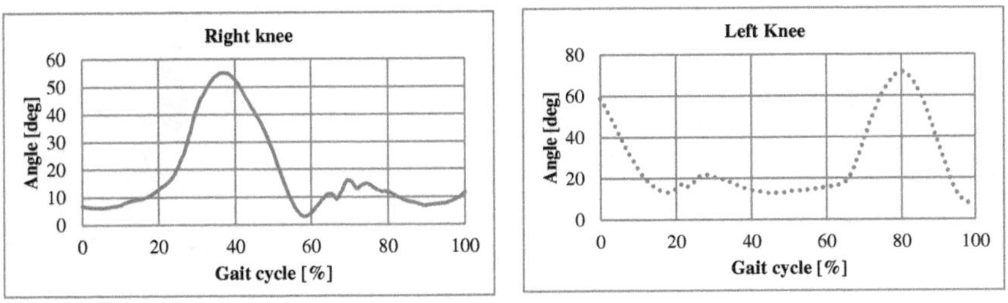

Fig. 3. Knee joints angle variation in time.

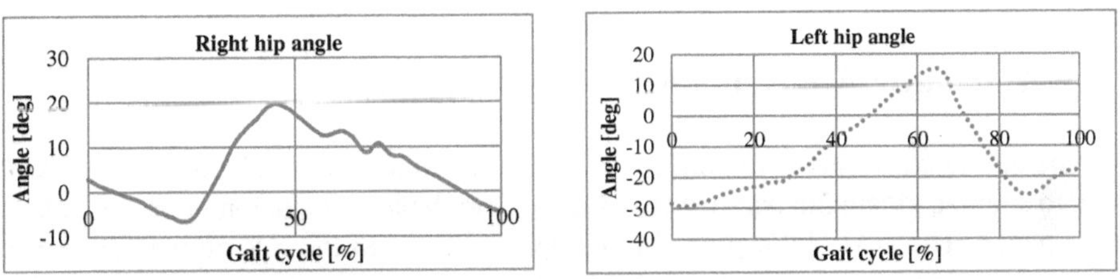

Fig. 4. Hip joints angle variations in time.

3 Leg Exoskeleton Linkage Design and Analysis

Considering previous achievements in this area, a planar closed loop linkage to design the legs of a motion assistance exoskeleton is proposed, as in Fig. 5. The proposed leg exoskeleton consists of a planar five links closed linkage. The links of the planar mechanism is note with digits from (1) to (5) and the joints with capital letters. The actuating motor of the leg is placed on the joint A. The mechanism reproduces the human leg structure, with the link (3) representing the thigh and link (5) materializing the shank. In addition, the knee joint is noted H and the hip joint is D. To adjust the mechanism on patients with different anthropometric data, the links (3) and (5) are designed with adjustable length in future developments. This construction is designed in order to achieve an ovoid trajectory of the ankle joint, M. Because is intended as a low cost solution, the exoskeleton leg realizes the mobility only for knee and hip joints. For an optimized design, to improve exoskeleton motion capabilities, an elliptical cam will replace the link (1).

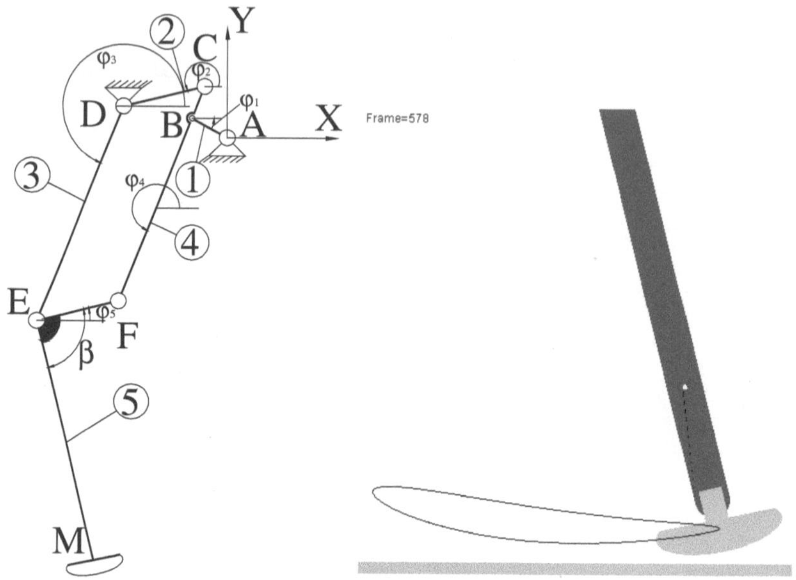

Fig. 5. A kinematic scheme of the leg linkage and achieved ankle ovoid computed trajectory.

4 A Kinematic Analysis of the Leg Mechanism

The kinematics equations of the linkage motion are deducted and solved in Matlab with plots, to analyze kinematic parameters variation and depict conclusion upon the linkage performance. The ankle joint point M position, reported to the XY global coordinate, Fig. 5, is computed as a function of the crank position input angle ϕ_1 and others dimensional parameters of the linkage. The position variation of the point C position, reported to the fixed coordinate system is compute with Eq. (1), while the point E position is compute with Eq. (2):

$$\begin{aligned} x_C &= x_B + l_{BC} \cdot \cos \varphi_4 = x_D + l_{DC} \cdot \cos \varphi_2 \\ y_C &= y_B + l_{BC} \cdot \sin \varphi_4 = y_c + l_{DC} \cdot \sin \varphi_2 \end{aligned} \tag{1}$$

$$\begin{aligned} x_E &= x_D + l_{DE} \cdot \cos \varphi_3 = x_F + l_{EF} \cdot \cos \varphi_5 \\ y_E &= y_D + l_{DE} \cdot \sin \varphi_3 = y_F + l_{EF} \cdot \sin \varphi_5 \end{aligned} \tag{2}$$

The velocity of the points C and E can be computed by derivation of Eqs. (1) to (2), function of time. The angles φ_i (i = 2–5) are computed resolving the closure loops Eq. (3) as a function of the crank position angle ϕ_1.

$$\varphi_i = 2\tan^{-1}\frac{A_i \pm \sqrt{A_i^2 + B_i^2 - C_i^2}}{B_i - C_i};\ i = \overline{2,5} \tag{3}$$

With:

$$\begin{aligned}
&A_2 = -2(y_D - y_B)l_{DC};\ B_2 = -2(x_D - x_B)l_{DC};\\
&C_3 = l_{BC}^2 - (x_D - x_B)^2 - (y_D - y_B)^2 - l_{DC}^2.\\
&A_3 = 2(y_F - y_D)l_{DE};\ B_3 = 2(x_F - x_D)l_{DE};\\
&C_3 = l_{EF}^2 - (x_F - x_D)^2 - (y_F - y_D)^2 - l_{DE}^2.\\
&A_4 = 2(y_D - y_B)l_{BC};\ B_4 = 2(x_D - x_B)l_{BC};\\
&C_4 = l_{DC}^2 - (x_D - x_B)^2 - (y_D - y_B)^2 - l_{BC}^2.\\
&A_5 = -2(y_F - y_D)l_{EF};\ B_5 = -2(x_F - x_D)l_{EF};\\
&C_5 = l_{DE}^2 - (x_F - x_D)^2 - (y_F - y_D)^2 - l_{EF}^2.
\end{aligned} \tag{4}$$

The presented equations are useful to determinate the mechanism position and orientation angles, when it's operating without ground interaction (contact). Resolving the above equations in Matlab, the kinematic parameters variation is established and is represented as plots, as, for example, orientation angle φ_3 (Fig. 6).

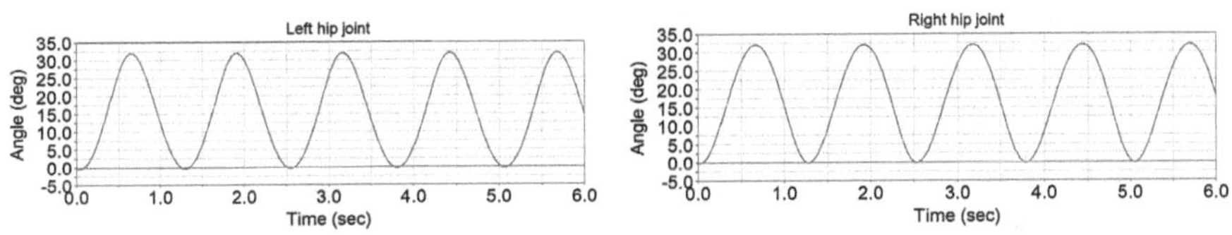

Fig. 6. Computed plot for the hip joints orientation angle.

The considered input date for the linkage geometric configuration are: l_{AB} = 17 mm; l_{CD} = 160 mm; l_{BC} = 65 mm; l_{BF} = 375 mm; l_{EF} = 160 mm; l_{DE} = 440 mm; l_{EM} = 450 mm; $\beta = 90°$. The computed hip joints angular variations presented in Fig. 6 are comparable with the angular variation obtained from experimental study of human gait. The angular amplitude for hip joint is 34° while for knee joint is 65°. Values computed for the angular amplitude of the exoskeleton hip and knee joints are close to experimental values obtained by using goniometers sensors.

5 Virtual Prototype Used to Dynamic Simulation

A mechanical prototype of a proposed exoskeleton is shown in Fig. 7. The links of the exoskeleton has to be designed to allow adjustments of lengths in accordance with different human body sizes. A dynamic simulation is developed considering the situation when the exoskeleton performs gait on ground. For that, proper joints, ground contact model, are defined with parameters indicated in Table 1, as used in ADAMS environment. Contact from foot and ground, contact stiffness, damping coefficients and friction force coefficients are specified considering the impact model in ADAMS, as mentioned in previous research [12].

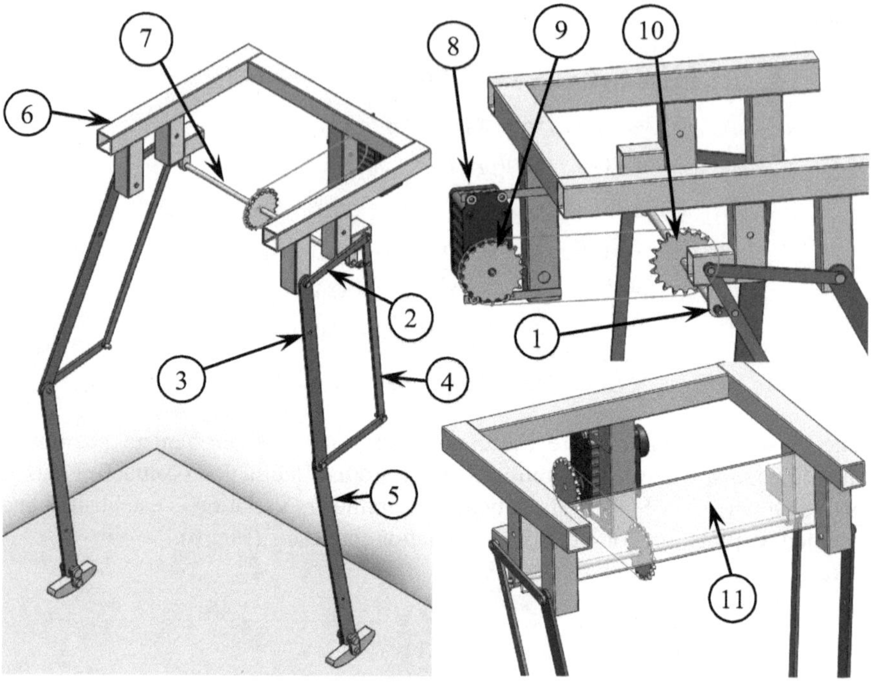

Fig. 7. Mechanical design in Solid Works of the proposed exoskeleton.

The exoskeleton consists of two mechanisms: for the left and right legs. The legs are composed of five elements connected by seven kinematic rotation joints. The electric motor with reducer (8), which is mounted on the upper frame (6), is used to drive the shaft (7) and the links (1). By using of a chain wheel drive, the motion reaches to the shaft (7), which is mounted with radial-axial ball bearings on the upper frame (10) made of welded rectangular beams. The driving links (1) of the two leg mechanisms are fixed to the drive shaft (7) and are positioned to 180°, to assure consecutive steps. On the upper frame, the links (2) and (3) are coupled with rotation joints (materialized with bolts, on joint D). Two segments (EF) and (EM) compose the link (7) which is manufactured (welded) at an angle of 90°. The designed exoskeleton has an anthropomorphic human leg-like structure: the joint (D) represents the human hip joint, the joint (E) represents the human knee joint, (DE) link represents the femur, and (EM) - the tibia.

ADAMS computed exoskeleton-walking sequences are shown in Fig. 8 and some kinematic results are presented in Fig. 9 (exoskeleton hip and knee joint angle). Analyzing simulation results, one can observe that the obtained values are similar if are compared with those from human gait experimental tests and with results of theoretical kinematic analysis (similar variations of parameters).

Table 1. ADAMS parameters used for dynamic simulation

Parameter	Value	Parameter	Value
Elasticity modulus	$2.1 \cdot 10^5$ N/mm^2	Friction force, μ_s	0.35
Density	7.8 E–6 kg/mm^3	Contact force exponent	1.4
Penetration depth	1 mm	Damping (Ns/mm) stiffness (N/mm)	400, 2200

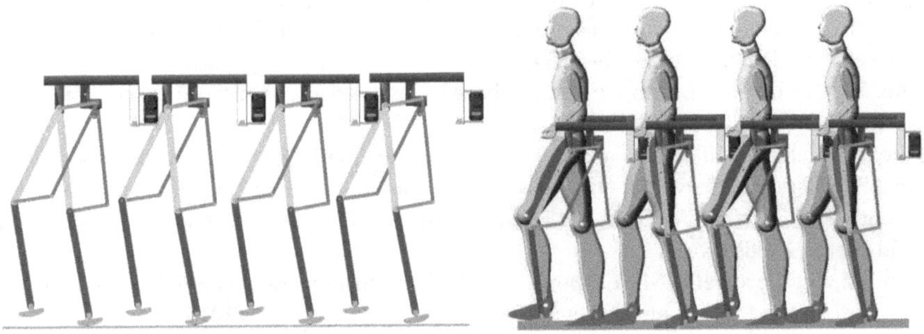

Fig. 8. A walking frames in ADAMS simulation of the exoskeleton.

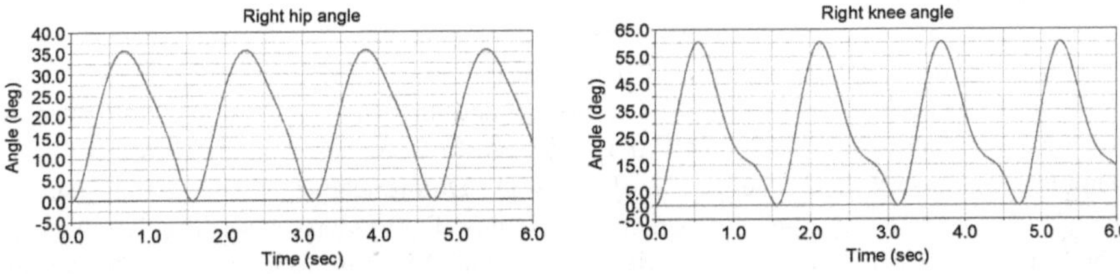

Fig. 9. Computed plot of the exoskeleton hip and knee angle joint variation.

6 Conclusions

A new prototype of an exoskeleton used for human motion assistance and rehabilitation with easy-operation features is proposed. Simulation results obtained with ADAMS computational model have revealed suitable performance for operation in walking rehabilitation, although its design may require additional improvements in future developments. The proposed design is low-cost, simple, wearable and light, with anthropomorphic structure. It operates with only one motor easy controlled as concerns the angular velocity. The exoskeleton purpose is to fully help or partially assist human walking and assure proper gait rehabilitation.

Acknowledgment. This work is supported by PN-III 239PED from 01/09/2017 grant of the Executive Agency for Higher Education, Research, Development and Innovation Funding (UEFISCDI).

References

1. Chen, B., et al.: Recent developments and challenges of lower extremity exoskeletons. J. Orthop. Transl. **5**, 26–37 (2016)
2. Anama, K., Al-Jumaily, A.A.: Active exoskeleton control systems: state of the art. Procedia Eng. **41**, 988–994 (2012)
3. Díaz, I., Gil, J.J., Sánchez, E.: Lower-limb robotic rehabilitation: literature review and challenges. J. Robot. **2011**, 11 (2011)
4. Yan, T., et al.: Review of assistive strategies in powered lower-limb orthoses and exoskeletons. Robot. Auton. Syst. **64**, 120–136 (2015)
5. Huo, W., et al.: Lower limb wearable robots for assistance and rehabilitation: a state of the art. IEEE Syst. J. **10**(3), 1068–1081 (2016)
6. Plitea, N., et al.: Kinematic analysis of an exoskeleton-based robot for elbow and wrist rehabilitation. In: Carvalho, J., Martins, D., Simoni, R., Simas, H. (eds.) Multibody Mechatronic Systems. MuSMe 2017. Mechanisms and Machine Science, vol. 54. Springer, Heidelberg (2018)
7. Copilusi, C., Ceccarelli, M., Carbone, G.: Design and numerical characterization of a new leg exoskeleton for motion assistance. Robotica **33**, 1147–1162 (2015)
8. Geonea, I., Ceccarelli, M., Copilusi, C.: New assistive device for people with motor disabilities. Appl. Mech. Mater. **772**, 574–579 (2015)
9. Copilusi, C., Ceccarelli, M., Dumitru, N., Carbone, G.: Design and simulation of a leg exoskeleton linkage for a human rehabilitation system. In: Visa, I. (ed.) The 11th IFToMM International Symposium on Science of Mechanisms and Machines. Mechanisms and Machine Science, vol. 18. Springer, Cham (2014)
10. Geonea, I., Ceccarelli, M., Carbone, G.: Design and analysis of an exoskeleton for people with motor disabilities. In: The 14th IFToMM World Congress, Taipei, Taiwan (2015)
11. Dumitru, N., Copilusi, C., Geonea, I., Tarnita, D., Dumitrache, I.: Dynamic analysis of an exoskeleton new ankle joint mechanism. In: Flores, P., Viadero, F. (eds.) New Trends in Mechanism and Machine Science. Mechanisms and Machine Science, vol. 24. Springer, Heidelberg (2015)
12. Geonea, I.D., Tarnita, D.: Design and evaluation of a new exoskeleton for gait rehabilitation. Mech. Sci. **8**, 307–321 (2017). https://doi.org/10.5194/ms-8-307-2017
13. Tarnită, D.: Wearable sensors used for human gait analysis. Rom. J. Morphol. Embryol. **57** (2), 373–382 (2016)
14. Tarnita, D., Geonea, I., Petcu, A., Tarnita, D.N.: Numerical simulations and experimental human gait analysis using wearable sensors. In: Husty, M., Hofbaur, M. (eds.) New Trends in Medical and Service Robots. Mechanisms and Machine Science, vol. 48. Springer, Heidelberg (2018)
15. Batayneh, W., et al.: Biomimetic design of a single DOF Stephenson III leg mechanism. Mech. Eng. Res. **3**(2), 43–50 (2013)
16. Tarnita, D., Tarnita, D.N., Bizdoaca, N., Popa, D.: Contributions on the dynamic simulation of the virtual model of the human knee joint. Mater Werkst **40**(1-2), 73–81 (2009). Materials Science and Engineering Technology, Special Edition Biomaterials, Willey-Vch

Numerical Simulation of a Leg Exoskeleton for Human Motion Assistance

C. Copilusi[1(✉)], S. Dumitru[2], I. Geonea[1], E. Rosu[1], and M. Ceccarelli[3]

[1] Faculty of Mechanics, University of Craiova, Craiova, Romania
cristache03@yahoo.co.uk, igeonea@yahoo.com, acasa@mectrat.ro

[2] Faculty of Automation, Computers and Electronics, University of Craiova, Craiova, Romania
sorindumitru83@yahoo.com

[3] LARM, University of Cassino and South Latium, Cassino, Italy
ceccarelli@unicas.it

Abstract. The objective of this paper is to validate the design of a leg exoskeleton used to aid persons with neuromotor disorders rehabilitation. The proposed design of the leg exoskeleton was performed in two stages. The first stage comprises the motion capture of 30 test subjects under normal walking conditions. An equipment for motion capture was used for this task, namely VICON. By using the obtained experimental data, a numerical simulation was computed in second stage for validating the leg exoskeleton proposed design. Finally, the results obtained from leg exoskeleton proposed design will allow improving rehabilitation therapies by reproducing different normal walking trajectories. The proposed solution shows a new mechanism design that combines capture motion data for synthesizing sprocket-mechanisms. The research also presents a new methodology to evaluate the design of a low-cost leg exoskeleton.

Keywords: Design · Numerical simulation · Leg exoskeleton Chain transmission

1 Introduction

Nowadays exoskeletons and assistive walking devices used for medical purposes are especially designed for human gait recovery or rehabilitation. The ones used for gait recovery are designed for superficial and temporal injuries like bone fractures, lower limb specific surgical interventions typically to tendon fixtures. From this category can be found the ones from [1, 2, 5] and these are designed by taking into account simplicity, low-cost and easy operation features criteria.

By referring to the ones used for gait rehabilitation, there are exoskeletons especially designed for persons with AVC disorders (Accident Vascular Cerebral). In this case, several exoskeletons types like [3, 4, 6, 8, 9] were designed by considering the emphasizing task for hip, knee and ankle joints during walking and these are complex

G. Carbone et al. (Eds.): *New Trends in Medical and Service Robotics*, MMS 65, pp. 101–108, 2019.
https://doi.org/10.1007/978-3-030-00329-6_12

ones and also expensive. Considering the one from [6] it has six DOF (one compact actuator for each human lower limb joint with a harmonic unit for increasing the power and torque parameters), a mechanical structure that supports the patient weight and exoskeleton, a complex hardware especially designed for coordinating and assisting human gait during rehabilitation programs. By taking into account the simplicity and price criteria, one major solution is to use actuator units from industrial domain. In this case there is a problem with the actuator unit dimensions and position for assuring the desired motion. Thus, each actuator unit cannot be placed directly at each joint level due to the size and weight. For this it is necessary to design specific actuating mechanisms (typical linkages, cam mechanisms) as well as mechanical transmissions (belt or chain transmissions). For solving this problem, a leg exoskeleton will be designed with a central actuation unit placed near of the exoskeleton mechanical structure. The motion parameters will be transmitted through proper chain transmissions and bar linkages. For validating the functionality principle it is necessary to have human locomotion system motion laws as input data and these are presented in second section of this research. These were obtained with VICON Equipment.

In the third section a structural scheme of the proposed leg exoskeleton solution was elaborated. This can be parameterized according with the human anthropometric data. In fourth part of this research, a numerical simulation of the proposed leg exoskeleton will be performed with the aid of SolidWorks and MSC Adams. Thus the proposed leg exoskeleton will be validated through a comparative analysis between the experimental results and the ones obtained through virtual simulations.

2 Human Walking Experimental Analysis

The experimental analysis was performed with a VICON Equipment existent in University of Craiova biomechanical laboratories [7, 10]. The experimental analysis workflow is schematized in Fig. 1(a). The analysis purpose was to obtain the input data for validating the exoskeleton solution through numerical processing. These data are characterized by each human joint motion law for one gait in case of hip, knee and ankle joints. The obtained results will be useful for creating the motion laws in a polynomial equations form for numerical processing of the proposed leg exoskeleton. Most important output of this analysis is represented by the knee and ankle joints angular amplitude obtained in radians.

On the other hand, it is necessary to design an exoskeleton solution with adjustable linkage for adapting this to any anthropometric person. Thus, a database was created which consists of 30 healthy human subjects (15 male and 15 female persons) used on walking activity recordings and it represent walking patterns for AVC persons rehabilitation. For this research stage, from the database, it was retained the obtained motion laws with a known anthropometric data for a female human subject (dimensional leg parameters, age 31, 61 kg, 1,70 m height). A snapshot during experimental analysis is represented in Fig. 1(b). The obtained results validation was done by considering the existent data in specialty literature reported in [7]. The time period for a complete gait analysis was 1.45 s and the same variable will be used on virtual

a.

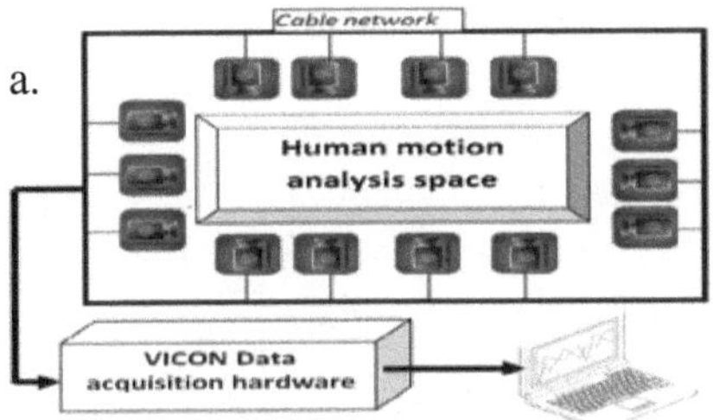

b.

Fig. 1. VICON equipment workflow [10] – (a) and a snapshot during experimental tests (b).

simulations. In this case the obtained results, represented through motion diagrams for each joint, are shown in Fig. 2.

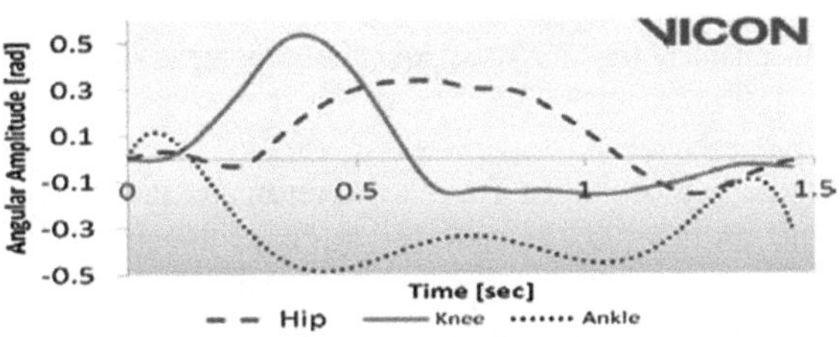

Fig. 2. Hip, knee and ankle joint angular variations during one gait vs. time

3 A Leg Exoskeleton Mechanism Solution

By taking into account leg mechanism solutions elaborated in [2, 4], a new leg exoskeleton solution will be elaborated for AVC persons. The main requirements for designing this are: high-torque and high inertia momentums for actuating the main human locomotion system joints; the possibility to actuate in an independent mode each joint; teaching functions in case of a manual manipulation for each human leg imposed by the physician; designing the exoskeleton in a test-bed frame especially created for rehabilitation; possibility to implement motion laws patterns for a large persons anthropometric data. Thus, for a new leg exoskeleton mechanism, a solution is presented in a structural scheme from Fig. 3(a). This has 3 DOF on each leg and the actuation was made through chain sprockets mechanical transmissions and bar linkages. A structural scheme of the proposed leg exoskeleton solution can be observed in Fig. 3(b), where the bar linkages, namely "I" and "II" are represented in a detailed form (each one has 1DOF).

Thus, in Fig. 3(a) are identified the following components: APS – servomotor for hip joint mechanism actuation; AGS – servomotor for knee joint mechanism actuation; ASS – servomotor for hip joint actuation through chain sprockets transmission; I – bar linkage for knee joint actuation; II – bar linkage for ankle joint actuation.

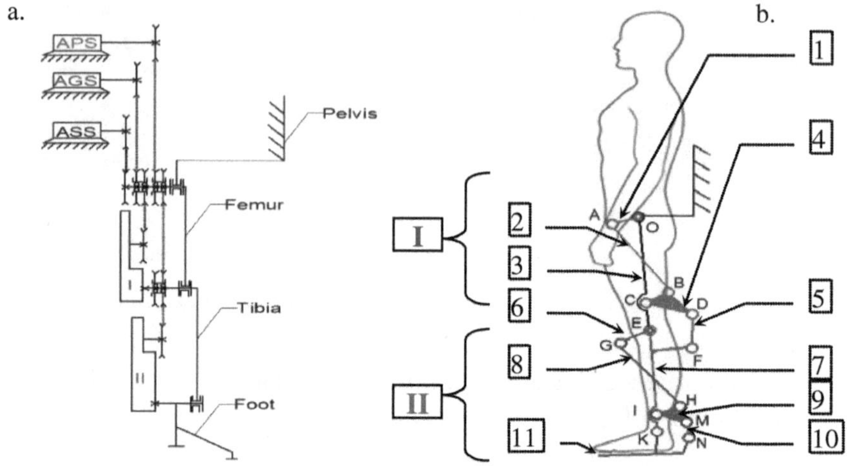

Fig. 3. Exoskeleton mechanical transmissions structural scheme (a) and bar linkages for a single leg actuation (b)

By considering the bar linkages for exoskeleton actuation from Fig. 3(b), in a parameterized form according with the analyzed human subject, it can be identified: OE – femur (3) equivalent segment with a length of 441 mm; EK – tibia equivalent segment with a length of 434,37 mm; OA – drive element (1) for knee actuation mechanism with a length of 22.13 mm; AB – connecting rod (2) from knee mechanism actuation mechanism with a length of 366 mm; EC – distance of 50 mm where C joint is placed; BC- element (4) with a length of 39.18 mm; CD – component of element (4) with a length of 63.8 mm; BD - component of element (4) with a length of 29.41 mm; DF - element (5) with a length of 90 mm; EG – drive element (6) for ankle actuation mechanism with a length of 30 mm; GH- connecting rod (8) from ankle actuation mechanism with a length of 322 mm; IH - element (9) with a length of 75 mm; HM – component of element (9) with a length of 40 mm; IM - component of element (9) with a length of 39.18 mm; IK – distance of 50 mm where I joint is situated on tibia equivalent segment; MN – element (10) with a length of 40 mm. The size identification of all components is mandatory for numerical processing in case of virtual simulations for the proposed leg exoskeleton mechanism. This solution has adjustable links for adapting to any person anthropometric parameters, respectively: femur (3); tibia (7); link AB and link GH.

The motor units will be placed in the centre of O and E joints, which these are considered double revolute joints and due to the chain sprockets transmissions these will be independent. These joints are identified in Fig. 3(b).

4 Modeling and Simulations

By considering the proposed exoskeleton solution from Fig. 3(b) and by having in sight the anthropometric data obtained through experimental analysis, a parameterized 3D model was designed with the SolidWorks aid. The purpose of these simulations is to obtain the motion laws developed by the equivalent motor units, namely O and E joints through the chain sprockets transmissions and bar linkage actuation mechanisms. These joints correspond only for equivalent knee and ankle joints. Hip joint will be actuated without a linkage through a chain transmission with a ratio equal to one. Thus the virtual simulations will be performed on two stages. First stage when the proposed leg exoskeleton model use as motor joints placed on O, E and K revolute joints and the bar linkages are driven. The input data for this will be considered the experimental motion laws for these joints. The target on this stage was to obtain the motion curves for OA and EG driven links according with Fig. 3(b). In second stage these links will be considered drive ones and will be actuated with the motion curves obtained from the first stage. Thus, in second stage the E and K revolute joints equivalent to knee and ankle will be driven ones. For both stages a model was imported in MSC ADAMS environment, according with [11] and there were defined proper kinematic joints and constraints. This is represented in Fig. 4(a), where the bar linkage actuation mechanisms and the chain sprockets can be remarked. Proper materials parameters were defined for the leg exoskeleton links, namely aluminum alloys and the structure of each link was a solid one. In this way, the leg exoskeleton weight reaches a value of 8.74 kg. The leg exoskeleton has 15 revolute joints and 3 revolute actuators namely O, E, and K joints, according with the structural scheme from Fig. 3(b). A detailed view of a knee joint exoskeleton actuating mechanism is represented in Fig. 4(b), where a chain sprocket is presented for actuating the ankle joint mechanism. The exoskeleton position corresponds with the one of the analyzed human subject. The friction forces were neglected and also the ground-foot contact was not considered. It was defined the chain transmission between the actuation units and driven sprockets with typical parameters such as: closed chain transmission with 2 sprockets type, teeth profile according with ISO 606 standard, ratio equal to 1, steel material for each component; damping equal to

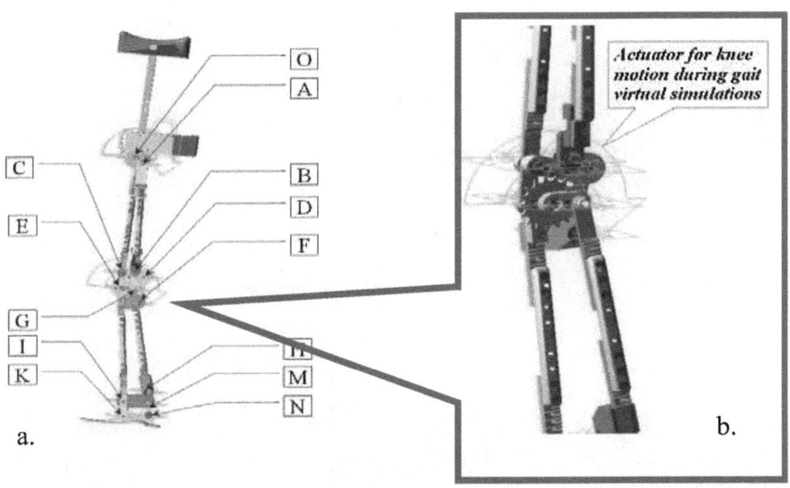

Fig. 4. Virtual model of the proposed exoskeleton (a) and knee joint detailed view (b)

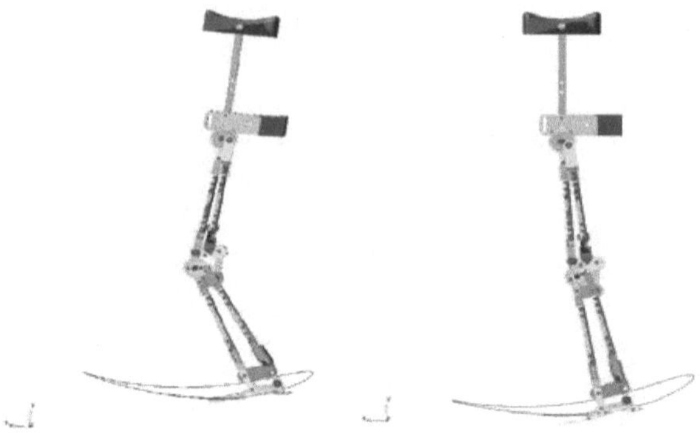

Fig. 5. Snapshots during virtual simulations of the proposed leg exoskeleton with ankle joint trajectory identification

100, stiffness of 1E+05 MPa and contact force exponent equal with 1.25. The simulation sequence was equal with 1.45 s and corresponds to a complete gait developed on the experimental analysis for both stages. The desired results for this research consist on motor motion laws developed through bar linkages. These will be further implemented for controlling the displacements based on computed trajectories of the prototype. Snapshots during simulations in case of second stage are shown in Fig. 5, were a simulated ankle trajectory was shown. Thus the obtained results, considered from second stage, were processed in a numerical form and these are shown as a comparative analysis in diagrams from Fig. 6. A comparison between the second stage results and experimental analysis was obtained and this is represented in Fig. 6.

Thus, by having in sight the curves path from Fig. 3 for human knee and ankle joints and comparing these with the ones from virtual simulations, it can be observed that these have similar paths for performing a single gait. In case of knee joint the

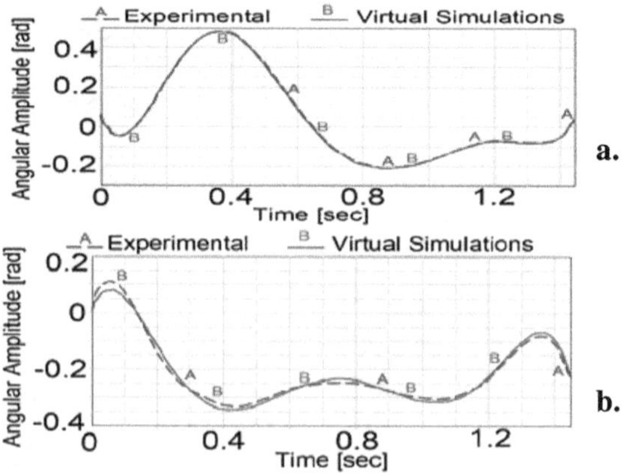

Fig. 6. Comparative study between experimental results and virtual simulations [radians] for a single gait vs. time [sec]: a-knee joint; b – ankle joint

maximum value is 0.4833 radians and the minimum one is 0.202 radians in both cases namely human subject and the developed motor joint obtained from virtual simulations as it can be seen in Fig. 6(a). In this case the bar linkage mechanism were designed properly. For ankle actuation mechanism the maximum value from virtual simulations was 0.1281 radians and a minimum one of –0.339 radians as reported in Fig. 6(b). These are close to the ones from experimental analysis and respect the standard accuracy of the VICON Equipment which is 0.36%.

5 Conclusions

A new design solution is proposed for hip, knee and ankle joints in a leg exoskeleton mechanism which is based on bar linkage actuation mechanisms combined through chain sprockets mechanical transmissions. The desired structure will be used on human neuromotor rehabilitation when the patient and exoskeleton will be suspended through cables.

Through this research, it was validated only the leg exoskeleton functionality, by taking into account as input data results of an experimental analysis and the obtained outputs was characterized by the leg exoskeleton motion laws for knee and ankle actuated joints. These will be further implemented on controlling the exoskeleton links displacements through computed trajectories.

Based on the performed experimental analysis, a database was created with modern equipment which will be used on further advanced research related to this work.

The obtained leg exoskeleton is a parameterized one and it has adjustable links in order to adapt to any anthropometric data of each human subject. Thus it can be obtained other variables for controlling the actuator units for different motion laws of the experimental database.

Also the virtual model gives the possibility to identify other parameters specific to particular kinematics or dynamic analyses for future research work.

On future work the obtained results will help for controlling the servomotor units. The suitable servomotor units will be Panasonic 750 W with planetary gearboxes characterized by high inertia momentum and torque values. These servo units can be placed at a reasonable distance, due to chain transmissions, on a special test bed and in this way the reduced size criteria imposed by the limit of each human natural joint will be eliminated. Also this will be equipped with safety devices and systems for patient protection during rehabilitation exercises such as program controlled protection and mechanical limitation on each actuated joint.

Acknowledgments. This work is supported by PN-III 239PED from 01/09/2017 grant of the Executive Agency for Higher Education, Research, Development and Innovation Funding (UEFISCDI).

References

1. Esquenazi, A., Talaty, M., Jayaraman, A.: Powered exoskeletons for walking assistance in persons with central nervous system injuries: a narrative review. PM R **9**, 46–62 (2017)
2. Copilusi, C., Ceccarelli, M., Carbone, G.: Design and numerical characterization of a new leg exoskeleton for motion assistance. Robot. Camb. J. **33**(5), 1147–1162 (2015)
3. Louie, D.R., Eng, J.J.: Powered robotic exoskeletons in post-stroke rehabilitation of gait: a scoping review. J. Neuroeng. Rehabil. **13**, 53 (2016)
4. Dumitru, N.A., Didu, C., Copilusi, P.C.: Mechatronic system for locomotion rehabilitation. Adv. Eng. Manag. 53–54 (2016). ISSN 2537-4443
5. Beil, J., Perner, G., Asfour, T.: Design and control of the lower limb exoskeleton KIT-EXO-1. In: Proceedings of IEEE International Conference on Rehabilitation Robotics (ICORR), pp. 119–124 (2015)
6. Bortole, M., Venkatakrishnan, A., Zhu, F., et al.: The H2 robotic exoskeleton for gait rehabilitation after stroke: early findings from a clinical study. J. NeuroEng. Rehabil. **12**(54), 1–14 (2015)
7. Keijsers, N.L.W., Stolwijk, N.M., Renzenbrink, G.J., Duysens, J.: Prediction of walking speed using single stance force or pressure measurements in healthy subjects. J. Gait Posture **43**, 93–95 (2016)
8. You, S.H.: Concurrent validity and test-retest reliability of the novel Walkbot-K system. J. Mech. Med. Biol. **16**, 1640029 (2016)
9. Federici, S., Meloni, F., Bracalenti, M., De Filippis, M.L.: The effectiveness of powered, active lower limb exoskeletons in neurorehabilitation: a systematic review. J. Neuro-Rehabil. **37**, 321–340 (2015)
10. Vicon Motion Analysis Equipment User Manual (2014)
11. MSC Adams User Manual. http://www.mscsoftware.com/sites/default/files/Book_Adams. Accessed (2017)

Exoskeletons as Mechatronic Design Example

Hannes Bleuler(✉), Tristan Vouga, Amalric Ortlieb, Romain Baud, Jemina Fasola, Jeremy Olivier, Solaiman Shokur, and Mohamed Bouri

EPFL Lausanne Robotic Systems Lab, LSRO, Lausanne, Switzerland
hannes.bleuler@epfl.ch

Abstract. Exoskeletons are a perfect example of a mechatronics product. They illustrate the close integration and interdependence of mechanical design, drive train, sensors, control strategy and user interface. Recent developments of our lab will be discussed in detail. Application examples include paraplegics, amputees, muscular dystrophy patients. The motivations of the users exoskeletons are as diverse as sporting challenge, life quality improvement for daily living, rehabilitation and social integration. Links to Cognitive Neurosciences will also be briefly discussed.

Keywords: Exoskeletons · Mechatronics · Assistive device · Haptics Human-machine interface

1 Introduction

Exoskeletons are a very lively research topic and first products have reached the market in recent years. However commercial success is still elusive for various reasons.

There are various reasons for this. We identify the absolute necessity to personalize exoskeletons in all aspects. This includes not only the engineering aspects (mechanical design, drive train, electronics), but also, more importantly, ergonomic issues and, essentially psychological and neuro-cognitive mechanisms such as acceptance, body consciousness, sense of agency, volition detection, to name the most salient topics. In this contribution, we will briefly report on three exoskeleton projects designed for very different types of users. We will concentrate on personalization of both, engineering and "psychological" aspects, prominently, of course, body consciousness. These projects have gone through a first design phase and are now in a testing phase, so that we are in a position to include first user feedback. We will also discuss the importance of integration of rich visuo-tactile feedback (virtual reality and haptic feedback) into these systems.

The three examples are: "TWIICE one": for a full paraplegic [4] (Fig. 1), "TWIICE H", for an amputee and "Autonomyo" for a muscular dystrophy patient.

G. Carbone et al. (Eds.): *New Trends in Medical and Service Robotics*, MMS 65, pp. 109–117, 2019.
https://doi.org/10.1007/978-3-030-00329-6_13

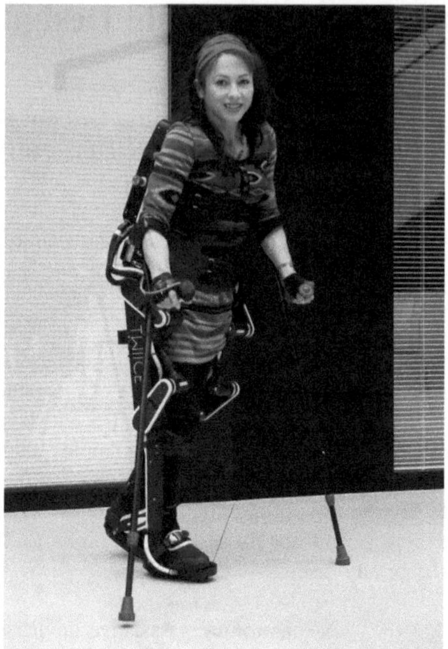

Fig. 1. The TWIICE exoskeleton with Silke Pan, an SCI- (ASIA A T10) full paraplegic who has never used an exoskelton (since her accident 10 years ago). TWIICE has 2 active and zero passive degrees of freedom per leg Photo: LSRO

2 The Twiice Exoskeleton

The project started three years ago (February 2015) in view of a sporting event for handicapped persons in the framework of the Pan-American games. It was proposed to design an exoskeleton for a teenager to race at the after-event showcasing pilots with disabilities. After the cancellation of this project with the Pan-American games, it was decided to register a candidacy for the Cybathlon scheduled for October 2016 [18]. The challenge here was a quite demanding series of obstacles (to be found on the Cybathlon web-site http://www.cybathlon.ethz.ch/.).

The obstacles included sitting down and standing up from a very low and soft living-room couch, going through a slalom, up a ramp, up and down standard stairs, opening and closing a door, planes inclined at various angles and irregularly spaced stepping stones.

The mechanical structure of the exoskeleton originally intended for the teen-ager at the American Para-Olympic games was already available. We were lucky to find an adult pilot fitting the physical restrictions of the teen-ager (size between 154 cm and 157 cm, weight less than ca. 55 kg. This "pilot" (Silke Pan) suffers from a Spinal Chord Injury (SCI). She is classified ASIA A (American Spinal chord Injury Assessment, "A" meaning complete loss of motor control AND sensation) at the vertebra T10 level (relatively high up, in the region of the diaphragm). She suffered a traumatic accident in 2009, chances of spontaneous recovery negligible.

2.1 Design, Basic Considerations

For such a patient, use of crutches should not be seen as a "technical inability" to provide balance. Actually, as all sensation is completely lost in the lower part of the body, such a patient will **want to feel physically the environment through crutches**. State of the art of exoskeletons without crutches is e.g. the New-Zealand designed "REX" exoskeleton, commercially available. It balances statically. This means very slow walking motion, complex multi-dof control and actuation, and, as a consequence, heavy weight and expensive price tag. Currently this type of exoskeleton is the only possibility for near-normal posture walking for tetraplegic patients (SCI at the level of the neck, i.e. loss of control & sensation on legs, arms and hands)

In our case (**paraplegic**, i.e. lower limbs only affected), balance in both degrees of freedom and walking direction changes are therefore **controlled by the crutches**. The legs need to be fully guided in predetermined sagittal trajectories.

Two to (max) three degrees of freedom per leg (hip, knee, and sometimes also ankle) are therefore enough. Motion will be **constrained in the sagittal plane**. Needless to say, it will not be possible for any user with full or limited control in his legs to walk in such a device: Lateral balance control is not possible (no abduction) without crutches.

2.2 Weight

Although the exoskeleton carries its own weight, light-weight is still important. This may not seem obvious, so let us briefly explain why: Low inertia especially at the fast moving body parts (foot, knee) is important for a fluid walking motion. Handling the exoskeleton when NOT using it should also not be underestimated: If it should be practical for daily use, it should be donned and doffed up to several times a day. Someone has to carry it, either the user himself (paraplegic!) or some attendant person (nurse, physio-therapeut, family member). It does make a big difference, if the device weighs 12 kg, 15 kg, 24 kg or even more! Moreover, even though the device should have enough power to climb stairs or a ramp, the user will tend to use a handrail if available. In this case, he or she will pull herself up by with the arms, so in the end it will be easier with a lighter device.

Finally, but by no means least, a lighter and less complex device will use less energy, and can be outfitted with smaller batteries, which will make longer operation possible before charging or exchanging the accumulators. This factor is actually of great importance in daily use. There are strict limits on Li-accumulator size in airlines and a handicapped person will also want to fly as anyone else.

2.3 Actuators: Two Active Degrees of Freedom Per Leg

For low weight, low inertia and simplicity, the choice was made not to actuate the ankle. The trade-off of potential advantages and drawbacks of a powered ankle in this situation clearly favours a passive ankle, although it is well known that a significant part of walking motion power is normally provided by the ankle joint. This choice implies careful design of the **rolling profile of the foot sole**. This important feature needs to be fine-tuned in close interaction with the user. It is obviously neither flat nor round, but essentially trapezoidal (short taper at heel, long one at toes), the exact geometric shape will depend on the pilot **AND** on the type of walking (daily life, sporting, smooth or natural terrain etc.)

Efficient electric motors rotate fast, so **reduction gearing** (in our case belt & harmonic drive) is necessary. The trade-off between walking speed, torque, lateral space and overall weight has determined the design of the TWIICE exoskeleton. The narrow lateral space requirement is needed for efficient use of the crutches and for moving at home around furniture, narrow doors etc.).

The low weight requirement can be met by using composite plates in a sandwich structure.

2.4 Control

Most publications on exoskeletons address the question of the control strategy. Basically the most cited control modes are either some kind of bio-signals, (EMG, Electromyograms, i.e. signals picked up at the muscles), body inclination (as e.g. Segway®, commercial exoskeletons like Rewalk® or Ekso®), or, very simply, but surprisingly little used, buttons. For a paraplegic as described above, lower limb EMGs are out of question (no signals at the level of muscles), except maybe some techniques of reinnervation or pickup at upper body muscles. Invasive or non-invasive brain-machine interfaces (BMI) have been proposed in preliminary studies (see review e.g. [19]. To our knowledge, the only attempt of an invasive BMI for locomotion of non-human primates has been published in 2016, [6]. For the time being, they seem not safe enough, but could become interesting in the future. A non- invasive BMI has recently been tested successfully by one of us with a group of paraplegic patients in the framework of a rehabilitation protocol [11]. While the results are promising for rehabilitation, it is at present too imprecise and not reliable enough for daily activity.

The simplest solution, push-buttons, in our case on the crutch, gives very good results and it is truly surprising that this method is not used more. Once volition is detected, the control strategy for paraplegics will be preprogramed trajectories. The motion mode and also the triggering of steps is controlled by these buttons. The user can choose between different operation modes such as slow or fast walking, going upstairs, downstairs, sit-to-stand transition and reverse etc. An interface screen (which could be the smartphone) displays the system status to the pilot.

The **Emergency button should block** all active links for a paraplegic, to prevent fall. For other users (as described in the next chapter), the emergency stop should on the contrary **restore full transparency**, so that the user still has a chance to restore balance with his residual motion capabilities.

3 Exoskeletons for Users with Full or Residual Control and Sensation in the Legs

3.1 Basic Considerations

In this case, the engineering task will start from completely different basic requirements as compared to the previous case (paraplegic). The users in this case could be from various populations, e.g. healthy persons needing assistance, elderly, most importantly stroke patients, or, more rare, sufferers of muscular dystrophy (myopathy) and similar, hemiplegics or others. As all these persons can at least partially walk and feel their feet & legs, the exoskeleton should be "transparent", i.e. not hinder normal walking. This alone is a huge technical challenge! It needs gravity compensation, full inertia compensation, friction compensation etc.

Once this is achieved, assistance should then be "added" gradually, similar to what is now familiar with most electric bicycles.

3.2 Mechanical Design

Active lateral stabilization has to be provided by the user, therefore the hip abduction movement, completely absent in case of TWIICE, is now needed. For such motions, low inertia is important for quick reaction. This results in the following design: No ankle actuation, three active degrees of freedom per leg (knee and hip flexion/extension, hip adduction/abduction), motors and reduction gear in the backpack, near the hip, with mechanical cable transmissions [8]. This allows assistance from zero up to almost 100% and low inertia at the legs. The mass is concentrated at the backpack where it contributes to balance control and does not impede quick motion of the extremities. (Figure 2)

Actuators (electric motors) are fast at low torque while the exoskeleton needs high torque at low speeds. This means reduction gear trains. As the motion is periodic, a lot of power is wasted in periodic acceleration and deceleration of the motor. One way to address this issue has been proposed by Jeremy Olivier: Using a combination of planetary gears, clutches and disk brakes to obtain changing output direction with a motor rotating always in the same direction, as presented in [17]. This would have clear advantages over an active inertia compensation through the feedback-loop. Disk brakes for example (as for bicycles) can be obtained off-the- shelf with low mass and very good torque control. There is room for interesting developments in mechatronics to obtain optimized exoskeleton designs.

3.3 Control, First Tests

Various sensing devices can be used for control: pressure transducers in the foot sole, proprioceptive angular sensors, possibly also force sensing, inertial measurement unit (IMU) at the backpack. Sometimes, electrodes on the muscles can be used for volition detection, as it is done e.g. for the HAL Cyberdyne commercial exoskeletons. Usually, force sensors will register the motion desired by the user and will then adapt, similar to torque detection in electric bicycles. We started testing this device with healthy users

Fig. 2. Exoskeleton « Autonomyo » for myopath. It has 6 d.o.f., incudes abduction. Motors at the back (photo LSRO)

and also, a few times up to now, with a patient suffering from muscular dystrophy. First results will be reported later.

The greatest challenge for such devices is balance control. Theoretically, such an exoskeleton should be able to walk almost autonomously (or at least close to it), i.e. like a humanoid robot. This means integration of the six components IMU, contact force-sensors at the feet and proprioceptive sensing. A user with weak muscular force, but still in possession of sensory feedback from his feet, will be able to control balance to a certain degree. At the same time, if the exoskeleton is equipped with its own active balancing system, there is a risk of conflict between the two balance control loops, the human's and the exoskeleton's. Especially if there is instability, each system (human and exoskeleton) can perceive the other one a disturbing factor. It will therefore be of prime importance, to implement control strategies that avoid this conflict. Achieving collaboration of the two control systems to improve the ability to recover from unbalance will be a prime challenge for future developments, especially for elderly, where fall prevention is a key issue.

4 Closing the Feedback Loop for Persons with Lower Limb Sensory Deficiency

Restoration of motor function is the main purpose of assistive exoskeletons. SCI patients usually also suffer from sensory deficiency, in this respect they are similar to amputees where the lack of sensory feedback is a major challenge. For many other potential exoskeleton users, sensory sensation is present, but motor function must be supplemented or fully provided by the device [9], Fig. 3.

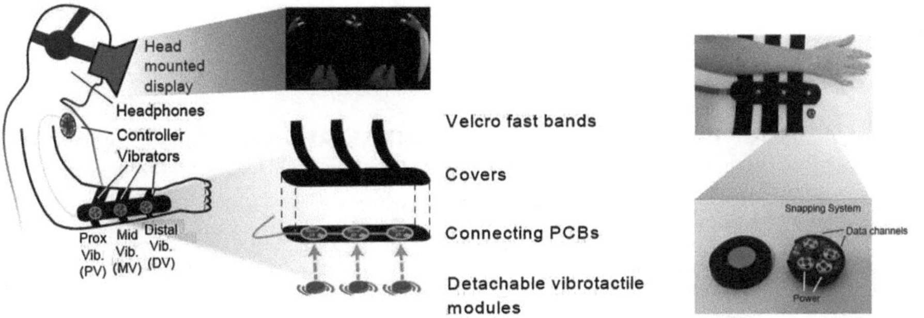

Fig. 3. Tactile feedback integrated with immersive virtual reality to study effect of use of tactile feedback for locomotion.

Recently, we developed and tested a strategy to help SCI patients recover missing lower-limb haptic information [10, 11] using a tactile shirt. It consists of arrays of vibrators on the patients' forearm displaying both the motion of the leg (combination of hip and knee angular position) and foot contact. Vibrations start e.g. at the distal spot on the forearm when ground contact is made and then progress towards the proximal vibrator as the hip-angle moves from front (less than 180° from vertical reference) to back (more than 180°).

Eight chronic complete SCI patients have been trained with this setup in a neuro-rehabilitation environment. After only a short of training, patients learn to perceive the position of the legs in space and, in several cases, even to "feel" the texture of floor on which they are walking, as they reported.

Integration of the tactile feedback into the locomotion is linked to the patients' perception of his own body [12]. It has been shown that following SCI [13] or amputation [14], the cortical representation of a patients' body changes. Such patients have a reduced sense of body ownership [15].

A psychometric assessment of body schema and embodiment has been implemented (cross congruency task [16]). Following a brief training in a virtual reality setting, the effect of embodiment of the virtual legs [10] can be observed. This effect is induced by the tactile feedback.

5 Conclusions

It has become evident that exoskeletons as mechatronic devices need to be personalized in all their aspects: Mechanical design, drive train, sensors and control, user learning strategy, medical follow-up. Emergency reaction of the device must be adapted for a specific user (transparency or the opposite, providing stability through blocking joints in case of any malfunction). Easy donning/doffing, device-body interface, overall weight, outdoor capability (also in rainy weather), and economic feasibility, certification, IP and legal aspects (product liability), acceptance by patients, therapists, doctors, health care insurance need to be considered carefully. All these aspects are interconnected. Development of exo-skeletons is therefore a particularly challenging mechatronic problem. But the potential benefits for large user populations are evident. The technical feasibility of efficient exoskeletons has been demonstrated by the impressing progress of various research projects on exoskeletons worldwide.

Acknowledgments. The authors wish to thank ASRIMM (Association Suisse Romande Intervenant contre les Maladies neuro-Musculaires), FSRMM (Fondation Suisse de Recherche sur les Maladies Musculaires), Fischer Connectors SA, Sonceboz SA, Lions Club Vevey for their support.

References

1. Olivier, J.: Development of Walk Assistive Orthoses for Elderly. PhD thesis No 6947, EPFL Lausanne (2016)
2. Shokur, S.: Virtual reality based Brain-Machine-Interface for Sensori-Motor and Social experiments with Primates (2013)
3. Young, A., Ferris, D.: State-of-the-art and future directions for robotic lower limb Exoskeletons. IEEE Trans. Neural Syst. Rehabil. Eng. **99**, 1–1 (2016)
4. Vouga, T., Baud, R., Fasola, J., Bouri, M., Bleuler, H.: TWIICE - A lightweight lower-limb exoskeleton for complete paraplegics. In 2017 International Conference on Rehabilitation Robotics (ICORR), pp. 1639–1645 (2017)
5. Tucker, M.R., et al.: Control strategies for active lower extremity prosthetics and orthotics: a review. J. NeuroEng. Rehabil. **12**(1), 1 (2015)
6. Vouga, T., et al.: EXiO - a brain-controlled Lower Limb Exoskeleton for Rhesus Macaques. IEEE Trans. Neural Syst. Rehabil. Eng. **25**(2), 131–141 (2017)
7. Ortlieb, A., Olivier, J., Bouri, M., Bleuler, H., Kuntzer, T.: From gait measurements to design of assistive orthoses for people with neuromuscular diseases. In 2015 IEEE International Conference on Rehabilitation Robotics (ICORR), pp. 368–373 (2015)
8. Ortlieb, A., Bouri, M., Bleuler, H.: AUTONOMYO: Design Challenges of Lower Limb Assistive Device for Elderly People, Multiple Sclerosis and Neuromuscular Diseases. Wearable Robotics: Challenges and Trends, pp. 439–443. Springer, Cham (2017)
9. Giuliani, C.A., Smith, J.L.: Stepping behaviors in chronic spinal cats with one hindlimb deafferented. J. Neurosci. **7**, 2537–2546 (1987)
10. Shokur, S., et al.: Assimilation of virtual legs and perception of floor texture by complete paraplegic patients receiving artificial tactile feedback. Sci. Rep. **6**, 32293 (2016)
11. Donati, A.R., et al.: Long-term training with a brain-machine interface-based gait protocol induces partial neurological recovery in paraplegic patients. Sci. Rep. **6**, 30383 (2016)
12. Iriki, A., Tanaka, M., Iwamura, Y.: Coding of modified body schema during tool use by macaque postcentral neurones. Neuroreport **7**(14), 2325–2330 (1996)
13. Fuentes, C.T., Pazzaglia, M., Longo, M.R., Scivoletto, G., Haggard, P.: Body image distortions following spinal cord injury. J. Neurol. Neurosurg. Psychiatry **82**, 201–207 (2013)
14. Canzoneri, E., Marzolla, M., Amoresano, A., Verni, G., Serino, A.: Amputation and prosthesis implantation shape body and peripersonal space representations. Sci. Rep. **3**, 1–8 (2013)
15. Lenggenhager, B., Pazzaglia, M., Scivoletto, G., Molinari, M., Aglioti, S.M.: The sense of the body in individuals with spinal cord injury. PLoS ONE **7**, e50757 (2012)

16. Maravita, A., Husain, M., Clarke, K., Driver, J.: Reaching with a tool extends visual–tactile interactions into far space: Evidence from cross-modal extinction. Neuropsychologia **39**(6), 580–585 (2001)
17. Olivier, J., Ortlieb, A., Bouri, M., Bleuler, H.: Mechanisms for actuated assistive hip orthoses. J. Robot. Auton. Syst. Special issue ``Wearable Robotics", Elsevier **73**, 59–67 (2014). https://doi.org/10.1016/j.robot.2014.10.002
18. Riener, R.: The Cybathlon promotes the development of assistive technology for people with physical disabilities. J. Neuroeng. Rehabil. **13**, 49 (2016)
19. He, Y., Eguren, D., Azorín, J.M., Grossman, R.G., Luu, T.P., Contreras-Vidal, J.-L.: Brain-machine interfaces for controlling lower-limb powered robotic systems. J. Neural Eng. **15**(8), 021004 (2018)

A Characterization of a Robotic Hand with Movable Palm

F. J. Espinosa García[1(✉)], M. Ceccarelli[2], M. Arias Montiel[3], G. Carbone[2], E. Lugo Gonzalez[3], and M. Russo[2]

[1] Postgraduate Division, Technological University of the Mixteca, 69000 Acatlima, Mexico
fjeg_1234@hotmail.com

[2] LARM: Laboratory of Robotics and Mechatronics, University of Cassino and South Latium, Via G. Dibiasio 43, 03043 Cassino, Italy
marco.ceccarelli@unicas.it

[3] Institute of Electronics and Mechatronics, Technological University of the Mixteca, 69000 Acatlima, Mexico
elugog@mixteco.utm.mx

Abstract. In this paper the numerical results for the characterization of a new design of an underactuated mechanism for a robotic hand are presented. The hand is composed by a palm and one finger. The finger is actuated by a four-bar mechanism and the palm is composed by two links which are actuated by a spring. The aim of the proposed mechanism is to improve the grasping performance of the LARM hand and avoid to increasing in the number of actuators. The numerical simulations have been carried out using the SolidWorks environment to evaluate the performance of the mechanism on grasping objects with different shapes, the obtained results show the optimal tracking of trajectories of this proposed solution.

Keywords: Robotic hands · Design · Finger mechanism · Movable palm

1 Introduction

Human hand is considered as a highly functional component, since it allows to successfully manipulate and interact with the environment. In recent years, in order to develop anthropomorphic finger mechanism, researchers have used two different approaches: namely complex mechanisms with a reduce number of degrees of freedom (DOFs) and actuators with reduced performance but a simplified device operation [1], and sophisticated designs with a full replica of human hand. The design of anthropomorphic hand must fulfill basic requirements for human-like motion and grasp, such as compact size actuation, lightweight, efficiency, position and force control, by respecting functional requirements of manipulative dexterity, grasp robustness, and human operability [2, 3].

By using underactuated mechanisms it is possible to achieve an adaptive grasp that mimics the human grasping action. Human hand can also be considered underactuated, since the distal phalanges of the fingers are not independently controllable [4].

G. Carbone et al. (Eds.): *New Trends in Medical and Service Robotics*, MMS 65, pp. 118–125, 2019.
https://doi.org/10.1007/978-3-030-00329-6_14

A mechanism is defined underactuated when its number of actuator is smaller than the number of degrees of freedom of the mechanism [1].

In this paper a characterization of a new underactuated artificial hand based on a movable palm is proposed for an improved the LARM hand grasping behavior as focused on easy operation and anthropomorphic grasp.

2 The Hand Design with Movable Palm

Most of robotic hands designs have been focused on the fingers mobility and they have a static palm. In those robotic hands, in contrast with the human hand, there is not a relative motion between fingers and the palm, which incurs limitation for manipulation and the finger usually operates on one plane only, and it is difficult to adapt the hands to the shape of the manipulated object [5, 6]. Based on a movable palm, a robotic hand has been developed overpassing those issues. The proposed mechanism is an innovative solution that is designed with an actuator which moves the palm and the finger to grasp an object. Figure 1 shows the kinematic scheme with link indications, while Fig. 2 gives a CAD design with the mechanical design in open and closed configurations of the palm.

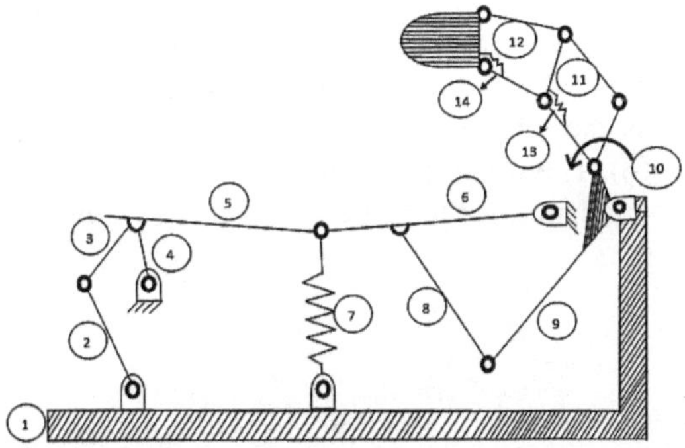

Fig. 1. A kinematic scheme of the proposed design of a new robotic hand.

Referring to Fig. 1, the mechanism for robotic hand is composed by the palm and one finger. The palm is composed by links (5) and (6). Link (5) is actuated by links (2, 3, 4). Palm link (6) is actuated by linkage (8, 9). Spring (7) acts on palm mechanism. The underactuated finger linkage is composed by links in the four bar linkages (11) and (12) and there are two springs (13, 14) which permit the finger to adapt its configuration to shape and size of the object under grasp. The finger can be moved by servo-motor (10). Figures 2(a) and (b) show an example of construction of the proposal mechanism for robotic hand illustrating the adaptive palm motion from a home planar configuration up to a wrapping configuration that is adjusted also by the spring action.

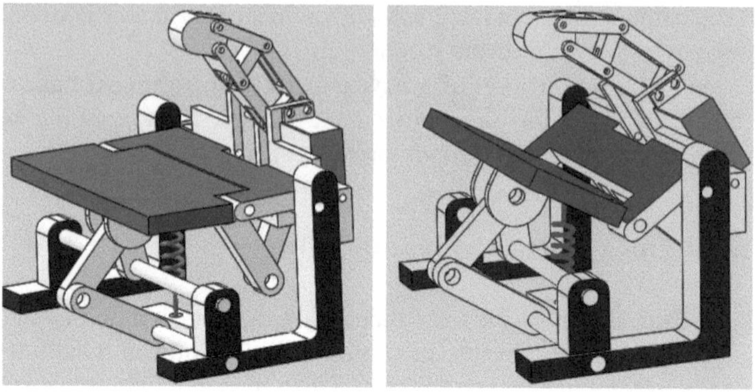

a) b)

Fig. 2. CAD model of the proposed robotic hand, (a) Open palm and (b) closed palm.

3 The Simulation Model

Simulations have been worked out using the SolidWorks environment. The simulations have been planned for three cases, namely without an object, grasping a cylindrical object and grasping an object with an irregular shape, the action of gravity is considered in all the cases. The aim of numerical simulations is to evaluate the parameters which are shown in Fig. 3. FP is the angular position of the fingertip body and its initial position is 221°. TM is the torque of the motor, PP1 and VP1 are the position and velocity on palm link (6). The initial position of the link 6 is 181°. LS is the length of the spring. In Fig. 4a simulation for the case in which a cylindrical object is gripped is

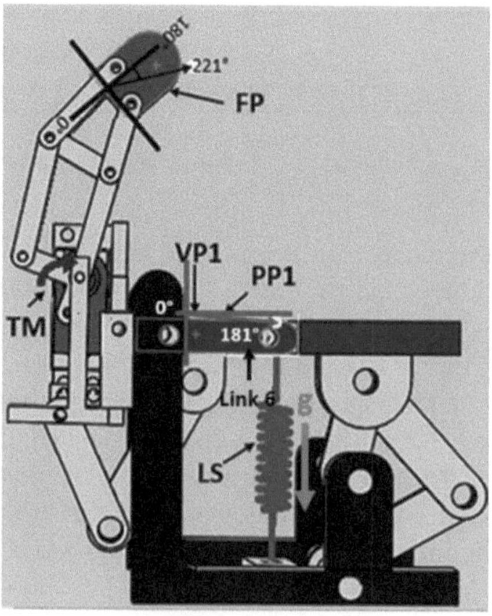

Fig. 3. Main parameters for simulation results.

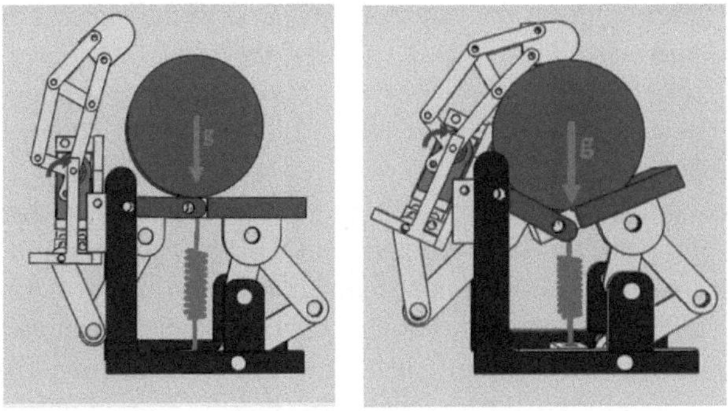

a) b)

Fig. 4. Example of simulation with a cylindrical object: (a) Initial position and (b) grasping.

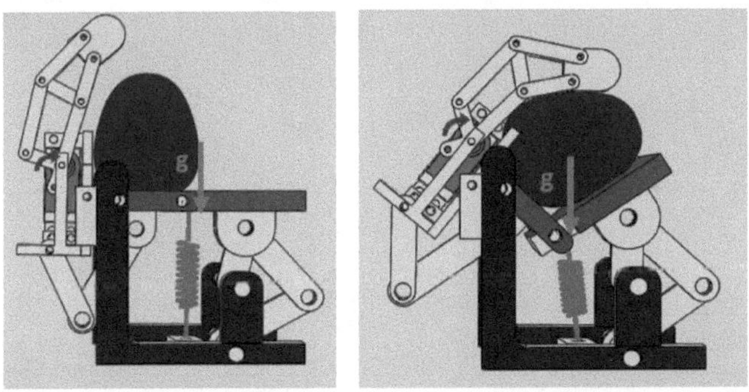

a) b)

Fig. 5. Example of simulation with an object with irregular shape: (a) Initial position and (b) grasping.

shown. This object is made of aluminum 7075, its weight is 1.81 kg and its body is 10 cm long with a diameter of 70 mm. In Fig. 5, an object with irregular shape is gripped. Its body is 8 cm long, its weight is 107.10 grams and it is made of plastic.

Referring to Figs. 4 and 5, the palm and the finger are on an initial position, after that, the palm and finger are actuated, so that the finger moves against the palm. With the movement of the palm the grasped object is closed also between palm links (5, 6). When the finger is in contact to the object, the four-bar linkages (13, 14) are actuated to get a wrapping grasp by the finger.

4 Numerical Evaluation

Numerical evaluations have been worked out in the SolidWorks environment by running the designed underactuated mechanism with design features and anthropomorphic motion with the characteristics that are previously presented. The results are reported in this section. Figures 6, 7 and 8 correspond to the simulation of a trajectory tracking without object and Figs. 9 and 10 show the results of grasping a cylindrical object. Finally, Figs. 11 and 12 correspond to simulation grasping an object with irregular shape. The time of simulations is 6 s due to that is enough to the palm and finger wrap an object.

In Fig. 6 the computed angular position of the fingertip is shown. The initial position is −41°, the position starts to change in 0.30 s. The spring is actuated by the weight of the mechanism in consequence finger starts to move until 4.80 s due the spring cannot be contracted anymore. After that time, the servomotor starts to rotate and the finger is actuated by it until final time of 6 s achieving a position of 16°. Figure 7 shows the angular position of the palm (link 6). The initial position is −179° and after 4.80 s, the palm achieves the maximum position which is −128°. The angular velocity of the palm is shown in Fig. 8. The maximum angular velocity of the palm is

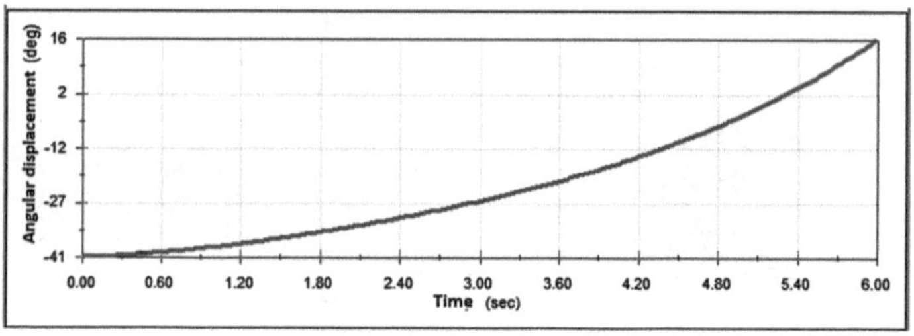

Fig. 6. Computed angular displacement of the fingertip during a motion without object.

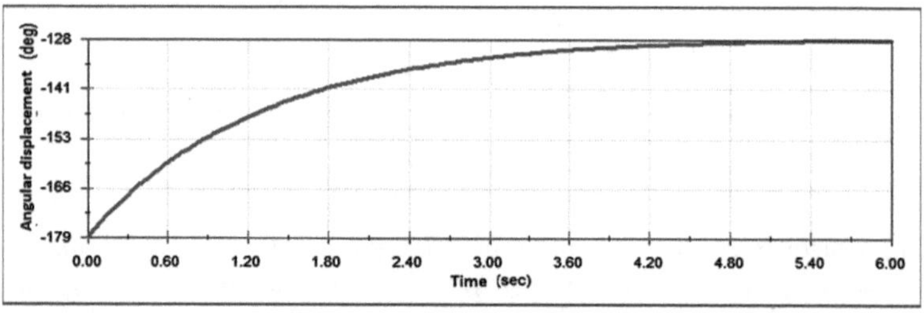

Fig. 7. Angular position of the palm.

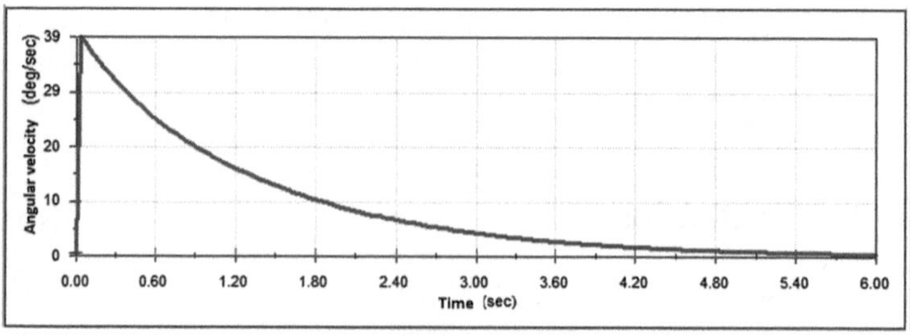

Fig. 8. Angular velocity of the palm.

39° per second near to the motion start. The palm is actuated by the spring and the movement is soft due to there is not an object that generates sudden movements.

Figure 9 shows the angular position of the fingertip for the cylindrical object grasp case. The initial position is −41°, from .60 s to 1 s the position is −58 due to the fingertip touches the cylindrical object. From 1.10 to 6 s, the finger achieves a position of −26°, in this period the finger is actuated by the spring and the position changes because the phalanxes touch the object and as result the fingertip is driven up and it stops touching the object. After that time, the palm cannot move and the servomotor starts to move the finger mechanism to the aim to adapt to the shape of the cylindrical object, as a consequence the fingertip achieves to touch the object again. The difference is that the finger has adapted to the shape of the object (Fig. 4b). Figure 10 shows the angular position of the link 6 (palm). From 0 to 1.20 s. link 6 changes its position until −151° and after that, the position is the same. It is due to the object is wrapped by the palm and a consequence, the palm cannot move any more.

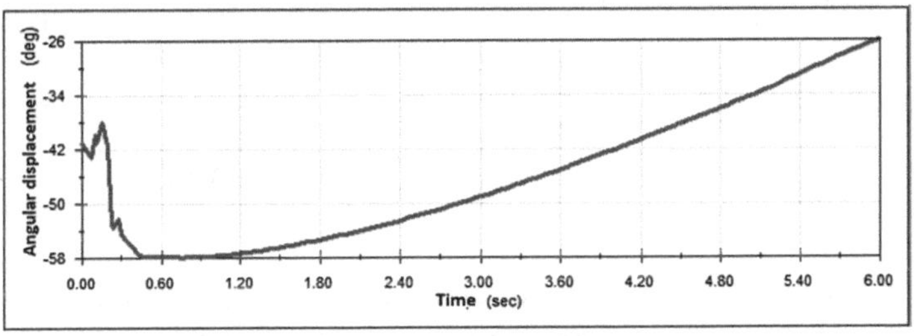

Fig. 9. Angular position of the fingertip.

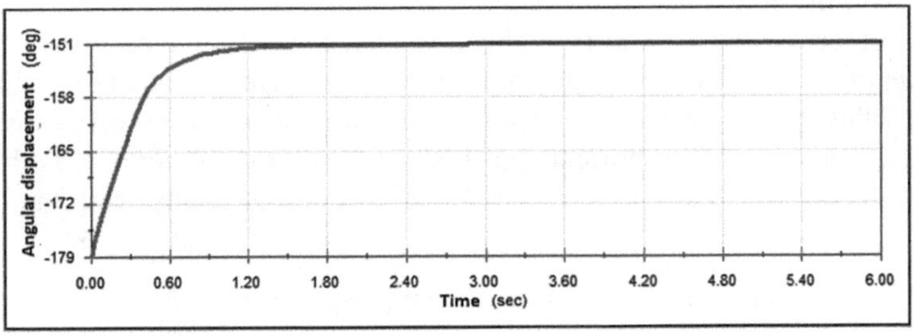

Fig. 10. Angular position of the palm (link 6).

Figure 11 shows the angular position of the fingertip in grasping with an irregular object. The initial position is −41°, from 0 s to 1.20 s the position is −69 due to the fingertip touches the irregular object. From 1.50 to 6 s, the finger achieves a position of 12°. The finger is actuated by the spring until 2.40 s. after that the servo-motor actuates the finger. Figure 12 shows the angular position of the link 6 (palm). From 0 to 2.40 s. link 6 changes its position until −136° and after that, the position is the same. It is due to the object is wrapped by the palm and a consequence, the palm cannot move any more.

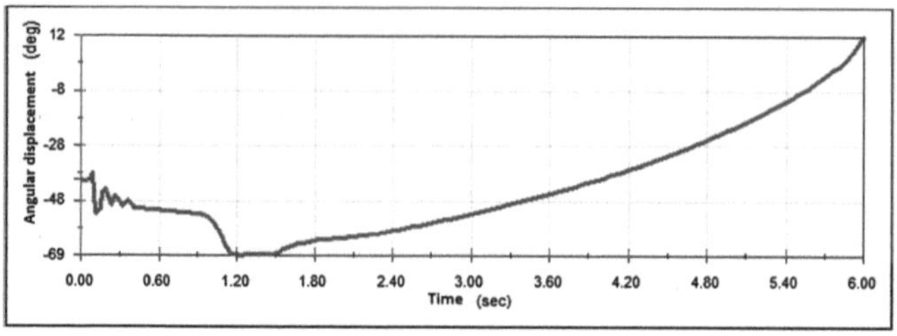

Fig. 11. Angular position of the fingertip.

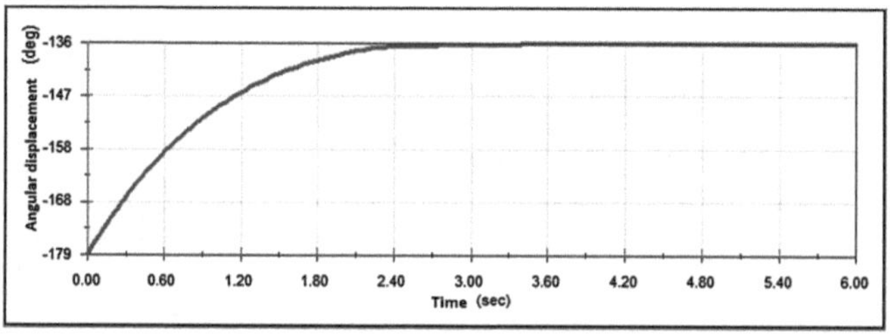

Fig. 12. Angular position of the palm (link 6).

Numerical results obtained from the simulation for the motor torque and grasping forces are listed in Table 1. It shows the values of maximum torque for grasping the two different objects. Based on the results, grasping an irregular object requires more torque power that when object has a regular shape. The forces applied by the palm and the fingertip to the objects are also indicated with different values. In the case to grasp a cylindrical object a higher force is applied. Finally, the deformation presented by the spring is reported from its initial value of 68 mm and a spring constant of 1 N/mm. The minimum value is 52 mm that is not due to the weight of the object because if the objects are compared, the cylindrical object is heavier than the other one. In this case is due to the shape as the palm requires to be closed more to totally wrap the object.

Table 1. Maximum values for grasping simulations

	Without object	Cylindrical object	Object with irregular shape
Torque [Nmm]	10.0	30.1	36.0
Finger force[N]	0	1.5	0.8
Palm force[N]	0	5.0	3.0
Spring length [mm]	55	56	52

5 Conclusion and Future Works

In this paper, the numerical characterization of a new mechanism for a robotic hand is presented with an active palm. The finger is actuated by one servomotor, in cooperation with a spring on the palm. The robotic hand mechanism has been simulated in the SolidWorks environment to grasp objects with regular and irregular shapes. The main advantage of this mechanism is that palm and finger can adapt their configuration to the shape and size of an object due to the movement of the palm. The undergrasp structure of the mechanism permit moves the palm and the finger only with one motor. Other advantages are the easy operation and simple design for a fairly simple manufacturing. This mechanism can be useful in the field of robot arms due to improve their grasp abilities. As a future works a dynamic study of the hand, the study of the influence of the spring constant in the gripping and an experimental study of the prototype are proposed.

Acknowledgements. First author is grateful to the National Council of Science and Technology (CONACYT) from Mexico for supporting his master studies by the national grants program and his research stay at the Laboratory of Robotics and Mechatronics (LARM) in the University of Cassino and South Latium, Italy during 2017–2018.

References

1. Ceccarelli, M.: Finger mechanism for robotic hands. In: Bai, S., Ceccarelli, M. (eds.) Recent Advances in Mechanism Design for Robotics, pp. 3–13. Springer (2015)
2. Krausz, N.E., Rorrer, R.A., Weir, R.F.: Design and fabrication of six degree-of-freedom open source spurce hand. IEEE Trans. Neural Syst. Rehabil. Eng. **24**(5), 562–572 (2016)
3. Ceccarelli, M., Rodriguez, N.E., Carbone, G.: Optimal design of driving mechanism in a 1-DOF anthropomorphic finger. Mech. Mach. Theor. **41**(8), 897–911 (2005)
4. Ceccarelli, M., Tavolieri, C., Lu, Z.: Design considerations for underactuated grasp with one D.O.F. anthropomorphic finger mechanism. In: International Conference on Intelligence Robots and Systems IROS, pp. 1611–1616 (2006)
5. Wei, G., Dai, J.S., Wang, S., Luo, H.: Kinematic analysis of a metamorphic anthropomorphic hand with a reconfigurable palm. Int. J. Humanoid Robot. **8**(3), 459–479 (2011)
6. Dai, J.S., Wang, D., Cui, L.: Orientation and workspace analysis of multifingered metamorphic hand-metahand. IEEE Trans. Robot. **25**(4), 942–947 (2009)

Wearable Working Assist Mechanism for Hemiplegics Capable of Changing Step Length and Walking Direction

Daisuke Matsuura[1(✉)], Yuta Chounan[2], Yusuke Sugahara[1], and Yukio Takeda[1]

[1] Tokyo Institute of Technology, Tokyo, Japan
matsuura@mech.titech.ac.jp
[2] ex-Tokyo Institute of Technology, 2-12-1 Ookayama, Meguro-ku, Tokyo 152-8552, Japan

Abstract. In the present paper, a systematic method useful in the structural design and motion synthesis of motion assist exoskeleton. To this end, a method for analyzing complex human motions with multiple degrees of freedom for different functions and for regenerating arbitrary motions from the analyzed data by adjusting parameters based on the principal component analysis is described. An illustrative example on the design of walking assist device for hemiplegics is presented.

Keywords: Walking assist machine · Joint input generation Design optimization · Principal component analysis

1 Introduction

Approximately 1.18 million patients suffer from stroke in Japan in the year 2014. Stroke is one of the major reasons for elderly to need nursing care. After stroke, patients may have possibility to become hemiplegia. In order to recover such situation, rehabilitation is quite important. In the development of practical and wearable assist device for walking rehabilitation, mechanism design of such device which can perform cooperative movement of limbs of human (patient) with allowable costs should be done. To this end, a systematic methodology to reflect the characteristics of human locomotion on the design of such mechanism is demanded.

In our previous works, walking assist devices using crutches for paraplegics and hemiplegics have been proposed, and structural design and motion synthesis for several gaits for step ascending and descending gaits as well as straight walking gaits on a horizontal floor, have been carried out [1–5].

As a design tool of such assistive devices, systematic procedure for structural design and motion synthesis using reference motions obtained from healthy subjects would be helpful. In this article, a method for analysis of motion obtained by motion capture experiments and regeneration based on the principal component analysis are introduced. As an example of the application of the systematic method, a design and motion

G. Carbone et al. (Eds.): *New Trends in Medical and Service Robotics*, MMS 65, pp. 126–133, 2019.
https://doi.org/10.1007/978-3-030-00329-6_15

synthesis of a wearable walking assist device for hemiplegics which has functions of changing step length and walking direction on the flat floor will be described.

2 Gait Analysis and Generation Based on Principal Component Analysis

As the first step of human gait analysis, motion of the gait should be digitized and corrected in computer. For this purpose, optical motion capture systems capable of tracking many target markers simultaneously are widely used these days because of their ease of use and scalability. In addition to the target marker's position, other measurement information such as ground reaction forces (GRF) measured by force plates can be corrected at the same time. Accuracy of typical motion capture systems capable of measuring human gate is in-between several millimeters to sub-millimeters.

Corrected information by the motion capture system such as time trajectory of target marker's position and ground reaction force can be analyzed by using mathematical analysis methods. Key point of such analysis methods is how effective and easy they are in finding proper explanatory variables. Since human's gait is a complex phenomenon, purpose of the measurements is valid such like extracting personal characteristics (according to one's age, sex, body height or weight, disease, etc.), motion pattern characteristics (walking, turning, running, etc.), change of motion due to the effects of wearing robotic devices, etc. In order to extract such characteristics effectively, the analysis method should be able to reduce the number of explanatory variables. In addition, the analysis method should also work reversely to generate the motion of robotic welfare devices.

Principal Component Analysis (PCA) is one of the latest and most effective methods satisfying above-mentioned requirements. Mathematically, it is a kind of orthogonal linear transformation to transform the database matrix Z into a new coordinate frame of which 1st axis is aligned to the most significant explanatory component. This can be achieved by obtaining a set of eigenvectors called *principal components* and eigenvalues called *factor loadings* (or *factor scores*). Figure 1 shows a basic scheme of data analysis and synthesis using PCA. First of all, a database matrix Z is constructed by arranging a number of N component vectors (say, N is a number of trials), $z_{n\ (n\,=\,1,\ldots,N)}$, which has p particular arguments. Since each of the p columns of Z can be different kind of information such as position, force, and so on, each of the n rows should be normalized (z-scored). After constructing the database matrix, order reduction, and extraction of explanatory vectors and variables are performed.

As the result, Z can be written as a dot product of principal component scores $\boldsymbol{t}_n$ and factor loadings $\boldsymbol{w}_m$ as $\boldsymbol{t}_n = z_n\ \boldsymbol{w}_m$, for each $n = 1,\ldots,N$, $m = 1,\ldots,M$, where M is the number of principal components. This first component can be found by solving the following maximizing problem:

$$\boldsymbol{w}_1 = \arg\max\left\{\sum_{n=1}^{N}(t_1)_n^2\right\} = \arg\max\left\{\boldsymbol{w}^{\mathrm{T}}Z^{\mathrm{T}}Z\boldsymbol{w}\right\},\quad \text{while } \|\boldsymbol{w}\| = 1. \tag{1}$$

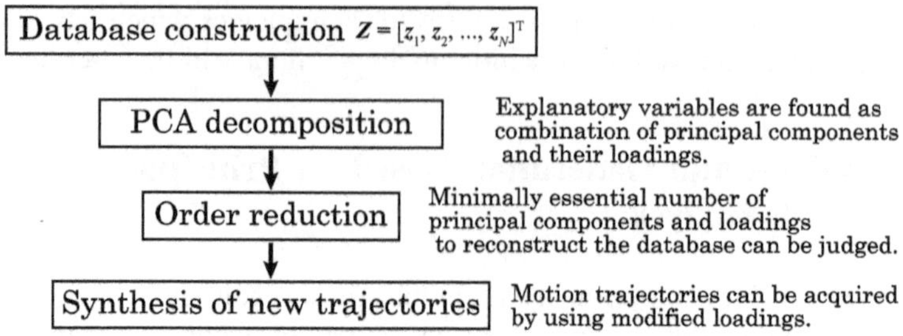

Fig. 1. Flowchart of principal component analysis

Since the constraint $\|\boldsymbol{w}\| = 1$, Eq. (1) can be written as

$$\boldsymbol{w}_1 = \arg\max\left\{\frac{\boldsymbol{w}^{\mathrm{T}} Z^{\mathrm{T}} Z \boldsymbol{w}}{\boldsymbol{w}^{\mathrm{T}} \boldsymbol{w}}\right\}. \tag{2}$$

A standard result for a positive semidefinite matrix, $Z^{\mathrm{T}}Z$, is that the quotient's maximum possible value is the largest eigenvalue of the matrix, which occurs when $\boldsymbol{w}$ is the corresponding eigenvector. In this way, first component $\boldsymbol{w}_1$ can be found. After that, further principal components can be found recursively. In this time, database $\mathbf{Z}$ should be rewritten as

$$\hat{Z}_m = Z - \sum_{s=1}^{m-1} Z \boldsymbol{w}_s \boldsymbol{w}_s^{\mathrm{T}}, \tag{3}$$

which represents the remaining part of the database independent from the first principal component. Further principal components can be obtained by using rewritten Eq. (1) as

$$\boldsymbol{w}_m = \arg\max\left\{\frac{\boldsymbol{w}^{\mathrm{T}} \hat{Z}_m^{\mathrm{T}} \hat{Z}_m \boldsymbol{w}}{\boldsymbol{w}^{\mathrm{T}} \boldsymbol{w}}\right\}. \tag{4}$$

As the result, full principal components decomposition can be given as

$$T = ZW, \tag{5}$$

where T and W are a factor loading matrix and a p by p matrix of which each column is an eigenvector of $Z^{\mathrm{T}}Z$, principal components namely, respectively. This decomposition can also reversely work as

$$Z = TW^{\mathrm{T}} \text{ (since } W \text{ is an orthogonal matrix, } W^{-1} = W^{\mathrm{T}}). \tag{6}$$

By modifying each component of W^{-1} from original ones, one can synthesize new (modified) data set. When a meaningful relationships between demanded features and principal components can be found, enhancement or decline of that feature can be done by modifying relating factor loadings.

3 Joint Input Trajectory Synthesis and Design Optimization of the Wearable Walking Assist Device

3.1 Mechanism Design of the Wearable Walking Assist Device

The walking assist device for hemiplegics proposed in this paper is aiming to assist whole lower limb motions of a user's disabled side, it should be able to drive one's hip, knee and ankle joints. Those three joints can be represented by a spherical, revolute and again spherical joint, respectively, as shown in the leftmost of Fig. 2. In order to design an assist device that can generate a whole-lower limb motion, and is allowed to be separated into several components, one can find out four different patterns as shown in Figs. 2(i)–(iv) (in each of the patterns, numbers in the ellipses identify the components). Among those candidates, pattern (ii) could be picked up as a suitable one, since it is generally said that among the disabled joints of hemiplegics, ones closer to trunk, which is hip joint in this case, tend to be recovered earlier than the others. Accordingly, hip joint assist unit tends to become unnecessary at first although it is the heaviest part among the device, and therefore it should be removable from the remaining components of the device.

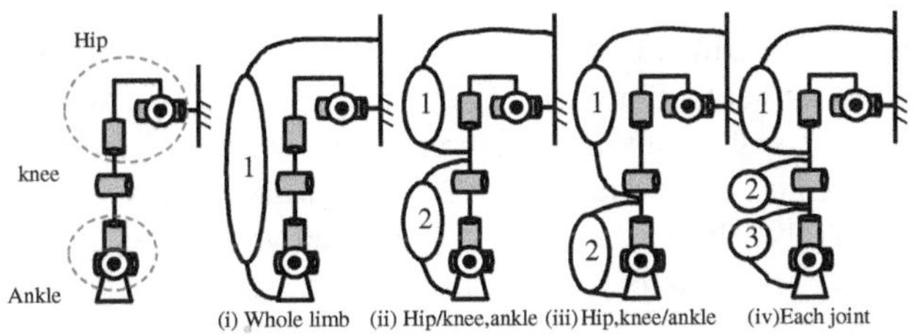

Fig. 2. A kinematic model of human lower limb (leftmost) and four possible separation patterns of the walking assist device (i)–(iv)

Based on the above consideration, mechanisms to assist hip joint, and knee and ankle joints were figured out as shown in Fig. 3. For ease of designing, those mechanisms were composed of revolute joints only. In addition, in the upper mechanism shown in Fig. 3(a), the first rotation axis, A, is parallelized to a user's hip joint's extension/flection axis. The Secon and fourth, B and E, and third and fourth axes, C and D, are in parallel each other, and the fifth axis, E, is parallelized to the inner/outer rotation axis. As the result, rotational motion around the axis perpendicular to both axes of D and E is always constrained while decoupling spatial translational movement of a thigh against a trunk. In the lower mechanism shown in Fig. 3(b), axes F and G, and H and I are parallelized to each other to decouple the spatial translational movement and rotation of a shank against a thigh. As the result, rotational motion around the axis perpendicular to both axes of G and H is always constrained. As the result, adduction and abduction of user's hip and ankle joints is naturally constrained by the two mechanisms, and remaining axes of the whole-lower limb can be driven. This feature can make the fabrication and motion synthesis/control easy. In order to make a decision where to attach minimally essential actuators and to synthesize input trajectories for them, PCA were applied.

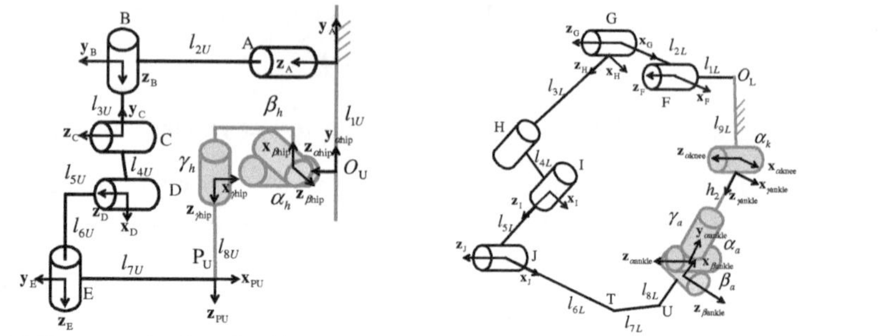

(a) Upper mechanism (for hip joint) (b) Lower mechanism (for knee and ankle joints)

Fig. 3. Joint compositions of each of two walking assist mechanism components.

3.2 Joint Input Trajectory Synthesis Based on PCA

In this section, joint input trajectories that can achieve straight gait with variable step length and turning gait on a horizontal ground are synthesized. First of all, database matrix was constructed by following steps.

(a) Motion capture experiments using a marker sets shown in Fig. 4(L) were carried out with seven subjects. For each of subjects, five-different walking patterns that were [i] Straight walk without specifying step length, [ii] Straight walk with as large a step length as possible, [iii] Straight walk with small a step length as possible, [iv] Walk while turning right, [v] Walk while turning left, were recorded (each of the patterns was performed five times).

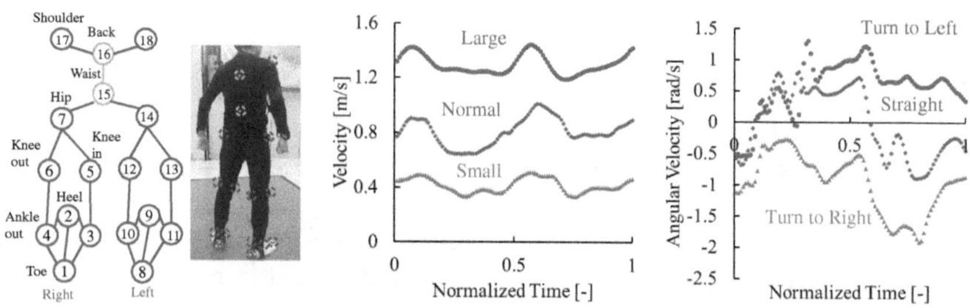

Fig. 4. Target marker positions (L), Measured velocity (Mid.) and Angular velocity (R).

(b) From acquired data, each joint angle, and translation and angular velocity of the trunk, v_{cx} and ω_{cz}, were calculated. Figures 4(Mid.) and (R) show one of the obtained trajectories of translation and angular velocities. From those trajectories, walking pattern evaluation measures were calculated as

$$v_{cx} = \sum_{\tilde{t}=1}^{100} v_{x,\tilde{t}}, \quad \omega_{cz} = \sum_{\tilde{t}=1}^{100} \omega_{z,\tilde{t}}, \tag{7}$$

that approximately correspond to translational and rotational displacement among entire trials.

(c) By using the acquired joint angles and evaluation measures in Eq. (5), database matrix Z was constructed with z-scored element vectors:

$$Z = \begin{bmatrix} \tilde{z}_1 \\ \vdots \\ \tilde{z}_N \end{bmatrix}, \text{ where } z_n = \begin{bmatrix} \boldsymbol{\alpha}_{1,n} & \boldsymbol{\beta}_{1,n} & \gamma_{1,n} & \cdots & \gamma_{6,n} & v_{cx,n} & \omega_{cz,n} & T_n \end{bmatrix}, \quad (8)$$

where $\boldsymbol{\alpha}$, $\boldsymbol{\beta}$, γ are angular displacement vectors on the axes of flexion and extension, adduction and abduction, and inner and outer rotation, and T is the length of gait cycle. Since each of angular displacement vectors includes 100 scalars, one z_n vector includes 1803 scalars in total.

After carrying out the PCA decomposition, factor loadings with respect to the three evaluation measures, v, ω, and T are obtained as shown in Fig. 5. As shown in Table 1, cumulative contribution ratio up to eighth principal component passed over 80%. It could thus be said that up to eighth principal components can sufficiently explain the experimental results. In addition, the fourth principal component has a strong positive relationship between the angular velocity, the fifth component has a strong negative relationship between the translational velocity, and also a mid-strong positive relationship between the gait cycles. These relationships can also be observed in the distribution of factor loadings plotted in Figs. 6(a) and (b). Circled numbers in each figure represent one of the above-mentioned five gait patterns. Factor loading against fourth component has significant peak on the pattern [iv] and [v], in which turning right and left had performed. On the other hand, that against fifth component has a peak at

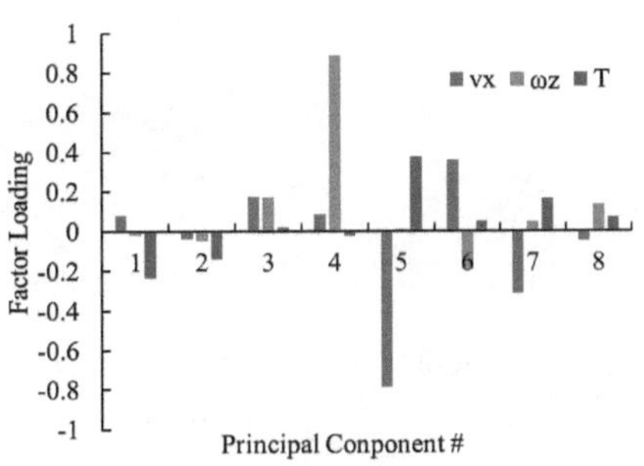

Fig. 5. Factor loadings with respect to each principal component.

Table 1. Contribution ratio of each principal component.

ID number of PC	#1	#2	#3	#4	#5	#6	#7	#8
Contribution ratio	20.6	19.3	15.4	7.6	6.2	5.5	3.8	2.8
Comulative contribution ratio	20.6	39.9	55.3	62.1	69.1	74.6	78.4	81.2

the pattern [iii] in which subjects had tried to walk with a smaller step length than the other patterns. From these results, it was predictable that input joint trajectories to achieve step length changing and turning can be synthesized by using reverse PCA decomposition with modified factor loadings in which fourth and fifth loadings were promptly changed. Figures 7 and 8 illustrate numerical examples. By changing the fourth factor loading in positive and negative direction, heading direction was changed to left and right, respectively, as shown in Fig. 7. On the other hand, by changing the fifth factor loading in positive and negative direction, step length became small and large step lengths, respectively, as shown in Fig. 8.

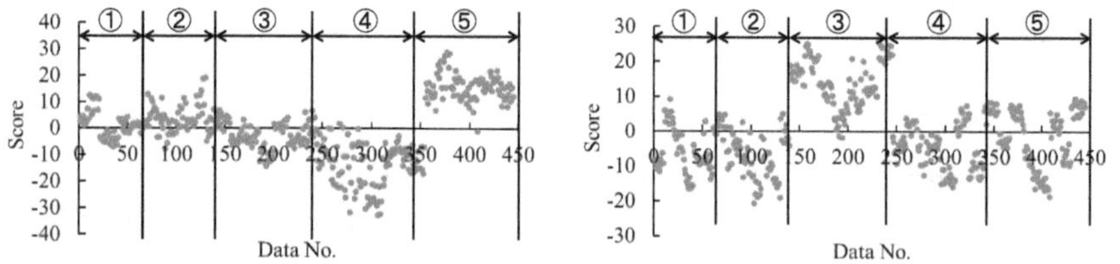

(a) Factor loadings against 4th component (b) Factor loadings against 5th component

Fig. 6. Factor loadings with respect to each data element.

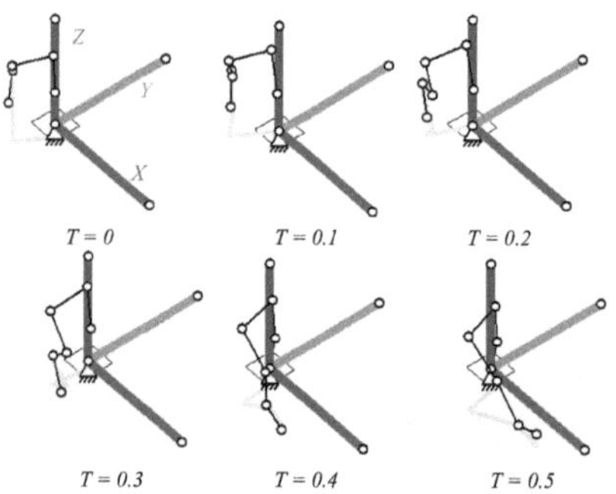

Fig. 7. Result of motion generation ($w4$ = 20).

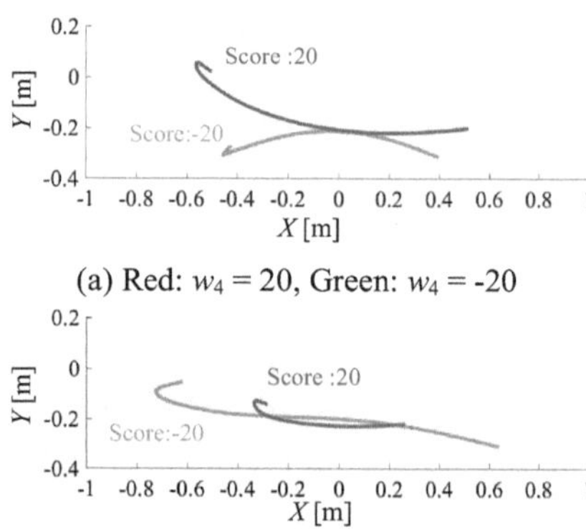

(a) Red: $w_4 = 20$, Green: $w_4 = -20$

(b) Red: $w_5 = 20$, Green: $w_5 = -20$

Fig. 8. Ankle trajectories with modified loadings.

3.3 Selection of Joints That Should Be Actively Driven

In the above section, synthesis of angular input trajectory for each joint was discussed. Remaining problem is which joint should become active to establish a fine driving capability. This problem can be solved by a consequence of PCA-based motion synthesis. It can be said that axes having a larger ROM than others are more sensitive to the key factor loadings, and it is thus meaningful to actuate those axes to achieve the target motion. By picking up axes having larger standard deviations of ROM, flexion/extension and inner/outer rotation of hip joint, adduction/abduction of knee joint, and flexion/extension and inner/outer rotation of ankle joint can be selected as significant ones. Joint trajectory generation using PCA can be again performed by using angular displacements around only above five axes. In the obtained result, new factor loading distribution is obtained in which cumulative contribution ratio up to seventh principal components passed over 80%. Fourth PC has a negative relationship to the translational velocity, and sixth PC has a positive relationship to the angular velocity. Just like the above case, turning left and right can be achieved by changing the sixth factor loading and step length change can be achieved by changing the fourth factor loading. The selected five axes are therefore sufficiently significant to achieve the desired gaits, and joints of the wearable assist device parallel to those axes, namely A, B, F, G, and H should be actuated.

4 Conclusions

This paper presented a systematic mechanism design method of a walking assist exoskeleton which is based on a degrees-of-freedom analysis, a kinematic analysis and a static analysis, and motion synthesis method. Throughout these methods, motion analysis of human gait including different kind of walking patterns was performed based on PCA, and regeneration of gait with variable step length and turning direction and speed was also achieved by reversely applying PCA with adjusting parameters.

References

1. Higuchi, M., Ogata, M., Sato, S., Takeda, Y.: Development of a walking assist machine using crutches (composition and basic experiments). J. Mech. Sci. Technol. **24**, 245–248 (2010)
2. Iwata, T., Takeda, Y., Ogata, M., Higuchi, M.: Development of a walking assist machine using crutches – motion for ascending and descending steps. In: New Trends in Mechanism Science – Analysis and Design, Mechanism and Machine Science, vol. 5, pp. 319–326. Springer (2010)
3. Matsuura, D., Funato, R., Ogata, M., Higuchi, M., Takeda, Y.: Efficiency improvement of walking assist machine using crutches based on gait-feasible region analysis. Mech. Mach. Theory **84**, 126–133 (2015)
4. Matsuura, D., Funato, R., Chounan, Y., Tharasrisuthi, K., Takeda, Y.: Synthesis and analysis of the mechanism for walking assist machine for persons with hemiplegia. In: De Vin, L.J., Solis, J. (eds.) Proceedings of the 14th Mechatronics Forum International Conference, Mechatronics 2014, 13–19 June 2014, Karlstad, Sweden, pp. 303–309 (2014)
5. Takeda, Y., Matsuura, D., Higuchi, M., Sato, S., Iwaya, T., Ogata, M., Funato, R.: Mechanism design and control of a simple and low-cost walking assist machine. In: De Vin, L.J., Solis, J. (eds.) Proceedings of the 14th Mechatronics Forum International Conference, Mechatronics 2014, 13–19 June 2014, Karlstad, Sweden, pp. 555–562 (2014)

Wearable Upper Limb Exoskeletons: Generations, Design Challenges and Task Oriented Synthesis

Sakshi Gupta(✉), Anupam Agrawal, and Ekta Singla

Indian Institute of Technology Ropar, Rupnagar 140001, Punjab, India
{2016mez0024,anupam,ekta}@iitrpr.ac.in

Abstract. This paper first summarizes the research work in the active and passive upper limb exoskeleton robotics that have been majorly testified in the area of assistive and rehabilitation robotics. Recent techniques have been classified with respect to the reported generations defined by scientific communities and design challenges have been retrieved out of the discussions. Further, an approach is proposed to follow natural upper limb motion of a human arm, and for this required human motion data is taken using a Kinect sensor. A two degrees-of-freedom wearable manipulator is synthesized for emulating the collected data. Simulations are shown while emulating natural motion of an arm, and a corresponding architectural concept design for an active solution is also presented.

1 Rehabilitation Robotics: Wearable Upper Limb Exoskeletons

Due to stroke, the chances of functionality failure in patient joint/limb in term of mobility and strength will increase. Thus, Rehabilitation robotics is designed to assist a patient in such a manner so that patient can recover the movement of affected limb/ joint. These stroke patients are called the paralyzed patient.

Rehabilitation robotics is a combination of biomedical engineering and human-robot interaction. To understand the concept of wearable upper limb exoskeleton, human arm anatomy is considered as basis which consists of shoulder complex, elbow complex and wrist complex. Shoulder complex possesses three bones: the clavicle, scapula and humerus, and four articulations: the glenohumeral, acromioclavicular, sternoclavicular and scapulothoracic, with the thorax (chest) as a base [1]. Glenohumeral joint (Shoulder joint) is a ball and socket joint between humerus and scapula which has 5-dof (degrees of freedom) with 3-rotational and 2-translational joints. Elbow complex has majorly 1-dof and wrist complex consists of 3-dof. Each dof has movement in both directions positive and negative (Fig. 1).

The classification of upper-limb robotic devices is shown in Fig. 3. The wearable exoskeleton robots come under the type of orthosis and are generally worn by the human beings. In addition, wearable exoskeleton robots have normally

G. Carbone et al. (Eds.): *New Trends in Medical and Service Robotics*, MMS 65, pp. 134–142, 2019.
https://doi.org/10.1007/978-3-030-00329-6_16

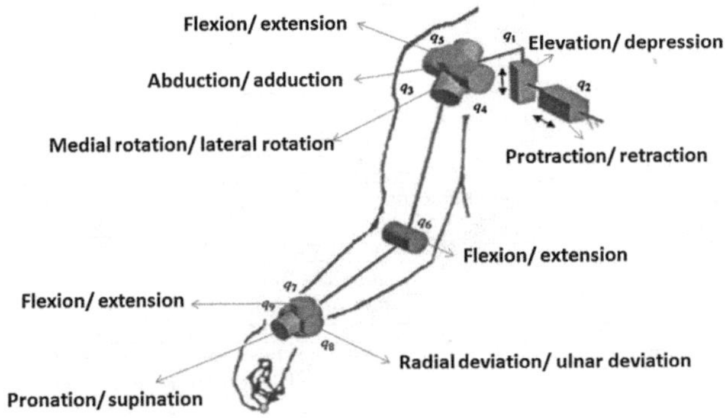

Fig. 1. Human upper limb motions, courtesy [2]

a larger range of motion compared to the end-effector type robots which permit an extensive variety of movements to be used in therapy exercises. Further, wearable exoskeletons are divided into four types: active exoskeleton, passive exoskeleton, Pseudo-passive exoskeleton and hybrid exoskeleton [3,4] (Fig. 2).

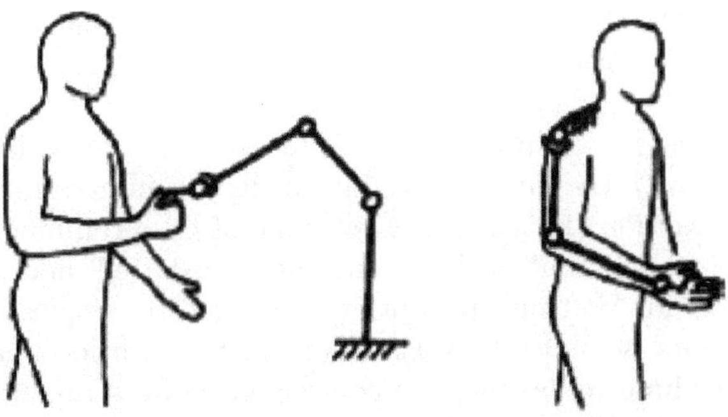

Fig. 2. (a) End-effector robot. (b) Exoskeleton robot courtesy: [5]

Active exoskeleton offers active gesture assistance and has at least one actuator to perform the task. Since it is not essential to put on active assistance to resist patient's movement or to increase patient's force, as an alternative *passive devices* may be used, which are equipped with components providing resistance force only (i.e. brakes, springs, cables etc.). Another *pseudo passive exoskeleton* consists of batteries viz. sensors and other electronic equipment and are not used to provide actuation and uses the electronics only to control a variable damper in the joint. However, *hybrid exoskeleton* possesses all the controllers and sensors of a powered exoskeleton and normally uses functional electrical stimulation (FES) of the muscles which act as an actuator.

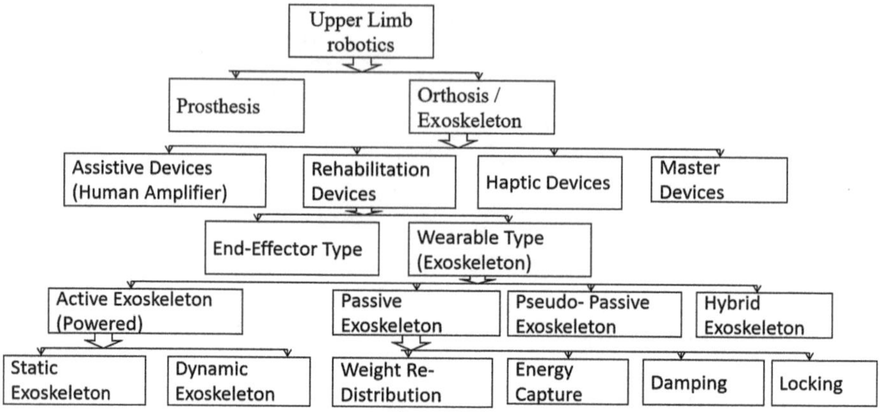

Fig. 3. Broadly classification of upper limb robotics

2 Generations of Upper Limb Exoskeletons

This section briefs the evolution through four major generations of upper limb exoskeletons. The concept of an exoskeleton and mobility was introduced in mid-fifties which considered as zeroth generation. The first generation era focusses at boosting the power of unfit human and normally several hydraulic actuators were in used [6,7].

Real development of an upper-limb exoskeleton robot was first appeared in second generation with the objective of increasing the human limb strength and human robot interaction (HRI). In this era, one of the prominent developments include 'Gentle exoskeleton' utilizing machine-mediated neuro-rehabilitation which is based on integrating appropriate haptic technologies to high-quality virtual environments, so as to deliver a challenging and meaningful therapies to people with upper limb impairment in consequence of a stroke [8,9].

After the second generation, Researchers inclination shifted towards high performance, increasing dof, and powered exoskeleton which was called as the third generation. Armin II has been designed in this generation. It is a semi-exoskeletal structure with seven active degrees of freedom (two of them coupled), five adjustable segments to fit in with different patient sizes, and is equipped with position and force sensors [10–12] (Fig. 4).

Recent research works are focussed at size and weight reduction by introducing cable-based wearable exoskeletons and series elastic actuators. Modular based custom-designed actuation, advanced control strategies for joints, cognitive human-robot interaction (cHRI) and physical human-robot interaction (pHRI) are major focus in recent era [13–15].

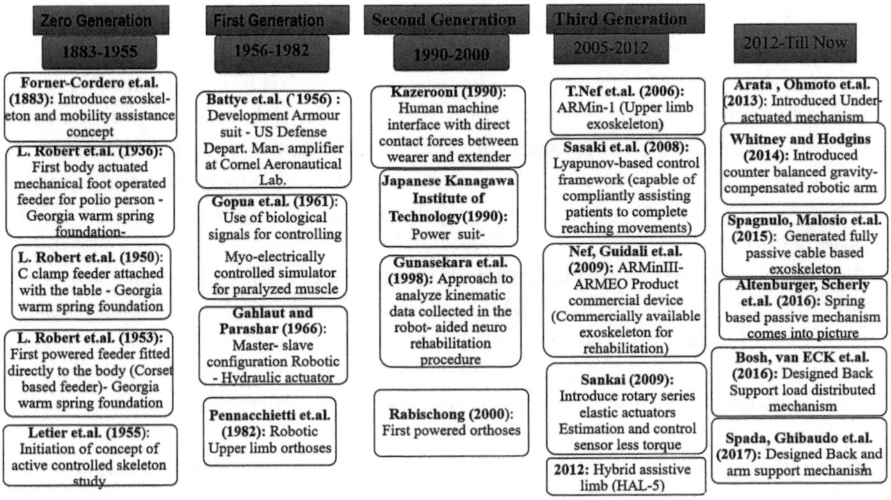

Fig. 4. Generation of upper limb exoskeleton

3 Design Challenges in Upper Limb Exoskeletons

The literature review highlights the major challenges worked upon by the researchers:

1. **Anatomical alignment:** The robot joints and links are in direct correspondence with the human joints and limbs respectively. The axis of the robot is expected to be aligned with the anatomical axes of the upper limb.
2. **Length adjustment:** It is very difficult to adapt different arm lengths, multiple interfaces is considered an effective way to fully determine the upper limb posture for an exoskeleton and apply controlled torques to each upper limb joint independently.
3. **Variation in center of rotation:** The design of a shoulder is considered critical with respect to of anatomy of upper-limb because there is a change in center of rotation with its motion. So it is difficult to set the position location of the shoulder joint to be same as the patient shoulder joint. Some designs [16,17] proposed as the solution of shoulder complex-ability. So far, in elbow complex case, the joint acts as a uniaxial hinge joint while it has three bones (ulna, radius, and humerus). It is not difficult to locate its rotational axis with the patient axis(London 1981). In case of the wrist, It has been considered that wrist design is generated through single axis instead of wrist radial and ulnar deviation [18,19].
4. **Singularity:** Another limitation is singular configurations. Singularity normally occurs when two rotary joints align with each other and a few dof's are lost.
5. **Back-drivability:** A property of mechanical design indicating that the patient is able to move the device, even when the device is in a passive

state. It increases patient safety because it does not constrain limb movements and keeps patients limb in a comfortable position.

6. **Power to weight (mass) ratio and bandwidth of actuator:** While the high power to weight ratio and high bandwidth type of actuators are demand-able, recent actuators cannot provide both high powers to weight ratio and high bandwidth. Motors do have sufficient bandwidth but may possess a poor power to weight ratio, while pneumatic muscle actuator (PMA) have vice versa.
7. **Achieve optimum performance parameter:** Exoskeletons have many varying parameters such as EMG signals convey mechanical properties of muscle and tissue while carrying the load at human-robot interaction. Optimum performance parameters need to be customized for particular patients.
8. **Carry more loads with minimum stress on human muscles:** A common target is to design an exoskeleton which can carry more loads with minimum stress on human muscles. These kinds of exoskeletons are not only helped to stroke affected persons but also enhance the capability of fit persons.
9. **Light weight:** The overall weight of exoskeleton should be light. In this respect, researchers have tried to focus at the light weighted material as well as minimize the use of actuators.
10. **High strength and stiffness:** Maintaining a trade-off between light weight and high strength (stiffness) is a general design challenge.
11. **Human safety and comfort:** Mostly exoskeletons are designed for physically unfit persons and sometimes the wearer has no sense of hurt. In any such scenario, human safety is another most important parameter. A number of sensors have been introduced viz. temperature sensor, blood pressure sensor, SO_2 measuring sensor, heart beat sensor for improving safety and comfort.

    Recent challenges include the possibility of modular design and developments for less expensive customized solutions.

4 Task Oriented Synthesis: Discussion and Results

For natural and comfortable human upper limb movements, a model is proposed which is based on the concept of emulating human upper limb articulations and movement. Considering safety of robot users and to provide assistance in performing essential activities in daily life (ADL) e.g., feeding, reaching tasks, grasping, combing, washing the body, preliminary studies on the anatomical range of the human upper limb movements have been carried out in order to choose a suitable movable range for the exoskeleton.

To emulate the natural gesture of a human arm in the field of assistance and rehabilitation robotics, human motion data has been taken through the Kinect sensor. Kinect sensor is a depth camera which provides shoulder joint, elbow joint and wrist joint co-ordinates, So that dimensional synthesis of upper limb exoskeleton and its configuration is achieved. For simplification of action, the task is divided into planer motions, so as to utilize synthesis of planar mechanism and keeping the total number of dof minimum. Considering vertical flexion/extension of the shoulder as joint 1 and corresponds to flexion/extension of the elbow joint as joint 2. A human motion data retrieved for up and down motion through Kinect camera as shown in Fig. 5. An algorithm is proposed to optimally synthesis and analyze the topology following the required motion.

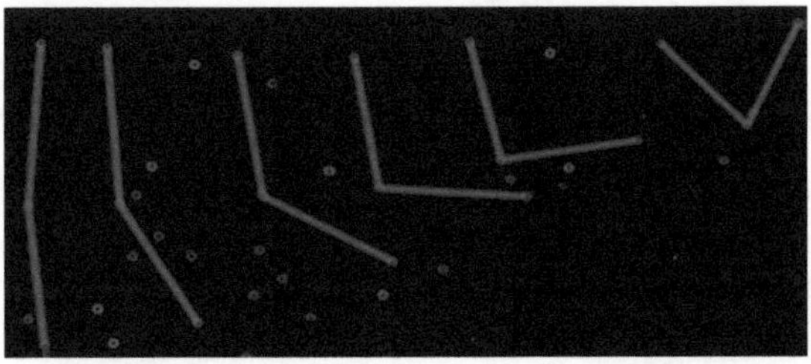

Fig. 5. Some Kinect camera pictures for up and down movement of human upper limb

Table 1. Data retrieved through Kinect sensor for up and down movement of human upper limb

Xshoulder	Yshoulder	Xelbow	Yelbow	Xwrist	Ywrist
0	0	−38.4	106.6	−19.2	174.2
0	0	−40.8	109.2	−76.8	166.4
0	0	−60	109.2	−96	171.6
0	0	−48	132.6	−45.6	239.2
0	0	−31.2	137.8	−16.8	257.4

Plots of the data retrieved through the Kinect sensor showcase the captured postures of the upper limb. Corresponding joint locations are given in Table 1. One of the shoulder joint motions are targeted to be tracked through four bar mechanism link with single bar mechanism. Beauty of using four bar mechanism with single bar mechanism is avoid the chance of variation of center of rotation. This is one of critical challenge faced by researchers. Consider upper arm length and lower arm length fixed based on the study of anthropometric data. To formulate this problem as optimization problem, all the link lengths and the joint angles are taken as design variables and the design objective is the best kinematic condition at all intermediate locations in joint configurations. Units of all the lengths are in mm and that of angles are in radians as shown in Fig. 6.

The formulated synthesis optimization problem has maximization of kinematic conditioning as objective function and kinematic equation as the equality constraints. An NLP (Non-linear problem) is formulated. The method of Genetic Algorithm is utilized for solving the problem. Resulting values of the link-lengths

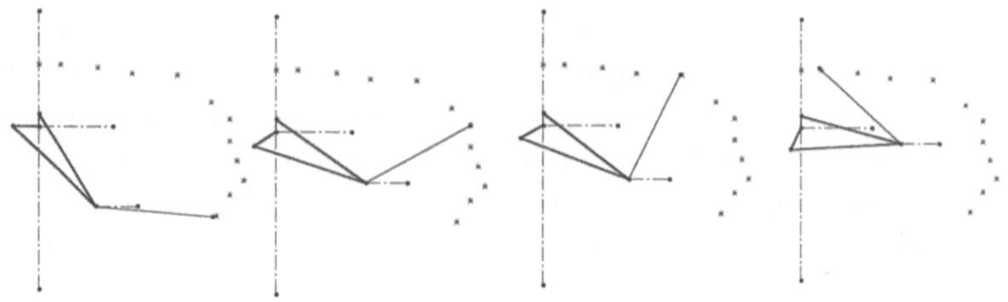

Fig. 6. Joint locations captured by Kinect sensor for data-driven synthesis for up and down upper limb movement, and corresponding synthesized results. The 2-dof manipulator is emulating the natural motion at its end effector

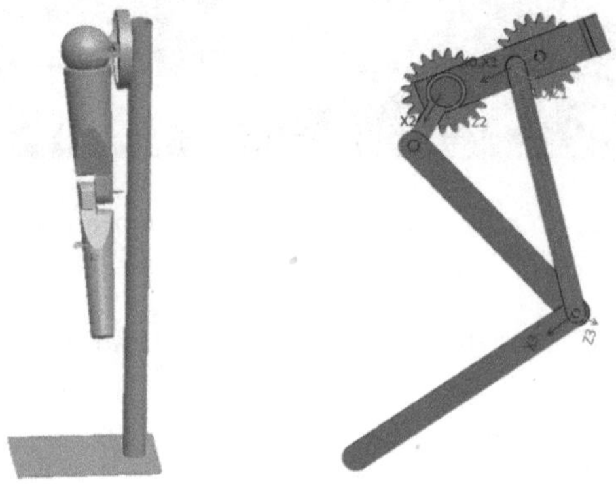

Fig. 7. Solid model architecture (a) Mannaquin with stand, (b) 2-DOF exoskeleton with frame

and the location of the pivots for tracking the targeted points. Solid model architecture of a test-rig is worked upon for the synthesized 2-dof manipulator, as shown in Fig. 7. Design is done by solid-works and considering resulting value as dimensions of proposed model. The test-rig consists of two major parts: one is the mannequin which is designed upon standard human data and another is the prototype of 2-dof active exoskeleton. The test-rig is a prototype of conceptual design of task-oriented configuration. In future, it will help to achieve more effective results by increasing more DOF.

5 Conclusions

The paper briefly presents a review of active and passive upper-limb exoskeletons and highlights design challenges in the field. To reduce the anatomical alignment feature and to work towards lesser weighed mechanism, a 2-dof planar manipulator is synthesized for emoluting natural motion of a standard human arm. Data is retrieved through Kinect Sensor. Synthesis problem is formulated as an optimization problem and utilized Genetic Algorithm for solutions. Proposed 2-dof wearable exoskeleton is an active manipulator with one actuator planned for vertical flexion/extension motion of shoulder and other for flexion/extension motion for elbow. For transferring motion from shoulder joint to elbow joint, a belt drive mechanism is used. This mechanism helps to reduce the weight of exoskeleton as well as it consist of four bar mechanism link with a single bar. The paper presents the concept design and synthesis results which are expected to be improved on accuracy.

Acknowledgement. The authors would like to thank Global Innovation and Technology Alliance (GST) and Department of Science and Technology (DST) for financial support of this work.

References

1. Gopura, R., Kiguchi, K.: Mechanical designs of active upper-limb exoskeleton robots: state-of-the-art and design difficulties. In: 2009 IEEE International Conference on Rehabilitation Robotics, ICORR 2009, pp. 178–187. IEEE (2009)
2. Xie, S.: Advanced Robotics for Medical Rehabilitation. Springer Tracts in Advanced Robotics, vol. 108, pp. 1–357. Springer, Switzerland (2016)
3. Galiana, I., Hammond, F.L., Howe, R.D., Popovic, M.B.: Wearable soft robotic device for post-stroke shoulder rehabilitation: identifying misalignments. In: 2012 IEEE/RSJ International Conference on Intelligent Robots and Systems (IROS), pp. 317–322. IEEE (2012)
4. Pehlivan, A.U., Rose, C., O'Malley, M.K.: System characterization of ricewrist-s: a forearm-wrist exoskeleton for upper extremity rehabilitation. In: 2013 IEEE International Conference on Rehabilitation Robotics (ICORR), pp. 1–6. IEEE (2013)
5. Frisoli, A., Borelli, L., Montagner, A., Marcheschi, S., Procopio, C., Salsedo, F., Bergamasco, M., Carboncini, M.C., Tolaini, M., Rossi, B.: Arm rehabilitation with a robotic exoskeleleton in virtual reality. In: 2007 IEEE 10th International Conference on Rehabilitation Robotics, ICORR 2007, pp. 631–642. IEEE (2007)
6. Gopura, R., Bandara, D., Kiguchi, K., Mann, G.K.: Developments in hardware systems of active upper-limb exoskeleton robots: a review. Robot. Auton. Syst. **75**, 203–220 (2016)
7. Battye, C., Nightingale, A., Whillis, J.: The use of myo-electric currents in the operation of prostheses. Bone Joint J. **37**(3), 506–510 (1955)
8. Loureiro, R., Amirabdollahian, F., Topping, M., Driessen, B., Harwin, W.: Upper limb robot mediated stroke therapygentle/s approach. Auton. Robots **15**(1), 35–51 (2003)
9. Kazerooni, H.: Human/robot interaction via the transfer of power and information signals. In: 1989 Proceedings of IEEE International Conference on Robotics and Automation, Part I Dynamics and Control Analysis, pp. 1632–1640. IEEE (1989)
10. Mihelj, M., Nef, T., Riener, R.: Armin ii-7 DOF rehabilitation robot: mechanics and kinematics. In: 2007 IEEE International Conference on Robotics and Automation, pp. 4120–4125. IEEE (2007)
11. Stienen, A.H., Hekman, E.E., Van Der Helm, F.C., Van Der Kooij, H.: Self-aligning exoskeleton axes through decoupling of joint rotations and translations. IEEE Trans. Robotics **25**(3), 628–633 (2009)
12. Sasaki, D., Noritsugu, T., Takaiwa, M.: Development of active support splint driven by pneumatic soft actuator (assist). In: 2005 Proceedings of the 2005 IEEE International Conference on Robotics and Automation, ICRA 2005, pp. 520–525. IEEE (2005)

13. Ball, S.J., Brown, I.E., Scott, S.H.: MEDARM: a rehabilitation robot with 5DOF at the shoulder complex. In: 2007 IEEE/ASME International Conference on Advanced Intelligent Mechatronics, pp. 1–6. IEEE (2007)
14. Kim, B., Deshpande, A.D.: Controls for the shoulder mechanism of an upper-body exoskeleton for promoting scapulohumeral rhythm. In: 2015 IEEE International Conference on Rehabilitation Robotics (ICORR), pp. 538–542. IEEE (2015)
15. Forner-Cordero, A., Pons, J.L., Turowska, E., Schiele, A., Baydal-Bertomeu, J., Garrido, D., Mollá, F., Belda-lois, J., Poveda, R., Barberá, R., et al.: Kinematics and dynamics of wearable robots. In: Wearable Robots: Biomechatronic Exoskeletons, pp. 47–85 (2008)
16. Kiguchi, K., Iwami, K., Yasuda, M., Watanabe, K., Fukuda, T.: An exoskeletal robot for human shoulder joint motion assist. IEEE/ASME Trans. Mech. **8**(1), 125–135 (2003)
17. Papadopoulos, E., Patsianis, G.: Design of an exoskeleton mechanism for the shoulder joint. In: 12th IFToMM World Congress, pp. 18–21. Citeseer (2007)
18. Perry, J.C., Rosen, J., Burns, S.: Upper-limb powered exoskeleton design. IEEE/ASME Trans. Mech. **12**(4), 408–417 (2007)
19. Gopura, R.A.R.C., Kiguchi, K.: An exoskeleton robot for human forearm and wrist motion assist. J. Adv. Mech. Des. Syst. Manuf. **2**(6), 1067–1083 (2008)

Human-Robot Interactions

Neurorobotic Investigation by Using a Basic Vestibulo-Ocular Reflex System Algorithm

A. Nussibaliyeva[1(✉)], A. Mussina[1,3(✉)], G. Carbone[2], B. Tultayev[3], and G. Balbayev[1,3]

[1] Almaty University of Power Engineering and Telecommunications, Almaty, Kazakhstan
kz.a.mussina@gmail.com, gani_b@mail.ru
[2] University of Cassino and South Latium, Cassino, Italy
{carbone, ceccarelli}@unicas.it
[3] Gylym Ordasy, Almaty, Kazakhstan

Abstract. Bioinspired and biomimetic approaches are widely researched in robotics in order to improve cognitive, autonomic functions of machines. Use of bioinspired adaptive control algorithms in machines stabilises the motor control. Gaze stabilisation system/vestibulo-ocular reflex (VOR) operates to counter-rotate the eyes to prevent retinal slip for maintaining the stable gaze. Cerebellum has a lot of functions, like a cognitive, motor, autonomic functions and probably, emotions.

Test cases of implementation and evaluation of bioinspired adaptive control algorithm in the control of a robot eye show promising performance. Model of cerebellar function, which is analogue to the VOR is analysed. Inverse oculo-motor plant model is simulated on Matlab/Simulink by using simple Vestibulo-Ocular Reflex model.

Keywords: Neurorobotic · Robot 'eye' · Vestibulo-Ocular Reflex model Adaptive filter model · Bioinspired control algorithm · Cerebellum

1 Introduction

All mobile agents, both natural and artificial, are dealing with disturbance issue regarding to their vision system while they are moving. Reafference problem arises when sensory signals are contaminated by movement-produced noise. The simple example could be, when we are observing typical shaking on video footage shot with a hand-held camera whilst walking. Nature comes up with an efficient way of solving this gaze stabilisation task by decoupling the vision system to eye and head. According to basic knowledge from biology almost all vertebrates, from goldfish to humans, have an ability to rotate their eyes relative to their head by using extraocular muscles. The mechanism that stabilises vision system with respect to rotations and translations of head is called vestibulo-ocular reflex (VOR) [1, 4, 9]. This reflex drives the eyes in the opposite direction to the head movement and consequently keeps the image in the centre of the retina. It is necessary to further investigate the neurorobotic adaptive control in order to validate and extend current research findings [5–8].

G. Carbone et al. (Eds.): *New Trends in Medical and Service Robotics*, MMS 65, pp. 145–152, 2019.
https://doi.org/10.1007/978-3-030-00329-6_17

2 Basic Circuit of VOR Control

Diagram of circuitry that mediates the horizontal basic VOR control is presented in the Fig. 1. Result of the moving of image extremely fast across retina is the vision degradation. The retinal slip would be produced by movements of the head, as it happens in locomotion. The VOR operates to counter-rotate the eyes to prevent retinal slip for maintaining the stable gaze. Cerebellar flocculus receives information about head velocity, eye movement commands and retinal slip.

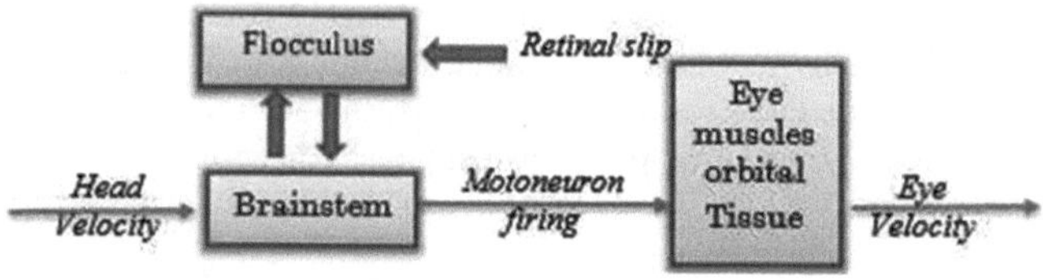

Fig. 1. Basic circuit of VOR control

In that structure there is no feedback control, which means that open-loop control is used. Here controller becomes an inverse model of the plant. The adaptive control is used to be sure that the inverse plant model is accurate. The information about system output has to be used for learning.

According to Figs. 2 and 3 training signal is retinal slip, which is sent to the flocculus, also consistent with flocculus being the adaptive part of the controller. Characteristics of basic VOR system model on Fig. 4 were chosen according to [3, 10, 11]. Transfer functions of the models and chosen values are the similar to [11] as paper presented the part of the research.

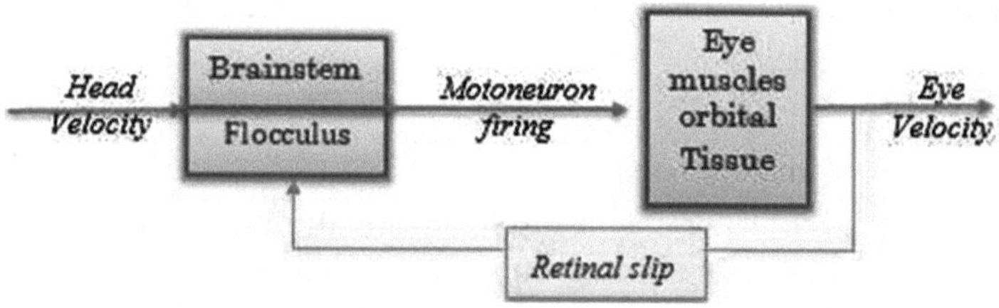

Fig. 2. Bio-control circuit

P – the first order dynamic model of oculomotor plant, with transfer function $P(s)$ between eye-in-head velocity $x(t)$ and motor command $u(t)$ as a Laplace transform in Eq. 1.

$$P(s) = \frac{ks}{s + 1/T_p}, \tag{1}$$

where $T_p = 0.2\,\mathrm{s}$ is time constant, $k = 1$ is a gain.

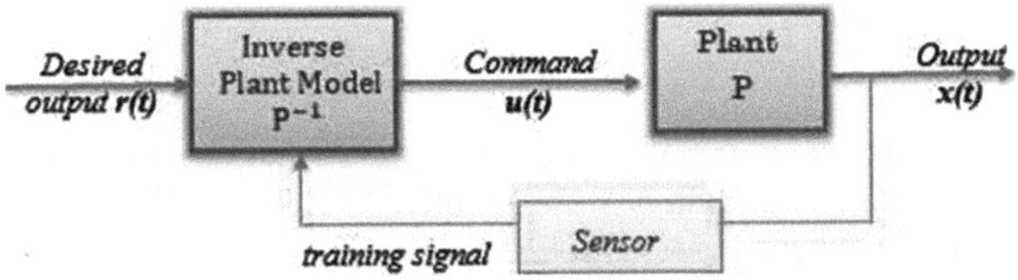

Fig. 3. Control circuit

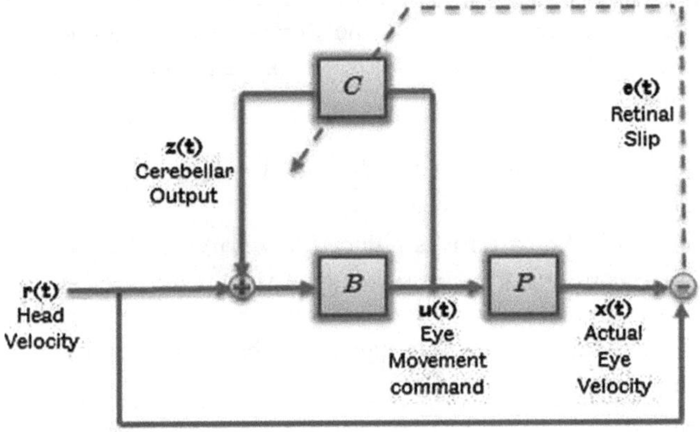

Fig. 4. The basic VOR system

B – brainstem, which is modelled as a first-order leaky integrator plus a pure dc gain. Transfer function of brainstem model $B(s)$ is presented as Laplace transform in Eq. 2. The output is a motor command $u(t)$; input is a conjunction of the vestibular system $r(t)$ and the cerebellar output $z(t)$.

$$B(s) = G_d + \frac{G_i}{s + 1/T_i}, \tag{2}$$

where $G_d = 1$ is the direct path gain and $G_i = 1/T_p = 5$ is the indirect path gain and $T_i = 0.5$ s is a time constant.

Further simulations were executed with different versions of the parameters T_p, G_d, G_i, T_i. It has to be noted, that the exact values are not important at this stage, but time constants should be in the 100 ms range [11].

Brainstem provides a control, which the cerebellum improves by adjusting the response via the filter weights. The perfect compensation of plant could be achieved by brainstem itself, when $T_i = \infty, G_d = 1, G_i = 1/T_p$.

The cerebellum is implemented as an adaptive FIR (finite impulse response) filter C, with output $z(t)$, which is given in Eq. 3.

$$z(t) = \sum_{i=0}^{L} w_i p(t - i \cdot \Delta T), \tag{3}$$

where input $u(t)$ to the adaptive filter C, which was splitted into number of L components $p_1(t), \ldots, p_n(t)$, with delays between them of $\Delta T = 0.02\,\text{s}$ (2 s in total); w_i – weight of the component p_i.

The rule for adjusting weights is provided below. For reducing the perceived visual slip from the input, the weights of adaptive filter have to be changed. As it is mentioned in [2] the weights are analogue to the parallel fibre Purkinje cell synapses of the cerebellar cortex. The inverse oculomotor plant is realised when the learning algorithm is successfully applied and adjustment of the weights is done in combination with the brainstem. As it could be seen in Fig. 4 the sensory error $e(t)$ is a direct result of the performance of the adaptive filter C.

$$e(t) = \left(P - B^{-1} + C\right) \cdot x(t) \tag{4}$$

As it can be seen in Eq. 4 error is reduced to zero when $\mathrm{C} = \mathrm{B}^{-1} - \mathrm{P}$.

Efficiently, this learning topology subsequently does not need a translation of the sensory error into a motor-command error. For that reason, the observed visual slip is expected to build the suitable teaching signal for the adaptation of the filter. Rule to adjust the weights is shown in Eq. 5.

$$\delta w_j = -\beta \langle p_j(t) \cdot \widehat{u}(t) \rangle, \tag{5}$$

where δw_j is the change in the j th weight w_j

β – a learning rate constant. The value of that is adjusted to give rapid learning without instability.

$\widehat{u}(t)$ – the value of retinal slip at time t.

$p_j(t)$ – the value of the j th filter signal at time t.

$\langle \rangle$ – denotes the expected value of the enclosed quantity over the time period used for training

The feature of learning rule is that it is identical to the LMS rule of adaptive control theory [11].

3 Simulation Results

Model architecture of the system, which is described above, was programmed by using Matlab and Simulink [3]. Experiments below were done for simple, first order plant, to prove the usefulness of the algorithm.

1. Pre-Training investigation

Behaviour of the system before training with a first-order plant is presented. Responses to the unit step input of both the oculomotor plant $P(1)$ and the brainstem model $B(2)$ are shown on Fig. 5.

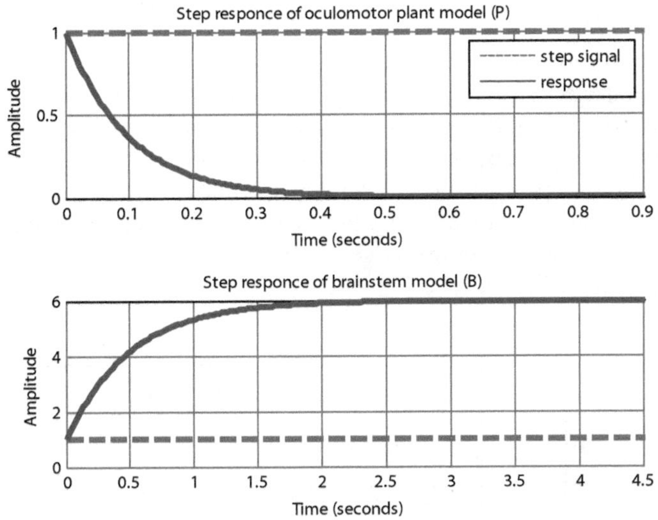

Fig. 5. Step responses of oculomotor plant P and brainstem B

Performance of the system to band limited white noise input, i.e. head velocity, gave rise to retinal slip with low frequency, which is expected because of the existing brainstem controller (Fig. 6a). According to these graphs it can be stated that it is unable to maintain eccentric gaze. The time course of the plant and the brainstem is directly effected on how fast eye position returns to initial value.

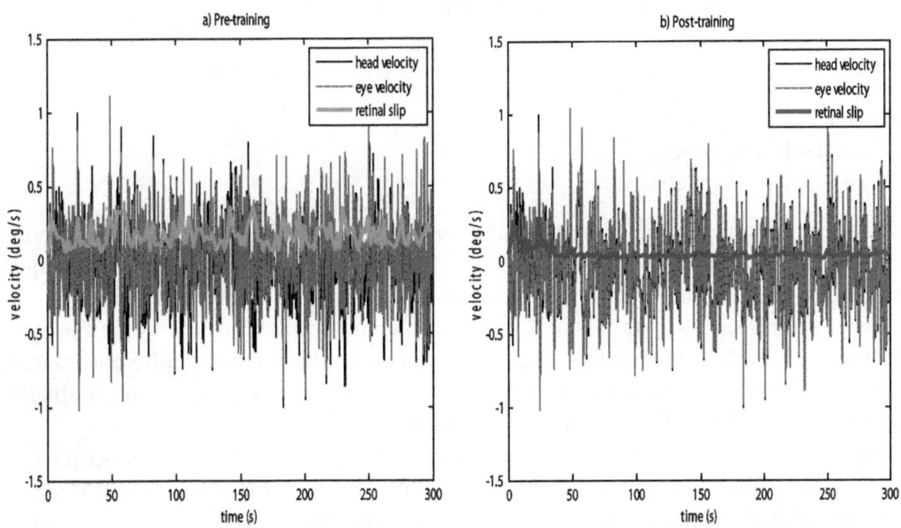

Fig. 6. Displacement tracking and retinal slip *a* - for pre-training, *b* - for post-training

2. Post-Training investigation

The performance of the model during and after training, with the first order plant *P* and brainstem controller *B* illustrated below. The characteristics of *P* and *B* are the same as in the previous experiment. Input for that system is a head-velocity signal modelled as a band limited white noise as before. Learning starts at 30 s. For testing how the value of learning rate effects on performance, each time the same set of signals are used as input *r*. The effect of training is shown in Fig. 6b. The difference between retinal slip of pre-training and post-training experiments gives better understanding how adaptive component reduces error (Fig. 7).

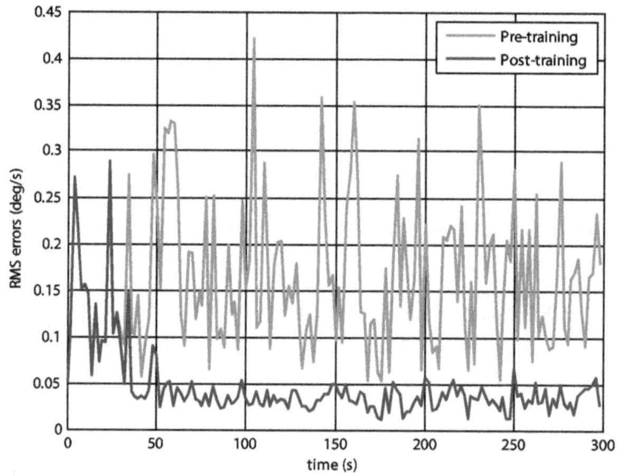

Fig. 7. Retinal slip before and after training (learning began at 30 s second)

Ensure that learning is slow enough to cope with following issues:

(a) plant disturbances;
(b) delays in responses;
(c) changes to plant dynamics.

The graph on Fig. 8 was plotted to analyse how the learning rate (β) affects to error. Actually, if the learning rate is changed, simulation time also has to be changed. Reason for this is that system needs appropriate time to apply that learning. For instance, if the learning rate is small it takes time and applying to the system is slow.

However, in the experiments simulation time was the same for different learning rates. According to graph on Fig. 8, especially for that set of input, error is smaller, when $\beta = 1$. For $\beta \gg 1$, the system becomes unstable.

Figure 9 shows how the RMS error between desired and actual eye-velocity over last 10 s for different numbers of filters. As it can be seen on graph that RMS error decreases with increase in number of filters; if the number of filters is enough, the error will be close to zero. It is concluded that bigger learning rates result in smaller RMS error. However higher learning rates ($\beta \ll 1$) results in instability hence complete loss of tracking.

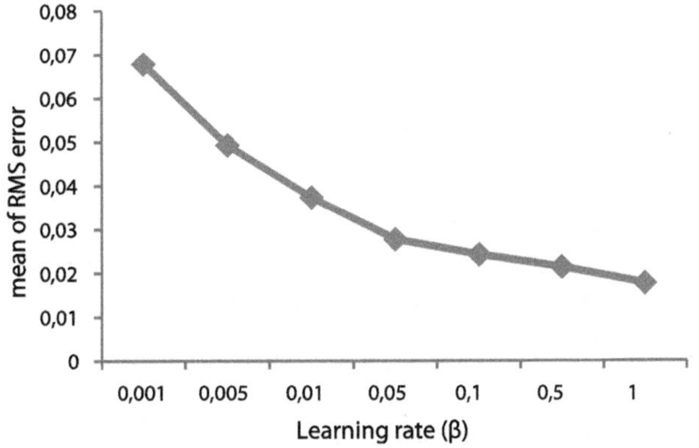

Fig. 8. Mean of RMS error against learning rate

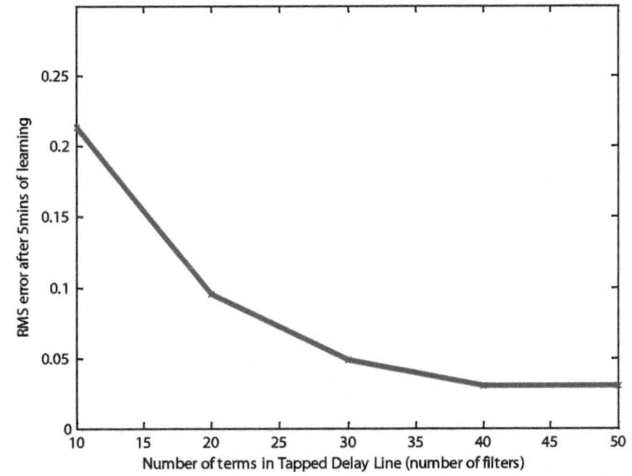

Fig. 9. Average RMS error over last 10 s for different numbers of filters

4 Conclusion

Pre-training and post-training investigation was done. An analysis of the RMS error after training revealed satisfactory performance.

Investigation of tracking on the dynamic model reveals that the adaptive controller is able to track even when changes occur in the plant model.

A study of cerebellar inspired adaptive control, which is augmented by reference model, provides a potential general solution for robotic control. Control of oculomotor plant is significantly improved by bioinspired adaptive filter control. The performance of the adaptive filter control is dependent on the learning rate; thus it is vital that for any design, an optimal choice of learning rate be sought and simulation on MATLAB/ Simulink was done successfully.

References

1. Lenz, A.: Cerebellar-inspired adaptive control of a robot eye actuated by pneumatic artificial muscles. IEEE Trans. Syst. Man Cybern. Part B Cybern. Publ. IEEE Syst. Man Cybern. Soc. **39**(6), 1420–1433 (2009)
2. Porrill, J., Dean, P., Stone, J.V.: Recurrent cerebellar architecture solves the motor-error problem. Proc. Biol. Sci./R. Soc. **271**(1541), 789–796 (2004)
3. Dean, P., Porrill, J., Stone, J.V.: Decorrelation control by the cerebellum achieves oculomotor plant compensation in simulated vestibulo-ocular reflex. Proc. Biol. Sci./R. Soc. **269**(1503), 1895–1904 (2002)
4. Anderson, S.R., Wilson, E.: SimulinkAdaptiveFilterCode_TDLs (2011). [Simulink block and Matlab scripts]
5. Wilson, E.D.: Developing the cerebellar chip as a general control module for autonomous systems, pp. 1–12 (2013)
6. Dean, P.: An adaptive filter model of cerebellar zone C3 as a basis for safe limb control? J. Physiol. **591**(Pt 22), 5459–5474 (2013)
7. Shibata, T., Schaal, S.: Biomimetic gaze stabilization based on feedback-error-learning with nonparametric regression networks. Neural Netw. Off. J. Int. Neural Netw. Soc. **14**(2), 201–216 (2001)
8. Dean, P.: The cerebellar microcircuit as an adaptive filter: experimental and computational evidence. Nat. Rev. Neurosci. **11**(1), 30–43 (2010)
9. Anderson, S.R.: Adaptive cancelation of self-generated sensory signals in a whisking robot. IEEE Trans. Robot. **26**(6), 1065–1076 (2010)
10. Porrill, J., Dean, P.: Cerebellar motor learning: when is cortical plasticity not enough? PLoS Comput. Biol. **3**(10), 1935–1950 (2007)
11. Mussina, A., Ceccarelli, M., Balbayev, G.: Neurorobotic investigation into the control of artificial eye movements. In: Mechanisms and Machine Science. vol. 57, pp. 211–221 (2018)

Kinect-Based Approach for Upper Body Movement Assessment in Stroke

S. Spasojević[1,2](✉), A. Rodić[1], and J. Santos-Victor[2]

[1] Mihailo Pupin Institute, University of Belgrade, Belgrade, Serbia
{sofija.spasojevic, aleksandar.rodic}@pupin.rs
[2] Institute for Systems and Robotics, Instituto Superior Técnico, University of Lisbon, Lisbon, Portugal
jasv@isr.tecnico.ulisboa.pt

Abstract. In this paper, we investigate how the objective movement assessment can support the clinical practice in the stroke treatment. The movement data are collected using vision-based, low-cost and marker-free Kinect sensor device. Sensor recordings are collected in the hospital settings for stroke outpatients with the supervision of medical doctors. We propose movement performance indicators, extracted from the sensor signals, to characterize the movements. The proposed approach for movement quantification is intended to support the clinical evaluations and to monitor the patients' state over time. The emphasis is on the verification of the proposed indicators and investigation of their importance for the stroke relevant clinical aspects.

Keywords: Kinect · Movement characterization · Stroke · Rehabilitation

1 Introduction

Stroke is a leading cause of motor disability, the second most common cause of death in general [1], and the third, taking into account the countries of the developed world [2]. Approximately 80% of individuals that survive stroke suffer the neurological damage that leads to impairments of the motor functions, long-term disability and limited every-day activities [3]. Recovery after stroke requires a lengthy and slow rehabilitation procedure that usually includes the neuro-stimulation and motor training organized in clinical centers.

Evaluation of the patients' performance during rehabilitation sessions is based on the specially designed clinical scales such as Jebsen-Taylor [4] and Fugl-Meyer [5] scale. However, both clinical scales are prone to subjective and imprecise ratings and intended to be carried out by a trained physiatrist. Hence, there is a clear need for introducing new sensing/processing techniques into the clinical practice, capable to support the clinical evaluations in stoke by introducing the objective approach for the movement assessment.

In this paper, we investigate how the data from the low-cost, marker-free and portable Kinect sensor can support the rehabilitation therapy and evaluation procedures in stroke. We examine large range upper body movements of stroke patients. A set of experimental exercises is defined according to clinical protocols. The special emphasis

G. Carbone et al. (Eds.): *New Trends in Medical and Service Robotics*, MMS 65, pp. 153–160, 2019.
https://doi.org/10.1007/978-3-030-00329-6_18

is on the design of quantitative descriptors (so-called *Movement Performance Indicators*), extracted from the sensor data and proposed for the objective movement assessment. This paper presents the continuation of our previous research on Kinect-based approach for progress monitoring of the stroke patients [6]. The extension refers to the verification of the proposed indicators and investigation of their importance for the clinically relevant aspects.

2 Related Work

Different types of sensor devices are used nowadays for the movement acquisition. Rough division addresses three main sensor groups: (i) vision-based with markers, (ii) vision-based without markers and (iii) wearable sensors.

The vision-based systems with markers (Marker based motion capture (MOCAP) systems) [7] involve the placement of the markers at particular body points and complex system of cameras for movement recording. Those systems deliver accurate measurements, but they are extremely costly and complex for use. In addition, MOCAP systems are not portable and the recordings need to be carried out in the specially designed environments. On the other side, low-cost and marker-free MOCAP systems such as the Kinect and Xtion [8] become very popular as a suitable alternative for expensive, complex and non-portable MOCAP systems. Using these new-generation devices, the movements can be acquired without markers, based on the inbuilt algorithms for skeleton tracking. The performance of lower-cost systems has been tested and the results report a satisfactory accuracy for the application in the rehabilitation therapy [9–12] and specifically in stroke rehabilitation [13]. However, their performance is quite lower in comparison to the advanced MOCAP systems and the readings are less robust to measurement noise. Still, they represent a good trade-off between the overall performance and cost. Some examples of Kinect-based rehabilitation systems are described in [14–17], but little attention has been devoted to the specific case of stroke [18–23].

The authors in [18] focus on the small range arm/hand movements (reaching tasks). Kinect is intended for the calculation of the shoulder and elbow angle, while the wrist angle is measured with the electro-goniometer and the data glove. The movement performance evaluation is limited only to those joint angles (shoulder, elbow and wrist). The study [19] proposes the game-based concept – tracking the patient's movements using Kinect and armband sensor. The study lacks the system validation through experiments with patients, as well as the signal processing, feature extraction and movement evaluation procedure behind the game interface. The authors in [20] perform the Kinect-based virtual reality training for motor functional recovery of upper limbs after stroke. However, the evaluation after the training is based only on the clinical assessment tool. The authors in [21] evaluate the food-related tasks as activities of daily living (ADL), intended for post-stroke patients. They use Kinect to measure joint positions and angular values and inertial sensors to measure the acceleration. The system was tested only for healthy subjects. The authors in [22] develop the system based on the 3D vision using Kinect, accompanied by virtual environment, ergonometric signals and a humanoid (Nao) for stroke rehabilitation. The study proposes a large set of potential quantitative measurements, resulting from the kinematics of the upper limbs, as well as the information based on the electromyography, goniometry, and inertial measurements. However, the study lacks the experimental verification with patients and evaluation of their performance based on the proposed set of quantitative measurements. The study [23] introduces the virtual rehabilitation system for stroke patients, composed of the Kinect device and haptic glove for tactile feedback. Kinect is used to track the upper limbs and to map the information to a virtual avatar. The study lacks the inclusion of sensor information in the performance evaluations and experimental verification based on the patients' data.

The overall conclusion is that the recent Kinect-based systems intended for stroke rehabilitation lack the following important aspects: (i) the movement evaluation procedure based on the sensor readings, (ii) experimental validation of the methods on patients' data and (iii) correlation of the proposed indicators with clinical scales.

3 The Proposed System Structure

The proposed system structure for the objective movement assessment from the Kinect sensor data is illustrated in Fig. 1. The first steps are the sensor calibration (placing the body in the particular position) and the movement data collection. As a second stage, the sensor signals are pre-processed using low-pass filters to reduce the measurement noise. A temporal segmentation algorithm is applied to the sensor signals since the movements are collected in the sequence, but each movement has to be analyzed separately.

The following step is the extraction of the movement performance indicators from the sensor signals. The indicators are clinically inspired and their design results from recommendations of medical domain experts. In the next steps, the proposed indicators are tested in terms of capability to support the relevant clinical aspects: (i) statistical investigation in terms of internal consistency and reliability; (ii) quantification of the differences between healthy and affected side and (iii) correlation with clinical (Fugl-Meyer) scale.

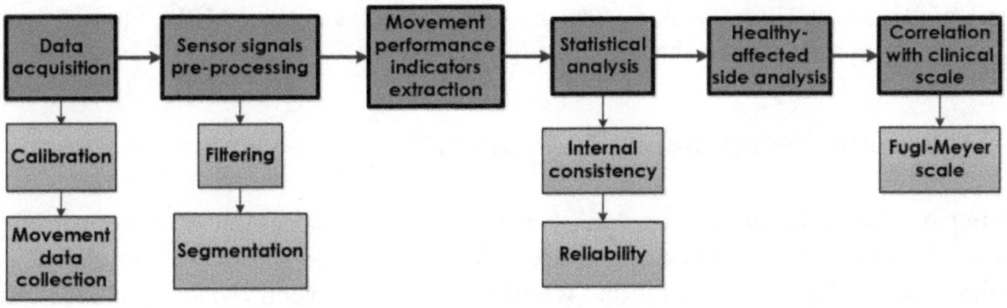

Fig. 1. The proposed system structure.

3.1 Sensor Data Acquisition

The experimental group consists of three stroke patients. All patients have been examined under the same conditions and they have performed five upper body movements, instructed by an experienced physiatrist. Illustration of the experimental movements is given in Fig. 2. The movement abbreviations listed in the caption of the Fig. 2 will be used from now on. The 3D coordinates of the characteristic skeleton joints are collected for every frame during the motion performance. Clinical measurements - Fugl-Meyer scale [5] are collected by an experienced physiatrist right before the sensor measurements.

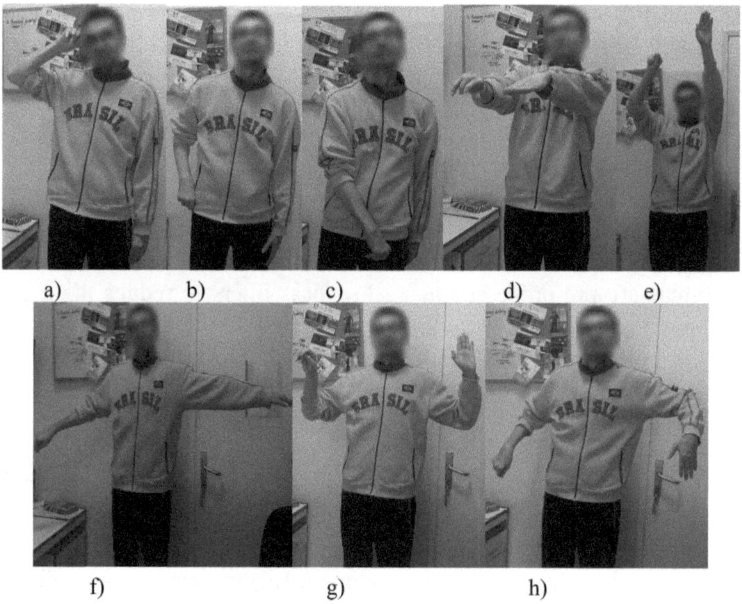

Fig. 2. **UB1:** Hand goes from the ear to the hip (same body side): (a) and (b); **UB2:** Hand goes from the ear to the hip (different body side - diagonal): (a) and (c); **UB3:** Shoulder flexion-extension: (d) and (e); **UB4:** Shoulder abduction-adduction: (f) and (e) and **UB5:** Elbow flexed at 90°: hands go up and down in shoulder joint (g) and (h)

4 Movement Performance Indicators

Relying on the collected sensor data, we have defined and extracted seven different movement performance indicators (**MPI**s). Those indicators refer to the elbow and shoulder angle (**A**)/range of motion (**ROM**), movement speed (**MS**), symmetry ratio (**SR**) among body sides and vertical distance (**VD**) between joints of interest – in our case between hands (**VDBH**, Fig. 3-a) and between elbow and shoulder (**VSED**, Fig. 3-b). The calculation of these indicators is explained in detail in our previous studies [6, 24]. An example of the indicator calculation is illustrated in Fig. 3. The figure presents the patient's skeleton in the X-Y plane, along with the joint position data, collected from Kinect. The vertical distance (**VD**) between joints of interest is a clinically-based and stroke-relevant indicator. It is defined as the difference in y-coordinate of joints at the final movement position. In the case of regular movements, those differences should be close to zero, while in the case of the impaired movements at stroke patients, they become significant (Fig. 3).

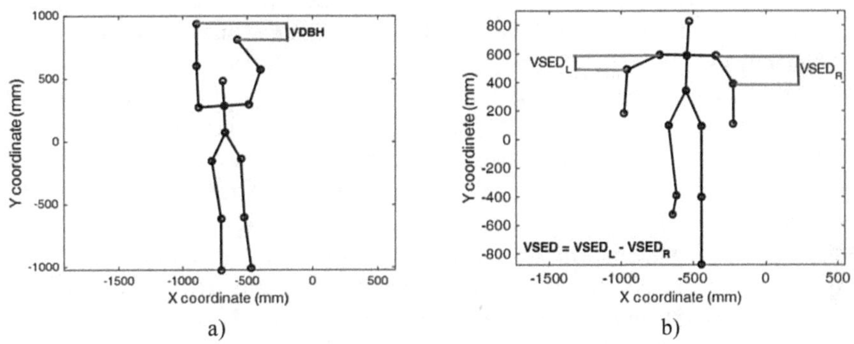

Fig. 3. Calculation of the **VD** between hands (**VDBH**) during the **UB4** movement (a) and **VD** between shoulder and elbow (**VSED**) during the **UB5** movement

The overview of the all extracted indicators across movements is presented in Table 1. The next step is the investigation of their capability to support the relevant clinical aspects, such as reliability, quantification of the differences between healthy and affected side and correlation with clinical Fugl-Meyer scale.

Table 1. Calculated Movement Performace Indicators across movements.

Movement*	Calculated Movement Performace Indicators (MPIs)				
UB1	**ROM** shoulder	**MS**	**SR**		
UB2	**ROM** shoulder	**MS**	**SR**		
UB3	**ROM** shoulder	**MS**	**SR**	**VD** hands	**ROM** elbow
UB4	**ROM** shoulder	**MS**	**SR**	**VD** hands	**ROM** elbow
UB5	**VD** shoulder-elbow	**MS**	**SR**	**VD** hands	mean shoulder **A**

* Movement notation is taken from the caption of Fig. 2

5 Results

We have proposed a set of 21 MPIs in total to characterize five upper body movements (Table 1). In this section, we conduct the analysis of the properties of these MPIs, to identify their importance in terms of the technical and stroke-related clinical aspects: (i) statistical analysis of the sensor measurements and indicators, (ii) healthy-affected side analysis and (iii) correlation with clinical scale.

Internal consistency of the sensor measurements is assessed using Cronbach's alpha parameter [25]. Obtained Cronbach's alpha values across sensor data for all movements have values within the range [0.95–0.99]. Values close to one indicate the high consistency of the Kinect sensor measurements. In order to investigate the reliability of the extracted MPIs, the test-retest method for the reliability analysis [25] has been applied. The test-retest method correlates the scores across repeated tests and the reliability is assessed using *Intraclass correlation coefficient* (ICC) [25]. ICC values across all indicators are within the range [0.72–0.97], whereby the values closer to 1 indicate higher reliability.

The progress monitoring of the affected hand is the most important task in the rehabilitation after stroke. Another important concept is the performance comparison between the healthy and affected hand. In order to complete this task, the statistical comparison has been performed. The two-sided Wilcoxon rank sum test is applied between the MPI values obtained from the healthy and affected hand. The corresponding MPI is considered as relevant if it satisfies the following conditions: (i) the difference between the MPI values for the healthy and affected hand is statistically significant ($p < 0.05$) and (ii) higher/lower MPI value in the case of healthy hand (depending on particular MPI). Only four upper body MPIs do not meet the statistical requirement in terms of the healthy-affected side analysis ($p > 0.05$). Those MPIs are **SR** for the movements **UB3**, **UB4** and **UB5** and **MS** for the **UB5** movement. Still, the remaining 17 MPIs turn out to be very relevant in distinguishing the performance of the healthy and affected hand, which is more than 80% of the proposed MPIs.

Table 2. Correlation between the proposed MPIs and Fugl-Meyer clinical scale expressed through the value of Pearson's correlation coefficient r

Movement	ROM S*	MS	SR	VDBH	VSED	ROM E*	MSA*
UB1	0.67	0.61	−0.97	/	/	/	/
UB2	0.72	0.78	−0.79	/	/	/	/
UB3	0.74	0.75	−0.68	−0.90	/	0.69	/
UB4	0.98	0.54	−0.90	−0.90	/	0.96	/
UB5	/	0.80	−0.69	−0.96	−0.80	/	0.92

* ROM S – ROM shoulder; ROM E – ROM elbow; MSA – mean shoulder angle

In order to verify whether the proposed MPIs evaluate the patients' movement performance in the same manner as the official clinical scales, we investigate the correlation between the proposed indicators and clinical Fugl-Meyer scale. The values of the Fugl-Meyer scale are obtained as a result of a physiatrist's evaluation right before the sensor measurements. Correlations were calculated using *Pearson's correlation coefficient* r (takes values between -1 and 0 for negative correlation and between 0 and 1 for positive correlation), along with the p-value and confidence intervals. Values of r closer to -1 in the case of negative correlation and closer to 1 in the case of positive correlation indicate a better correlation between the variables. Absolute values of the Pearson's correlation coefficient r across all MPIs and tested movements are inside the range [0.54–0.98], Table 2. MPIs that have demonstrated very high correlation with the clinical scale (0.80 < | r | < 1) are **VD** (both between hands and elbow-shoulder) and **MSA**. The remaining MPIs have shown a high correlation only for particular movements (Table 2).

6 Conclusions

Statistical analysis of the repeated sensor measurements confirmed the internal consistency of the sensor measurements and high test-retest reliability of the extracted MPIs. The results of the healthy-affected side analysis report the high percentage of the MPIs relevant for assessing the differences between the healthy and affected hand - more than 80% (17 out of 21 MPIs). The correlation analysis emphasizes the good correlation between the proposed MPIs and clinical (Fugl-Meyer) scale (Table 2). Based on the obtained results, we conclude that the majority of our proposed MPIs can be included in the stroke treatment protocols and used to support the clinical evaluations by medical doctors.

Acknowledgments. This work is partially funded by the Ministry of Science, Republic of Serbia under the contracts TR-35003, III-44008 and III-44004. This work was partially funded by the EU Project POETICON++ and the Portuguese FCT Project [UID/EEA/50009/2013].

References

1. Lozano, R., Naghavi, M., Foreman, K., et al.: Global and regional mortality from 235 causes of death for 20 age groups in 1990 and 2010: a systematic analysis for the global burden of disease study 2010. Lancet **380**(9859), 2095–2128 (2013)
2. Cheeran, B., Cohen, L., Dobkin, B., et al.: The future of restorative neurosciences in stroke: driving the translational research pipeline from basic science to rehabilitation of people after stroke. Neurorehabilitation Neural Repair **23**(2), 97–107 (2009)
3. Langhorne, P., Coupar, F., Pollock, A.: Motor recovery after stroke: a systematic review. Lancet Neurol. **8**(8), 741–754 (2009)
4. Jebsen, R.H., Taylor, N., Trieschmann, R.B., et al.: An objective and standardized test of hand function. Arch. Phys. Med. Rehabil. **50**(6), 311–319 (1969)
5. Fugl-Meyer, A.R., Jaasko, L., Leyman, I., et al.: The post-stroke hemiplegic patient. 1. A method for evaluation of physical performance. Scand. J. Rehabil. Med. **7**(1), 13–31 (1974)
6. Spasojević, S., Ilić, N.V., Rodić, A., Santos-Victor, J.: Kinect-based application for progress monitoring of the stroke patients. In: Proceedings of IcETRAN Conference, vol. ROI2.6, pp. 1–5, June 2017
7. Zhou, H., Hu, H.: Human motion tracking for rehabilitation – a survey. Biomed. Signal Process. Control **3**(1), 1–18 (2008)
8. Gonzalez-Jorge, H., Riveiro, B., Vazquez-Fernandez, E., et al.: Metrological evaluation of Microsoft Kinect and Asus Xtion sensors. Measurement **46**(6), 1800–1806 (2013)
9. Khoshelham, K., Elberink, S.: Accuracy and resolution of kinect depth data for indoor mapping applications. Sensors **12**(2), 1437–1454 (2012)

10. Clark, R., Pu, Y., Fortina, K., et al.: Validity of the Microsoft Kinect for assessment of postural control. Gait Posture **36**(3), 372–377 (2012)
11. Chang, C., Lange, B., Zhang, M., et al.: Towards pervasive physical rehabilitation using Microsoft Kinect. In: 6th International Conference on Pervasive Computing Technologies for Healthcare (PervasiveHealth), pp. 159–162, May 2012
12. Fernandez-Baena, A., Susin, A., Lligadas, X.: Biomechanical validation of upper-body and lower-body joint movements of kinectmotion capture data for rehabilitation treatments. In: 4th International Conference on Intelligent Networking and Collaborative Systems (INCoS), pp. 656–661. IEEE, September 2012
13. Webster, D., Ozkan, C.: Experimental evaluation of Microsoft Kinect's accuracy and capture rate for stroke rehabilitation applications. In: Haptics Symposium (HAPTICS), pp. 455–460. IEEE, February 2014
14. Chang, Y., Han, W., Tsai, Y.: A kinect-based upper limb rehabilitation system to assist people with cerebral palsy. Res. Dev. Disabil. **34**(11), 3654–3659 (2013)
15. Chang, Y., Chen, S., Huang, J.: A kinect-based system for physical rehabilitation: a pilot study for young adults with motor disabilities. Res. Dev. Disabil. **32**(6), 2566–2570 (2011)
16. Gama, A., Chaves, T., Figueiredo, L., et al.: Guidance and movement correction based on therapeutics movements for motor rehabilitation support systems. In: 14th Symposium on Virtual and Augmented Reality, pp. 191–200. IEEE, May 2012
17. Calin, A., Cantea, A., Dascalu, A., et al.: Mira – upper limb rehabilitation system using Microsoft Kinect. Studia Univ. Babes-Bolyai Informatica **56**(4) (2011)
18. Exell, T., Freeman, C., Meadmore, K., et al.: Goal orientated stroke rehabilitation utilising electrical stimulation, iterative learning and Microsoft Kinect. In: International Conference on Rehabilitation Robotics (ICORR), pp. 1–6. IEEE, June 2013
19. Esfahlani, S.S., Thompson, T.: Intelligent physiotherapy through procedural content generation. In: 12th Artificial Intelligence and Interactive Digital Entertainment Conference, September 2016
20. Bao, X., Mao, Y., Lin, Q., et al.: Mechanism of kinect-based virtual reality training for motor functional recovery of upper limbs after subacute stroke. Neural Regen. Res. **8**(31), 2904 (2013)
21. Hondori, H.M., Khademi, M., Lopes, C.V.: Monitoring intake gestures using sensor fusion (microsoft kinect and inertial sensors) for smart home tele-rehab setting. In: 1st Annual Healthcare Innovation Conference. IEEE, November 2012
22. Zannatha, J.M., Tamayo, A.J., Sánchez, Á.D., et al.: Development of a system based on 3D vision, interactive virtual environments, ergonometric signals and a humanoid for stroke rehabilitation. Comput. Methods Programs Biomed. **112**(2), 239–249 (2013)
23. Sadihov, D., Migge, B., Gassert, R., et al.: Prototype of a VR upper-limb rehabilitation system enhanced with motion-based tactile feedback. In: World Haptics Conference (WHC), pp. 449–454. IEEE (2013)
24. Spasojević, S., Ilić, T.V., Milanović, S., Potkonjak, V., Rodić, A., Santos-Victor, J.: Combined vision and wearable sensors-based system for movement analysis in rehabilitation. Methods Inf. Med. **56**(2), 95–111 (2017)
25. Field, A.: Discovering Statistics Using SPSS. Sage Publications, London (2009)

Analytical Study for the Capability Implementation of an Underactuated Three-Finger Hand

Chiara Cosenza(✉), Vincenzo Niola, and Sergio Savino

Department of Industrial Engineering, University of Naples "Federico II", Naples, Italy
chiara.cosenza@unina.it

Abstract. Combining the simple design with the implementation capability of underactuated hands is a challenging task. In this context, it is necessary to find a compromise between the number of actuators, the grasping abilities and the control strategy. Here, an analytical study is proposed to improve the precision grasping performance together with a simple control strategy of a three-finger underactuated hand.

Keywords: Underactuated robotic hand · Multibody simulations
Control strategy · Precision grasping

1 Introduction

There is a growing interest in the development of underactuated mechanical hands [1]. These devices have fewer actuators than the number of the degrees of freedom. Despite this design principle, these hands should possess manipulative dexterity and perform several grasping tasks for different applications. Moreover, the complexity design reduction must ensure the control robustness of these devices.

In previous works, the authors of this paper have already developed a five-finger mechanical hand (the Federica hand [2]) that is strongly underactuated. The hand has a tendon driven mechanism in which the traction and the antagonist tendons allow the fingers closing and reopening. Three phalanges compose each finger: the proximal, the medial and the distal. A differential system, obtained with a pulleys mechanism, allows the hand to be a self-adaptive mechanism in gripping tasks. Through previous multi-body model simulations, the authors have studied the influence of different types of antagonist tendons on the finger dynamics [3]. Furthermore, the grab ability has been studied in deep through theoretical and experimental activities [4].

The object of this current study is the implementation of the Federica hand capability. The hand has been divided in two subsystems: a three-finger one with the thumb, the index and the middle and a two-finger subsystem with the ring and the little. The attention has been focused on the three-finger mechanism that is mainly involved in the hand tasks. The three-finger mechanism can be thought both for prosthetic applications (like the previous prototype) and/or as a precision gripper.

G. Carbone et al. (Eds.): *New Trends in Medical and Service Robotics*, MMS 65, pp. 161–168, 2019.
https://doi.org/10.1007/978-3-030-00329-6_19

An analytical study has been implemented by means of the multibody simulation to develop an improvement strategy of the system capability. In this context, the studies of postural synergies have interested many researchers working on control theory and mechanical design of anthropomorphic hands [5, 6].

A critical issue has been the evaluation of the minimal number of actuators, with respect to degrees of freedom, needed to perform precision grasping tasks. Thus, to perform precision tasks the only simultaneous movement of the three fingers is not enough. An autonomous movement of the thumb with respect of the other two fingers has been supposed as a solution study.

2 The Three-Finger Gripper

We have considered a three-finger gripper composed by two identical fingers (the middle and the index) and a third finger in opposition (the thumb). In the thumb, the rotation axis of the proximal phalanx is rotated of 90° with respect to the other two phalanges. In this way, the thumb can realize two different rotations during the closing sequence. All three fingers are actuated by means of two tendons that are moved by the same actuation system. The two tendons work in opposition: the actuator tendon allows the closure while the antagonist tendon allows the reopening of the finger. With the aim to implement the capability of the underactuated system in its gripping tasks, in this study, our attention has been focused in tracking the fingertip distances.

3 Multibody Model Simulations

Simulink (MathWorks©) software has been employed to realize the multibody simulation model; the blocks adopted for the finger model parameterization are the Body block, the Rigid Transform block and the Revolute Joint block. In this current work, the multibody model is obtained from the 3D CAD model for the middle and the index (Fig. 1A) and for the thumb (Fig. 1B), that are joined to realize the three-fingers model, (Fig. 1C).

Fig. 1. The multibody model of the three mechanical fingers.

Each finger scheme consists in three main subsystems: the distal, the medial and the proximal phalanges. Three revolute joints provide the rotation of the phalanges; they can rotate of 90° due to a Hard stop subsystem that limits their rotations. To simulate the tendon action inside the fingers, forces were applied at the entry and the exit points of the tendon guides. A detailed description of the multibody model is reported in previous studies [7].

3.1 Analytical Expression for the Phalanx Rotations

For each finger, simulations have been performed to obtain the mathematical expression describing the rotations of the phalanges as a function of the tendon shortening.

Moreover, simulations have been also carried out at different closing velocities to study possible effects due to the decrease of the actuation time. The simulation data were analyzed to identify the functions describing the trend of the three-phalanx rotations in relation to the actuator tendon displacement. The mathematical expression was obtained by means of a multibody simulation model to consider the dynamical behaviour of the mechanical system. Finding an analytical model for the finger dynamical behaviour would have been much more complicated. Moreover, using the multibody simulation model, it would be also possible to improve the dynamical behaviour with the real phenomena that characterize the system (e.g. friction) and include these phenomena in the mathematical expression describing the rotation of the phalanges.

We choose the discontinuous piecewise function to best describe the data:

$$y = f + \left\{1 - \left[\frac{(d-x) + |d-x|}{2(d-x)}\right]\right\}\left\{b + \left[\frac{a(c-x) + |a(c-x)|}{2}\right]\right\} \tag{1}$$

where x is the tendon displacement and a, b, c, d, f are five parameters. The choice of a discontinuous piecewise function is justified from the fact that the multibody model does not consider the friction phenomena. Given this hypothesis, a three-section function best describes the trend of the phalanx rotations calculated through the model.

As an example, we have reported the rotation of the medial phalanx of the thumb as a function of the tendon shortening (Fig. 2). It is worth to notice that the function (1) describes the data well.

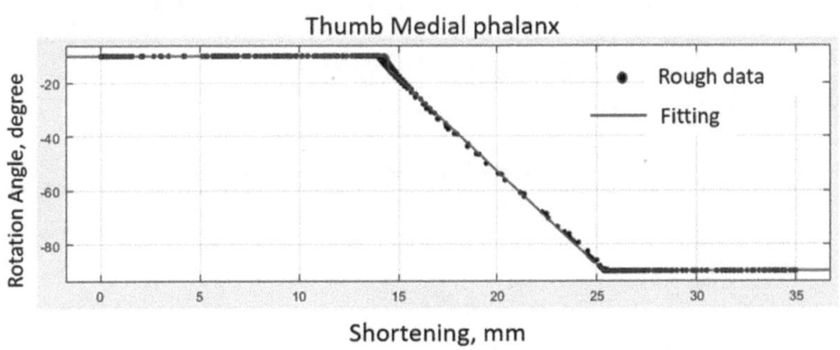

Fig. 2. Phalanx rotation as a function of tendon shortening.

4 One-Actuator System

Through the analytical expression (1) for each phalanx, it is possible to study the kinematics of the whole three-finger model. As a first step, only one actuator drives the gripper movement. The three-finger model, with the refence systems placed at the basis and in the fingertips of each finger, is showed in Fig. 3A.

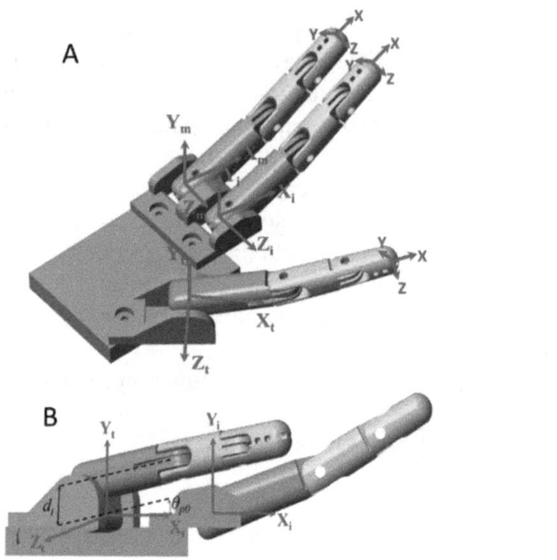

Fig. 3. Coordinate reference systems for fingertip displacement calculation.

For the middle and the index fingers, the ending points coordinates in the reference systems [X_i Y_i Z_i] and [X_m Y_m Z_m] are:

$$\begin{gathered} x = l_p \cdot \cos(\theta_p) + l_m \cdot \cos(\theta_m + \theta_p) + l_d \cdot \cos(\theta_d + \theta_m + \theta_p) \\ y = l_p \cdot \sin(\theta_p) + l_m \cdot \sin(\theta_m + \theta_p) + l_d \cdot \sin(\theta_d + \theta_m + \theta_p) \\ z = 0 \end{gathered} \tag{2}$$

For the thumb:

$$\begin{gathered} x_1 = l_m \cdot \cos(\theta_m) + l_d \cdot \cos(\theta_d + \theta_m) \\ y_1 = l_m \cdot \sin(\theta_m) + l_d \cdot \sin(\theta_d + \theta_m) \\ x = (x_1 + l_p) \cdot \cos(\theta_p) - d_t \cdot \sin(\theta_p) \\ y = (x_1 + l_p) \cdot \sin(\theta_p) + d_t \cdot \cos(\theta_p) \\ z = -y_1 \end{gathered} \tag{3}$$

Where l_p, l_m, l_d are proximal, medial and distal phalanx lengths; θ_p, θ_m, θ_d are the phalanx rotation angles described through the function found in the previous paragraph and d_t is the offset between the rotation axis, Z_t, and the middle axis (See Fig. 3B) of the thumb proximal phalanx.

The fingertip displacements in the coordinate refence system [X_i Y_i Z_i] are showed in Fig. 4.

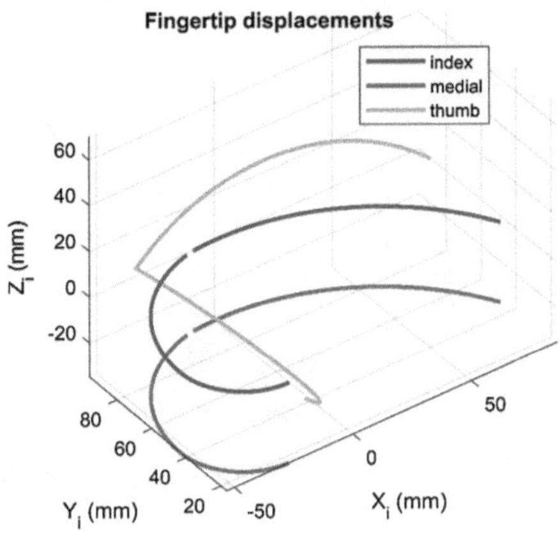

Fig. 4. Fingertip trajectories in the 3D space.

Since one actuator drives the fingers motion, it is possible to define the relations describing the thumb-index and the thumb-middle distances as a function of the tendon shortening (Fig. 5).

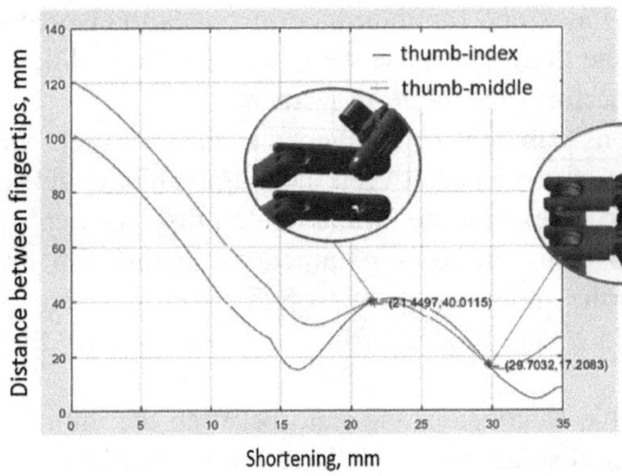

Fig. 5. Fingertip distances as a function of the tendon shortening for one actuator control.

There are two points in which the distances are equal, but they are both not feasible. These two configurations correspond to two tendon shortening values in which the fingers are completely closed, and/or the index should go through the thumb (insets in Fig. 6). This result demonstrates that is not possible to obtain a symmetric grasping, in which the thumb fingertip is located at the same distance from the index and the middle fingertips, with only one actuator.

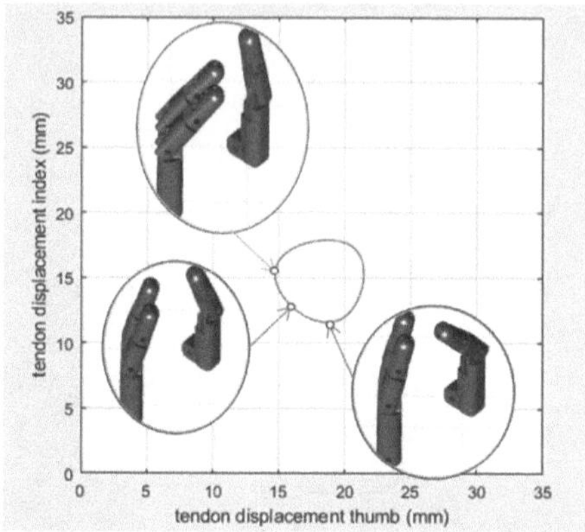

Fig. 6. Possible finger configurations realizing a 29 mm fingertip distance between the thumb and the index.

5 Two-Actuator System

The three-finger system has been modified, at this point, adding a second actuator: one to move the thumb and a second actuator to move the index and the middle fingers. In this case, the fingertip distances depend on two different tendon shortenings: one relative to the index and the middle and another relative to the thumb. The control on the distance between the index and the thumb corresponds to also control the distance between the middle and the thumb because the index and the middle are moved by the same actuator and then perform the same movement.

Multiple configurations can realize a given distance between the fingertips. For instance, if we set a fingertip distance between the thumb and the index, all the possible combinations between the index and the thumb tendon displacements are described by the blue curve in Fig. 6. This curve is obtained evaluating the fingertip distances between the thumb and the index by means of Eqs. (1), (2) and (3) in which two independent variables are introduced for the thumb and the index tendon displacements.

Moreover, it is possible to consider the ratio between the thumb tendon displacement and the index tendon displacement. We have reported all the curves associated to this ratio for multiple fingertip distances in Fig. 7A. Each curve can be well described with the analytical expression:

$$f = a_1 \cdot \sin(b_1 \cdot x + c_1) + a_2 \cdot \sin(b_2 \cdot x + c_2) + a_3 \cdot \sin(b_3 \cdot x + c_3) \tag{4}$$

Where a_1, b_1, c_1, a_2, b_2, c_2, a_3, b_3, c_3 are nine parameters and x the fingertip distance (Fig. 7B). Through the expression (2), it is possible to define a new control

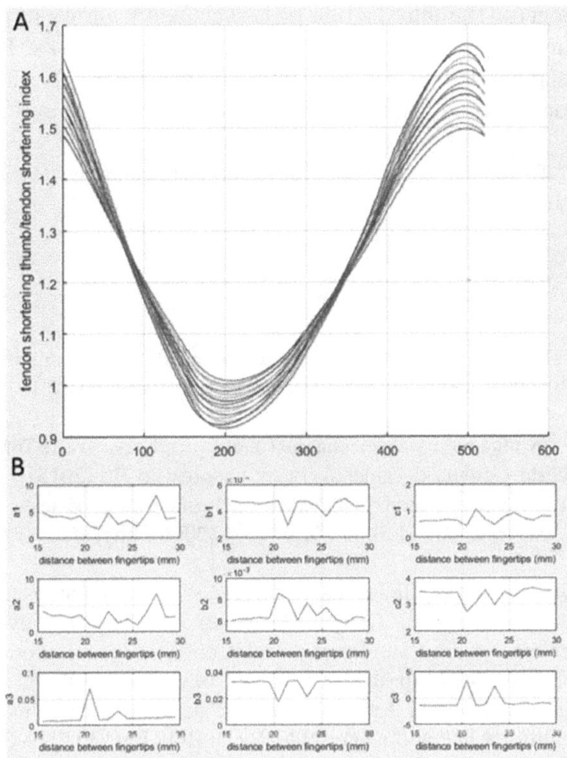

Fig. 7. Ratio between the two-tendon shortenings (A) and functions describing the nine parameters in the expression 4 (B).

strategy. If we set a given fingertip distance, we can compute the ratio between the tendon displacements and, therefore, obtain the possible shortening values that realize the desired configurations.

Equation (4) allows to define one tendon displacement, if the other one is assigned. In this way, the three-finger system has more capabilities to execute gripping tasks, but, even if it is necessary to control two actuators, the command is based on only one tendon displacement. In the insets in Fig. 6, we have reported three possible configurations able to achieve a 29 mm fingertip distance due to the proposed control strategy as an application case.

6 Conclusions

In this paper, the possibility to employ an underactuated three-finger gripper for precision grasping has been evaluated. The gripper with a single actuator is not able to realize precision grasping tasks. Conversely, the two-actuator gripper leads to a more precise control of the system due to the autonomous motion of the thumb with respect to the other two fingers. The introduction of two actuators allows to perform different control strategies and represents a solution for pincer grasp tasks. Finally, a simple control strategy is presented to achieve multiple finger configurations to carry out symmetric precision grasping tasks.

Acknowledgments. The authors thank Antonio Lombardo to contribute to this work during the master's degree thesis.

References

1. Tavakoli, M., et al.: Underactuated anthropomorphic hands: actuation strategies for a better functionality. Rob. Auton. Syst. **74**, 267–282 (2015). https://doi.org/10.1016/J.ROBOT.2015.08.011
2. Niola, V., et al.: An underactuated mechanical hand prosthesys By IFToMM Italy. In: The 14th IFToMM World Congress, Taipei, Taiwan, October 25–30 (2015)
3. Niola, V., Rossi, C., Savino, S.: Dynamical model and prototype tests of a self-adaptive mechanical hand. Int. Rev. Model Simul. **9**, 97–104 (2016). https://doi.org/10.15866/iremos.v9i2.8068
4. Niola, V., Rossi, C., Savino, S.: Influence of the tendon design on the behavior of an underactuated finger. In: Ferraresi, C., Quaglia, G. (eds.) Advances in Service and Industrial Robotics: Proceedings of the 26th International Conference on Robotics in Alpe-Adria-Danube Region, RAAD 2017. Springer, Cham, pp. 1033–1042 (2018)
5. Palli, G., et al.: The DEXMART hand: mechatronic design and experimental evaluation of synergy-based control for human-like grasping. Int. J. Rob. Res. **33**, 799–824 (2014). https://doi.org/10.1177/0278364913519897
6. Xiong, C.H., et al.: Design and implementation of an anthropomorphic hand for replicating human grasping functions. IEEE Trans. Robot. **32**, 652–671 (2016). https://doi.org/10.1109/TRO.2016.2558193
7. Niola, V., Rossi, C., Savino, S., Timpone, F.: Study of an underactuated mechanical finger driven by tendons. Int. J. Autom. Technol. **11**, 344–354 (2017). https://doi.org/10.20965/ijat.2017.p0344

IBM Cloud Services Enhance Automatic Cognitive Assessment via Human-Robot Interaction

Simone Varrasi[1], Alexander Lucas[1], Alessandro Soranzo[1], John McNamara[2], and Alessandro Di Nuovo[1(✉)]

[1] Sheffield Hallam University, Sheffield S1 1WB, UK
a.dinuovo@shu.ac.uk
[2] IBM Hursley Lab, Winchester SO21 2JN, UK

Abstract. Thanks to recent developments in artificial intelligence and social robotics, Human-Robot Interaction (HRI) can be used as a non-invasive screening tool for the assessment of cognitive decline. In this scenario, the robot manages the assessment by providing the instructions to the patient, registering his/her answers and objectively calculating the final score. This service can help to save time and reach a wider population. From the technical point of view, a challenge is to achieve a highly reliable speech and visual recognition as required for a valid scoring of cognitive performance that can support the assessment from a professional.

In this article, we evaluate a system for cognitive assessment that makes use of the IBM AI Cloud services embodied in one of the most popular platforms for social robotics: the SoftBank Pepper. Results of a pilot study with 16 human participants shows that IBM Cloud services for speech and visual recognition can improve the system performance in comparison with standard interfaces. Importantly, the improvement allows achieving a significant correlation with one of the most used paper-and-pencil tests and, therefore, the study demonstrates the validity of the robotic approach for cognitive assessment.

1 Introduction

Social robots have been increasingly studied for clinical applications [4]. Human-Robot Interaction (HRI), indeed, is a valid mean to provide patients with valuable services, even in the field of mental healthcare [8]. In the last period, initial evidence has been collected not only on robot-assisted treatments but also on the viability of robotic assessments, for instance in the measurement of Patient Reported Outcome [1] and in early diagnosis of Autistic Spectrum Disorder [2, 7]. When an assessment is managed by a robot, in fact, many advantages allow a more reliable scoring of performances: assessor neutrality, standardization of the interaction, better acceptance of the robotic platform than a non-embodied computer [3, 9]. These features are particularly relevant in the case of a robotic administration of screening tests for early detection of dementia, as robots could guarantee automatic large-scale screening exams for the elderly population.

G. Carbone et al. (Eds.): *New Trends in Medical and Service Robotics*, MMS 65, pp. 169–176, 2019.
https://doi.org/10.1007/978-3-030-00329-6_20

However, social robots will be able to be integrated into standard evaluation procedures only when the scores provided by them will be valid enough to represent an aid for clinicians, who will be, therefore, supported by the technology for some of their daily tasks. A careful development of robotic interfaces and artificial intelligence is needed to fulfil such a vision. However, in one of our pilot tests on cognitive assessment with the SoftBank humanoid social robot *Pepper* [10], it was found that the score automatically calculated using the robot software often failed to provide the correct score because of the failures of the embedded speech and object recognition interfaces. We concluded that for clinical validity was necessary the revision of the score by a professional supervisor.

This preliminary result prompted this follow-up study on alternative technologies to enhance the reliability of scoring and, therefore, a reliable cognitive assessment that could be used without necessarily involving a professional. An interesting solution is represented by cloud services offered by the major corporations, for instance Microsoft, Google, and IBM. In fact, these cloud services can be considered the state-of-the-art in artificial intelligence and provide a standardized and easily reproducible environment for development and testing of HRI applications. Even if these services were just recently introduced, they were already used for some studies in HRI [6].

After thorough consideration, we selected for further analysis the IBM AI Cloud services ("Watson"), which provide a comprehensive set of easy-to-use tools for speech recognition (speech-to-text) and production (text-to-speech) and object recognition from pictures that met the requirements for our application.

In this paper, we present an exploratory evaluation of the IBM Watson services in comparison with default Pepper's software with the aim to identify which approach can provide an automatic score closer to the one calculated by a professional supervisor, which represents the benchmark in our analysis. In addition, we administered and considered in the comparison a widely used paper-and-pencil test in order to provide also external validation to the robotic instrument.

2 Materials and Methods

2.1 Participants

A sample of 16 healthy adults was enrolled in our study (*M*-age = 31.5 years, range = 19–61, *SD* = 14.15). They were all proficient in British English, but only 2 of them were native speakers. The education level was usually higher (*M* = 19.5, *SD* = 4.07). All participants gave informed consent to use their data and pictures.

2.2 The Robotic Assessment Procedure (Data Collection)

The robotic cognitive assessment was inspired by one of the most important and widely used psychometric screening tools for Mild Cognitive Impairment, the Montreal Cognitive Assessment test (MoCA) [5], which is freely available from the official website. We programmed Pepper to administer and score a MoCA-like psychometric assessment, in which the robot provided the instructions (Fig. 1). The robotic cognitive test measures the same areas of the MoCAs, hence the subtests have the same names. In practice, speech recognition was used to score: Naming, Attention, Language, Abstraction, Delayed Recall, and Orientation. The visual recognition, instead, was used to assess Visuospatial/Executive skills by asking the user to draw a cube and a clock on two sheets of paper, which were automatically processed by robot's device.

All the administration was audio-recorded by the robot's microphones. The robot took pictures of the user drawings for the visuospatial/executive skills.

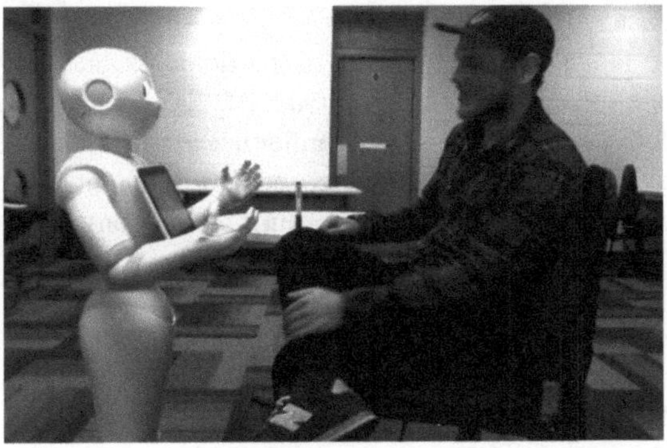

Fig. 1. Example of the human-robot interaction during the robotic administration

More detail on the experimental procedure, the assessment protocol, and the robotic software can be found in [10]. The direct result of the robotic administration was twofold: (i) a *Pepper score*, calculated by the robot thanks to its embedded AI software, and (ii) a *Supervised score*, which was the actual score achieved by the user during the robotic administration, obtained by a professional re-scoring the performance through the recordings available.

Furthermore, in order to provide external validity, the participants were also administered an alternative version of the MoCA test by a human psychologist, who calculated the *Paper and Pencil MoCA score*.

2.3 Data Processing via IBM Cloud AI Services

The results of the previous pilot experiment [10] show that the automatic processing must be improved, in order to have a reliable score that can be used for automatic screening. To this end, we used the IBM Cloud solutions for post-processing of the data collected from the previous experiment, with the aim to establish if these services can be used to improve the automatic assessment.

In particular, we used Watson Speech to Text to perform speech recognition and Watson Visual Recognition to perform the analysis of the hand-drawn images.

2.4 Watson Speech to Text

Pepper's default voice recognition system is tailored for recognizing individual words, which is not suitable for some parts of the test where long sentences must be recognized, e.g. the Language assessment. In addition, Pepper's microphones pick up constant noise coming from the cooling fans located inside the head. This has an impact on the quality of the audio source, which in turn negatively affects speech recognition. We planned to overcome these problems by using Watson speech to text service, which makes use of longer sentences for better context analysis and recognition and allows customizing the model to embed noise.

Considering the language background of participants and the characteristics of audios, we used the model *en-GB_BroadbandModel*. Because of the explorative character of this study, we did not set a confidence threshold under which the transcriptions were discarded. For the same reason, we took into account the 10 best alternatives of recognition transcription according to the confidence level.

By processing the audio with standard parameters, we obtained the *basic Watson score*, which is in fact the simplest version.

Also a *customized Watson score* was calculated thanks to the language customization service, which allows training the speech-to-text model for specific recognition requests. Indeed, we created three customized models by adding specific corpora. One model was trained for the recognition of numbers, months and weekdays (Attention and Orientation tasks), one for the recognition of all the words starting with *B*, *F* and *S* (Fluency task), and the last one for all the remaining tasks. The customization weight was set to 0.9.

For both basic and customized, scores were calculated following two scoring approaches: the *exact approach* allowed assigning the points only when the target word had the best confidence level and the string was fully contained in the transcriptions; the *flexible approach*, instead, accepted a certain percentage of error, so that the points were given if the transcription fitted for at least the 70% the target word or string.

2.5 Watson Visual Recognition

During the administration, the participants produced hand-drawing of a cube and a clock showing a specific time. They would then present these pictures so that the robot could take a photo. The photos were sent to Watson Visual Recognition for analysis, generating a general class and individual subclass of the object/setting the recognition system deemed as central on the picture. We reprocessed the drawings of cubes and clocks made by participants with the default visual recognition, without any kind of previous training. The system gave a score if the class/subclass contained the object the person was drawing.

The points were given when the cube was recognized as *polyhedron* and the clock as *clock* or *wall clock*. For the hours, points were given if numbers from 1 to 12 were recognized.

3 Experimental Results

The statistical analyses presented in this section were performed with the SPSS software (version 24). The descriptive statistics were calculated for the various automatic scores (Pepper, Watson), the supervised (benchmark) and the Paper-and-Pencil MoCA score: mean (*M*), standard deviation (*SD*), minimum (*Min*) and maximum (*Max*). The scores are the sum of the subtests scores. Then, Spearman correlations – chosen for the shape of the distribution and the typology of data – were calculated to explore the relationship among each test with the supervised score as well as with the paper and pencil MoCA score, which represents the external validity for the test. Finally, we conducted a repeated measure ANOVA on the differences between each test and the benchmark (supervised score).

3.1 Descriptive Analyses

The mean Pepper score is 12.69 (Min = 6; Max = 23; *SD* = 4.61), the mean Supervised score is 18.63 (Min = 10; Max = 27; *SD* = 4.83), and mean Paper-and-Pencil MoCA score is 25 (Min = 21; Max = 28; *SD* = 2.07).

In the case of Watson, the mean Exact basic score is 10.69 (Min = 2; Max = 17; *SD* = 4.71), the mean Flexible basic score is 11.44 (Min = 3; Max = 20; *SD* = 4.95), the mean Exact customized score is 15.50 (Min = 8; Max = 25; *SD* = 5.16), and the mean Flexible customized score is 16.75 (Min = 9; Max = 25; *SD* = 5.47).

3.2 Correlations

Spearman correlations were calculated among the global scores. Results show (Table 1) that the Pepper score does not correlate significantly either with the Supervised score nor with the Paper-and-Pencil MoCA score. The basic Watson score, instead, strongly and significantly correlates with the Supervised score, both in the Exact and Flexible versions. A stronger correlation can be found between the Exact and Flexible customized Watson scores, which also significantly correlates with the Pepper score and the Paper and Pencil score.

Table 1. Spearman correlations (Significant results in bold)

	Supervised score	Paper and Pencil MoCA score
Pepper score	0.38	0.01
Exact basic Watson	**0.738****	0.476
Flexible basic Watson	**0.737****	0.442
Strict customized Watson	**0.834****	**0.542***
Flexible customized Watson	**0.831****	**0.515***

** $p < .01$; * $p < .05$

3.3 Repeated Measures ANOVA

A Kolmogorov–Smirnov test for the normality of the raw data showed no significant deviation from normality. A repeated measure ANOVA on the differences between each test and the benchmark revealed a significant effect of the test [$F(5,75) = 54.2$; $p < 0.001$]. A post hoc analysis with Bonferroni correction revealed a significant difference between each test and the benchmark ($p < 0.001$) except for the Flexible customized Watson score. Figure 2 shows the departures of each test from the benchmark. As can be seen, the test that deviates the least from the benchmark is the Watson Customized Flex score.

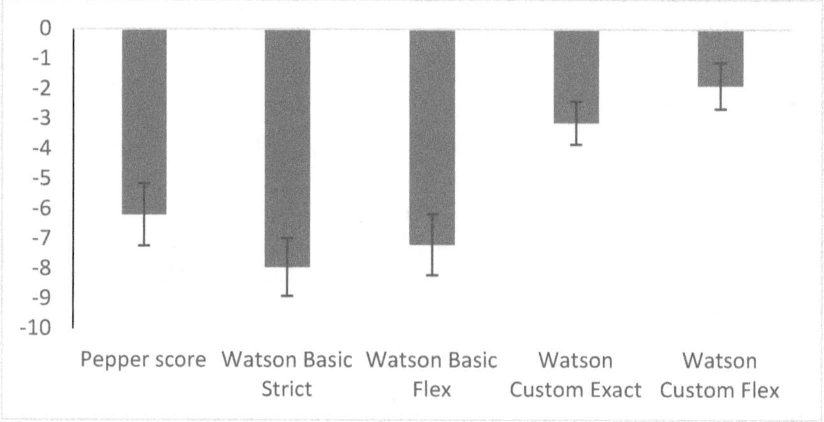

Fig. 2. Average departures of each approach from the benchmark.

For a more detailed analysis, Fig. 3 presents the average absolute deviations from the benchmark for each subtest. Watson versions strongly improve Visuo-spatial, Attention and Abstraction, while the basic version struggles with Naming, Delayed Recall and Orientation. However, after the customizations, Watson is always performing better than or as good as the default Pepper algorithms.

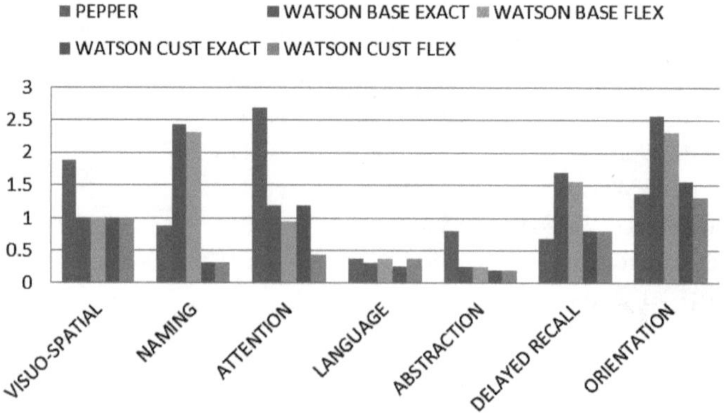

Fig. 3. Average absolute departures of each subtest from the benchmark.

4 Discussion and Conclusion

In this paper, we presented an exploratory study on the use of IBM Cloud AI services "Watson" for supporting the scoring of a MoCA-inspired psychometric test on a robot. To this end, we applied a new system on the collected data from 16 administrations and calculated Watson scores, which were compared with the score calculated using the default robot AI, and a benchmark *Supervised score* obtained by a professional review of the robot's recordings. We also included in the evaluation the paper-and-pencil MoCA score for testing external validity.

The analysis of the results shows that the score calculated by the robot with its embedded software does not fit the benchmark and lacks external validity. Vice versa, Watson improves the validity of the scoring, showing a strong and significant correlation with the *Supervised score*. The best result is obtained with the Watson customized version, which strongly correlates with the *Supervised score* and shows initial clinical validity as it is the only one that significantly correlates with the paper-and-pencil MoCa. The Flexible approach gives the best results by relaxing the confidence thresholds along with the customizations; in fact, this is the only case that doesn't significantly deviate from the benchmark.

We believe that the customizations are made necessary to overcome a structural problem of the Pepper head design, which places the cooling fan close to the rear microphones and the fan noise is always present in the background.

We conclude that the IBM Watson services can contribute in creating a reliable automatic scoring system that can be embedded on a physical device, such as a social robotic platform, and be used for screening of cognitive impairments in order to provide early treatment or for continuous assessment for personalized care. The overall results are very promising and represent a first step towards the development of artificial agents' contribution to psychological assessment.

However, there is still space for improvement and the present study has some of the limitations typical of preliminary studies, including a small sample, which is not balanced because of the majority of not native speakers, and a low quality of audio recordings. Our future work will focus on: (i) testing the performance of US Broadband model; (ii) further training and audio post-processing to enhance the quality of speech transcriptions; (iii) training the visual recognition with different kind of polyhedrons and to reliably assign intermediate points to the clock details; (iv) testing with people with pronunciation defects (such as stroke patients) or with low education level.

References

1. Boumans, R., et al.: Proof of concept of a social robot for patient reported outcome measurements in elderly persons. In: HRI 2018 Companion: Conference on ACM/IEEE International Conference on Human-Robot Interaction, 5–8 March 2018, Chicago, IL, USA, p. 2 (2018)
2. Conti, D., et al.: Robots in education and care of children with developmental disabilities: a study on acceptance by experienced and future professionals. Int. J. Soc. Robot. **9**, 51–62 (2017)

3. Feingold Polak, R., et al.: Differences between young and old users when interacting with a humanoid robot: a qualitative usability study. In: HRI 2018 Companion: Conference on ACM/IEEE International Conference on Human-Robot Interaction, 5–8 March 2018, Chicago, IL, USA, p. 2 (2018)
4. Iroju, O., et al.: State of the art: a study of human-robot interaction in healthcare. Int. J. Inf. Eng. Electron. Bus. **3**(3), 43–55 (2017)
5. Nasreddine, Z.S., et al.: The Montreal cognitive assessment, MoCA: a brief screening tool for mild cognitive impairment. J. Am. Geriatr. Soc. **53**(4), 695–699 (2005)
6. Novoa, J., et al.: DNN-HMM based automatic speech recognition for HRI scenarios. In: Proceedings of the 2018 ACM/IEEE International Conference on Human-Robot Interaction, pp. 150–159. ACM (2018)
7. Petric, F., Kovačić, Z.: No data? No problem! expert system approach to designing a POMDP framework for robot-assisted ASD diagnostics. In: HRI 2018 Companion: 2018 ACM/IEEE International Conference on Human-Robot Interaction Companion, 5–8 March 2018, Chicago, IL, USA, p. 2 (2018)
8. Rabbitt, S.M., et al.: Integrating socially assistive robotics into mental healthcare interventions: applications and recommendations for expanded use. Clin. Psychol. Rev. **35**, 35–46 (2015)
9. Varrasi, S., et al.: A social robot for cognitive assessment. In: HRI 2018 Companion: Conference on ACM/IEEE International Conference on Human-Robot Interaction, 5–8 March 2018, Chicago, IL, USA, p. 2 (2018)
10. Varrasi, S., et al.: Social robots as psychometric tools for cognitive assessment: a pilot test. In: Human Friendly Robotics, pp. 99–112. Springer, Cham (2019)

Concept Design and Dynamic Modelling of a Fibre-Based Continuum Robot for Early Cancer Detection Using DT-TMM

Mathias Bærenholdt[1], Lidai Wang[2], and Xuping Zhang[1(✉)]

[1] Department of Engineering, Aarhus University, Aarhus, Denmark
xuzh@eng.au.dk

[2] Department of Mechanical and Biomedical Engineering, City University of Hong Kong, Kowloon Tong, China

Abstract. This paper presents the dynamic modelling of a fibre-based continuum robot for early cancer detection using the discrete-time transfer matrix method (DT-TMM) with high computational efficiency. A brief introduction to the concept of the fibre-based robot is given in this paper. The principles of the DT-TMM are reviewed, and the dynamic modelling methodology is analysed and established for the fibre-based robot using the DT-TMM. Numerical simulations are performed to demonstrate the dynamic modelling method and the feasibility of using the robot for early cancer detection.

Keywords: Fibre-based robot · Deformation · Dynamic modelling
Transfer matrix

1 Introduction

Optical imaging plays a significant role in many biomedical applications. The urgent need for enhanced biophotonics in clinical or preclinical applications include early cancer detection, non-invasive neuron imaging, single-neuron manipulation and many others. The existing biophotonic technologies cannot efficiently focus in deep biological tissue due to the highly dynamic light scattering properties of living tissue [1]. To address this challenge, the fibre-based continuum robot, as shown in Fig. 1, offers a possible solution in combination with photoacoustically guided wavefront shaping [1], as it can provide better light focusing capabilities of optical imaging systems for imaging in deep biological tissue and thus improve numerous biomedical applications, such as physiological and pathological disease diagnosis, staging and treatment.

Continuum robots have distributed deformation with infinite degrees of freedom (DOF). This leads to a hyper-redundant configuration space in which the robot can attain an infinite number of shapes and configurations. To control a flexible continuum robot, a dynamic model with high computational efficiency needs to be established. In the past decades, significant efforts have been made towards the dynamic modelling of manipulators and mechanical systems with flexible components [2]. Different discretization techniques, such as the finite element method (FEM) [3, 4], the assumed mode method (AMM) [5, 6], and the lumped parameter method (LPM) [7, 8], have

G. Carbone et al. (Eds.): *New Trends in Medical and Service Robotics*, MMS 65, pp. 177–185, 2019.
https://doi.org/10.1007/978-3-030-00329-6_21

been reported extensively for modelling the dynamics of flexible robot manipulators. However, the computation of current dynamic modelling methods is still a big concern in relation to real-time control. An alternative method, the discrete time transfer matrix method (DT-TMM) [9, 10], features high computational efficiency because the global dynamics is avoided, and the matrix size of the robot dynamics does not increase with the DOF of a robot manipulator or mechanical system in DT-TMM. The aim of the present study is the dynamic modelling of a fibre-based continuum robot using DT-TMM and to provide insights into the design and control of the continuum robotic system for early cancer detection.

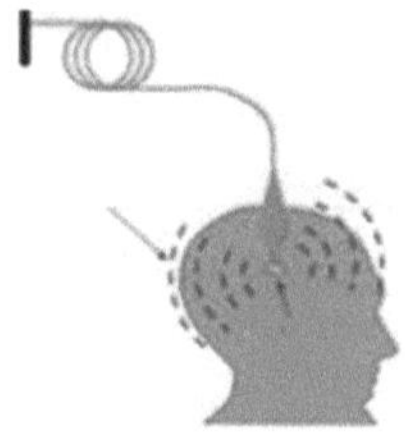

Fig. 1. Fibre-based Robot

2 Concept of a Fibre-Based Continuum Robot

A fibre-based continuum robot has been proposed to manipulate a single mode optical fibre as shown in Fig. 2, and hence to improve the photoacoustically guided wavefront shaping for imaging in deep biological tissue. The manipulation of light in an optical fibre is achieved by bending and/or squeezing the fibre. If only bending is applied to the fibre, the displacements should be at least 2 millimetres. However, if both squeezing and bending of the fibre is implemented, displacements in the order of micrometres are acceptable as long as the squeezing forces are large enough. In the proposed system, multiple piezoelectric actuators (PEA's) are employed to squeeze or change the shape of the optical fibre. The springs in the system are used to add squeezing forces to the fibre. The fibre is fixed in one end, and the other end is free so that the fibre is allowed to move in the axial direction.

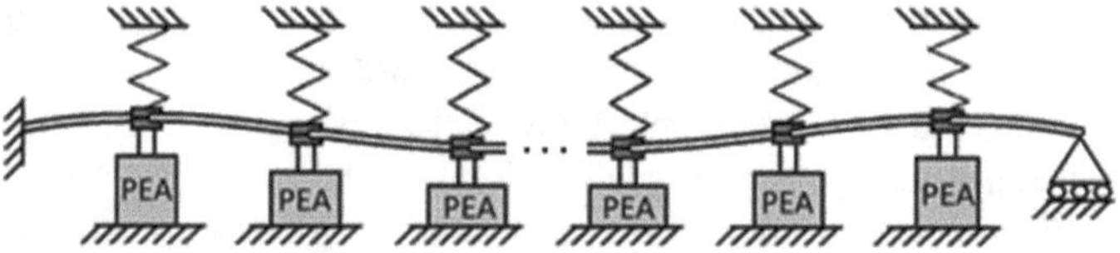

Fig. 2. Schematic illustration of a continuum robot concept

3 Dynamic Modelling with DT-TMM

The basic procedure of the DT-TMM consists of four main steps: (1) Decomposing a system into subsystems and elements. (2) Deriving the transfer matrix to transform the inboard state vector to the outboard sate vector for each element based on its linearized kinematics and dynamics. (3) Concatenating all the elements to obtain the system transfer matrix that transforms the system inboard state vector to the system outboard state vector. (4) Compute the state vectors based on the boundary conditions at the inboard and the outboard state vector.

The concept of DT-TMM can be seen in $z_O = U \cdot z_I$, where the outboard state vector, z_O, is calculated from the transfer matrix, U, and the inboard state vector, z_I. This formulation works on both elements and chain systems.

In this work, the state vector at any location can be expressed as $z = [x \quad y \quad \theta \quad m \quad q_x \quad q_y \quad 1]^T$, where x and y denote the position coordinates along the x and y-axis respectively, θ is the orientation angle, q_x and q_y are internal forces along x and y-axis respectively, and m is the internal torque.

To better present the principle and procedure of the dynamic modelling in this work, the fibre-based robot is considered as a continuum chain driven by only three piezoelectric actuators as shown in Fig. 3. The detailed procedure is: (1) The fibre-based robot is divided into four sub-chains; one at the left end, two in the middle, and one at the right end. (2) Each sub-chain is discretised further into a number of rigid bodies connected by spring hinges using the finite segment method. The springs represent the flexibility of the fibre. (3) The linearized dynamics and kinematics are established for rigid bodies and springs, and the transfer matrix is derived for each type of element. (4) The transfer matrix of each sub-chain is obtained by concatenating its rigid body and spring elements. (5) The system equation is established by concatenating the four sub-chains as $z_0 = U_4 U_3 U_2 U_1 z_I$. (6) The unknown variables of the inboard state vector of the system is determined from the boundary conditions. (7) The state vectors of each element are determined from pre-multiplication of the transfer matrices and the inboard state vector, z_I.

The two basic elements used to establish the dynamics of the fibre segments are a rigid body and a spring hinge. The derivation of the transfer matrix for each of these element types is briefly presented as the space is limited in this paper.

In DT-TMM, the transfer matrix for a basic element is derived based on the relationship between the outboard state vector and the inboard state vector. This is typically determined from the element's geometric properties and Newton-Euler equations. For example, the outboard position of a rigid body in the x-direction, as shown in Fig. 4, can be expressed in terms of the inboard position and orientation and the geometric properties of the rigid body:

$$x_O = x_I + b_1 \cdot \cos(\theta) - b_2 \cdot \sin(\theta)$$

Similarly, based on the kinematics and Newton-Euler dynamics, the other five equations for y, θ, q_x, q_y, m between the inboard and the outboard point can be established. Note that the six equations can consist of nonlinear (trigonometric, square, velocity, acceleration) terms with variables from the inboard state vector. These nonlinear terms need to be linearized in terms of the inboard variables to get a linear transformation from the inboard to the outboard. The linearization and dynamic derivations are not detailed in this work, but readers can refer to the related publications

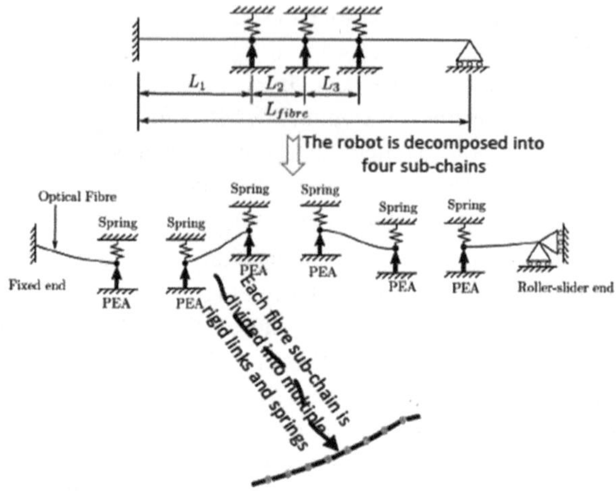

Fig. 3. Schematic diagram of dynamic modelling of the fibre robot using DT-TMM

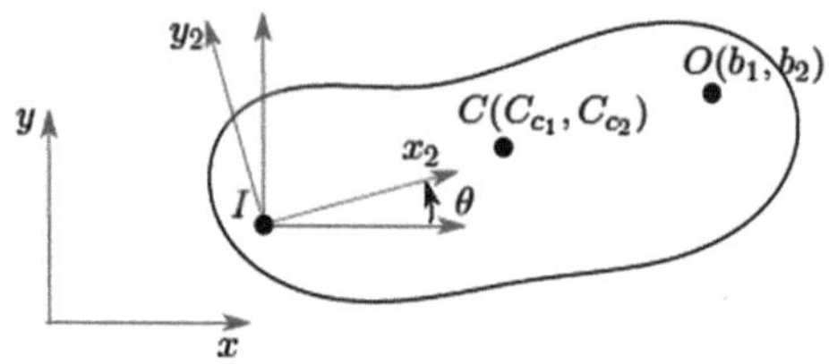

Fig. 4. Rigid body moving in plane. The inboard point is denoted by I, the outboard point is denoted by O and the centre of gravity is located at point C. The inboard point is the origin of the local coordinate system, x_2-y_2, of the body.

on DT-TMM [9, 10]. Finally, the linearized kinematic and dynamic equations $Z_O^{RigidBody} = U_{RigidBody} Z_I^{RigidBody}$ can be used to establish the transfer matrix of a rigid body element as seen below:

$$U_{RididBody} = \begin{bmatrix} 1 & 0 & u_{1,3} & 0 & 0 & 0 & u_{1,7} \\ 0 & 1 & u_{2,3} & 0 & 0 & 0 & u_{2,7} \\ 0 & 0 & 1 & 0 & 0 & 0 & 0 \\ u_{4,1} & u_{4,2} & u_{4,3} & 1 & u_{4,5} & u_{4,6} & u_{4,7} \\ u_{5,1} & 0 & u_{5,3} & 0 & 1 & 0 & u_{5,7} \\ 0 & u_{6,2} & u_{6,3} & 0 & 0 & 1 & u_{6,7} \\ 0 & 0 & 0 & 0 & 0 & 0 & 1 \end{bmatrix} \tag{1}$$

$$\bar{s} = \sin(\theta(t_{i-1})) \cdot \left(1 - \frac{1}{2} \cdot (\dot{\theta}(t_{i-1}) \cdot \Delta T)^2\right) + \cos(\theta(t_{i-1})) \cdot \left(\dot{\theta}(t_{i-1}) \cdot \Delta T + \frac{1}{2} \cdot \ddot{\theta}(t_{i-1}) \cdot \Delta T^2\right)$$

$$\overline{\mathrm{c}} = \cos(\theta(t_{i-1})) \cdot \left(1 - \frac{1}{2} \cdot (\dot{\theta}(t_{i-1}) \cdot \Delta T)^2\right) - \sin(\theta(t_{i-1})) \cdot \left(\dot{\theta}(t_{i-1}) \cdot \Delta T + \frac{1}{2} \cdot \ddot{\theta}(t_{i-1}) \cdot \Delta T^2\right)$$

$$\overline{\mathrm{S}} = \sin(\theta(t_{i-1})) - \theta(t_{i-1}) - \frac{1}{2} \cdot sin(\theta(t_{i-1})) \cdot (\dot{\theta}(t_{i-1}) \cdot \Delta T)^2$$

$$\overline{\mathrm{C}} = \cos(\theta(t_{i-1})) + \theta(t_{i-1}) - \frac{1}{2} \cdot cos(\theta(t_{i-1})) \cdot (\dot{\theta}(t_{i-1}) \cdot \Delta T)^2$$

$$x_{IO} = b_1 \cdot \cos(\theta) - b_2 \cdot \sin(\theta) \approx b_1 \cdot \overline{\mathrm{c}} - b_2 \cdot \overline{\mathrm{s}}$$

$$y_{IO} = b_1 \cdot \sin(\theta) + b_2 \cdot \cos(\theta) \approx b_1 \cdot \overline{\mathrm{s}} + b_2 \cdot \overline{\mathrm{c}}$$

$$x_{IC} = C_{c1} \cdot \cos(\theta) - C_{c2} \cdot \sin(\theta) \approx C_{c1} \cdot \overline{\mathrm{c}} - C_{c2} \cdot \overline{\mathrm{s}}$$

$$y_{IC} = C_{c1} \cdot \sin(\theta) + C_{c2} \cdot \cos(\theta) \approx C_{c1} \cdot \overline{\mathrm{s}} + C_{c2} \cdot \overline{\mathrm{c}}$$

$$u_{1,3} = -y_{IO}(t_{i-1})$$

$$u_{1,7} = b_1 \cdot \overline{\mathrm{C}} - b_2 \cdot \overline{\mathrm{S}}$$

$$u_{2,3} = x_{IO}(t_{i-1})$$

$$u_{2,7} = b_1 \cdot \overline{\mathrm{S}} + b_2 \cdot \overline{\mathrm{C}}$$

$$u_{4,1} = m \cdot \chi_1 \cdot (y_{IO} - y_{IC})$$

$$u_{4,2} = m \cdot \chi_1 \cdot (x_{IC} - x_{IO})$$

$$u_{4,3} = J_I \cdot \chi_1 - m \cdot \chi_1 \cdot (y_{IC} \cdot y_{IO} + x_{IC} \cdot x_{IO}) + 2 \cdot m \cdot \dot{\theta}(t_{i-1}) \cdot \chi_3 \cdot (y_{IC} \cdot x_{IO} - x_{IC} \cdot y_{IO})$$

$$u_{4,5} = -y_{IO}$$

$$u_{4,6} = x_{IO}$$

$$\begin{aligned} u_{4,7} = & J_I \cdot \chi_{2,\theta} + m \cdot x_{IC} \cdot \chi_{2,y_I} - m \cdot y_{IC} \cdot \chi_{2,x_I} \\ & - m \cdot \left(-\chi_{2,x_I} + (2 \cdot \dot{\theta}(t_{i-1}) \cdot \chi_{4.\theta} - \dot{\theta}(t_{i-1})^2 + \ddot{\theta}(t_{i-1})^2 \cdot \Delta T^2\right) \cdot x_{IC} \\ & + \chi_{2.\theta} \cdot y_{IC}) \cdot y_{IO} - m \cdot (\chi_{2,y_I} - (2 \cdot \dot{\theta}(t_{i-1}) \cdot \chi_{4.\theta} + \dot{\theta}(t_{i-1})^2 \\ & - \ddot{\theta}(t_{i-1})^2 \cdot \Delta T^2) \cdot y_{IC} + \chi_{2.\theta} \cdot x_{IC}) \cdot x_{IO} + f_{x,C} \cdot (y_{IC} - y_{IO}) \\ & + f_{y,C} \cdot (x_{IO} - x_{IC}) - m_C \end{aligned}$$

$$u_{5,1} = -m \cdot \chi_1$$

$$u_{5,3} = 2 \cdot m \cdot \dot{\theta}(t_{i-1}) \cdot \chi_3 \cdot x_{IC} + m \cdot \chi_1 \cdot y_{IC}$$

$$u_{5,7} = f_{y,C} + m \cdot \left(-\chi_{2,x_I} + \left(2 \cdot \dot{\theta}(t_{i-1})\right) \cdot \chi_{4.\theta} \cdot x_{IC} - \dot{\theta}(t_{i-1})^2 \cdot x_{IC} + \ddot{\theta}(t_{i-1})^2 \cdot \Delta T^2 \cdot x_{IC} + \chi_{2.\theta} \cdot y_{IC}\right)$$

$$u_{6,2} = -m \cdot \chi_1$$

$$u_{6,3} = 2 \cdot m \cdot \dot{\theta}(t_{i-1}) \cdot \chi_3 \cdot y_{IC} - m \cdot \chi_1 \cdot x_{IC}$$

$$u_{6,7} = f_{y,C} - m \cdot \left(-\chi_{2,y_I} + 2 \cdot \dot{\theta}(t_{i-1}) \cdot \chi_{4.\theta} \cdot y_{IC} + \dot{\theta}(t_{i-1})^2 \cdot y_{IC} - \ddot{\theta}(t_{i-1})^2 \cdot \Delta T^2 \cdot y_{IC} + \chi_{2.\theta} \cdot x_{IC}\right)$$

The transfer matrix of a rigid body moving in plane is a 7 × 7 matrix, where the χ's refer to the factors from the Newmark-β method, which is used for linearization of velocity and acceleration. $f_{x,C}$, $f_{y,C}$ and m_C are the externally applied forces and moment, and J_I is the mass moment of inertia about the inboard point.

The transfer matrix of a spring element can be derived using kinematics and dynamics as:

$$U_{spring} = \begin{bmatrix} 1 & 0 & 0 & 0 & u_{1,5} & 0 & u_{1,7} \\ 0 & 1 & 0 & 0 & 0 & u_{2,6} & u_{2,7} \\ 0 & 0 & 1 & u_{3,4} & 0 & 0 & u_{3,7} \\ 0 & 0 & 0 & 1 & 0 & 0 & 0 \\ 0 & 0 & 0 & 0 & 1 & 0 & 0 \\ 0 & 0 & 0 & 0 & 0 & 1 & 0 \\ 0 & 0 & 0 & 0 & 0 & 0 & 1 \end{bmatrix} \tag{2}$$

$$u_{1,5} = \frac{-1}{k_x + c_x \cdot \chi_3}$$

$$u_{1,7} = -c_x \cdot \frac{\chi_{4,x_O} - \chi_{4,x_I}}{k_x + c_x \cdot \chi_3}$$

$$u_{2,6} = \frac{-1}{k_y + c_y \cdot \chi_3}$$

$$u_{2,7} = -c_y \cdot \frac{\chi_{4,y_O} - \chi_{4,y_I}}{k_y + c_y \cdot \chi_3}$$

$$u_{3,4} = \frac{1}{k_\theta + c_\theta \cdot \chi_3}$$

$$u_{3,7} = -c_\theta \cdot \frac{\chi_{4,\theta} - \chi_{4,\theta}}{k_\theta + c_\theta \cdot \chi_3}$$

In the spring transfer matrix, k and c denotes spring constant and damping coefficient respectively. The coefficients for the bending stiffness as well as the axial stiffness are determined from the equations as.

$$k_x = \frac{E \cdot \pi \cdot \left(\frac{d}{2}\right)^2}{b_1} \tag{3}$$

$$k_\theta = \frac{n \cdot E \cdot I}{L} \tag{4}$$

In these equations, E is the modulus of elasticity, d is the diameter, L is the length of the fibre, I is the area moment of inertia, and n is the number of segments that the fibre is discretised into. The compliance and damping of the optical fibre in the radial direction are neglected in this work. The actuations are implemented in the model by adding them as position changes in the boundary conditions of the sub-systems. The final dynamic equation can be expressed as $Z_0 = U_4 U_3 U_2 U_1 Z_I$, where $U_j = U_j^{rigidBody1} U_j^{Spring1} \cdots U_j^{rigidBodyi} U_j^{Springi} \cdots U_j^{rigidBodyn} U_j^{Springn} (j = 1, 2, 3, 4)$. The matrix size of the final equation is 7×7. The size does not increase with the number of discretized rigid body elements and springs, and it does not increase with the number of decomposed sub-chains. This property means that the method has high computational efficiency.

4 Numerical Simulations

The numerical simulations are conducted with the proposed robot as shown in Fig. 3. The total length of the fibre is 124 [mm], L_1, is 47 [mm], L_2 is 15 [mm] and L_3 is 15 [mm]. The fibre consists of a core with a diameter of 9 [µm] and a density of 2210 [kg/m^3]. Around the core is a cladding with a diameter of 125 [µm] and density of 2210 [kg/m^3]. The coating of the fibre has a diameter of 245 [µm] and density of 1000 [kg/m^3]. The modulus of elasticity of the cladding is 73.6 [MPa]. The modulus of elasticity of the coating and the core are negligible. In the simulation, the fibre is discretised into 100 rigid body elements connected by springs. The actuation provided by the piezoelectric actuators are assumed in the equations as below.

$$PEA_1 = 10 \cdot 10^{-6} \cdot \sin\left(\frac{0.2 \cdot t}{\pi \cdot dt}\right) \tag{5}$$

$$PEA_2 = -10 \cdot 10^{-6} \cdot \sin\left(0.5 + \frac{0.1 \cdot t}{\pi \cdot dt}\right) \tag{6}$$

$$PEA_3 = -5 \cdot 10^{-6} \cdot \sin\left(\frac{2 \cdot t}{\pi \cdot dt}\right) \tag{7}$$

The simulation is run with time steps of 0.01 [s] for a total simulation time of 10 [s]. In Fig. 5, the shape of the fibre from the simulation is shown after the simulation has run for 2.5 s.

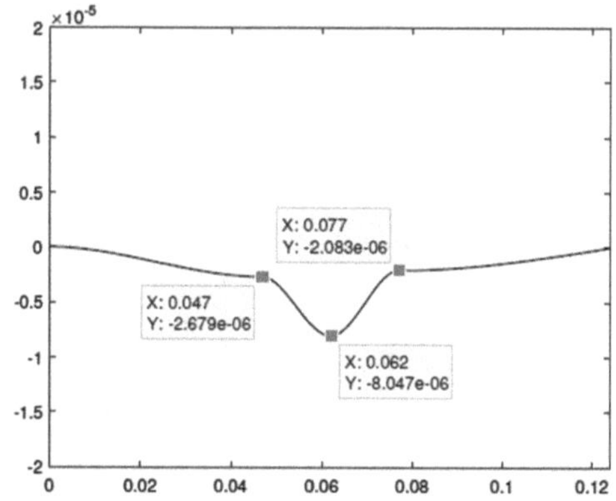

Fig. 5. Simulation results for the optical fibre after 2.5 s

The motions and shapes of the fibre that the simulation renders are in great accordance with the motions that one would expect the fibre to undergo. It can thus be concluded, that it is feasible to use DT-TMM to investigate the shape of the optical fibre based on the actuation provided by the piezoelectric actuators. The next step for this research is to build a prototype for experimental testing, and to investigate the 3-dimensional situation.

5 Conclusions and Discussions

In this work, the concept design of a fibre-based robot is proposed for biophotonic applications used in early cancer detection, and the dynamic modelling has been established for the fibre-based continuum robot using the discrete-time transfer matrix method (DT-TMM), so that its computational efficiency can be utilised to improve the existing photoacoustically guided wavefront shaping techniques. Numerical simulations have been conducted to demonstrate the feasibility of dynamic modelling using DT-TMM and manipulating the fibre-based robot for improved cancer detection. No experimental results are available to verify the simulation results, but the results seem reasonable, as the shape of the system looks as expected from the given actuation of the continuum robotic system. Our ongoing research is to develop a prototype and conduct experiments to validate the feasibility of the system concept.

References

1. Lai, P., Wang, L., Tay, J.W., Wang, L.V.: Photoacoustically guided wavefront shaping for enhanced optical focusing in scattering media. Nat. Photonics **3**, 126–132 (2015)
2. Dwivedy, S.K., Eberhard, P.: Dynamic analysis of flexible manipulators-a literature review. Mech. Mach. Theory **41**, 749–777 (2006)
3. Zhang, X., Yu, Y.: A new spatial rotor beam element for modeling spatial manipulators with moint and link flexibility. Mech. Mach. Theory **35**(3), 403–421 (2000)
4. Usoro, P.B., Nadira, R., Mahil, S.S.: A finite element/Lagrangian approach to modeling lightweight flexible manipulators. ASME J. Dyn. Syst. Meas. Control **108**, 198205 (1986)
5. Book, W.J.: Recursive Lagrangian dynamics of flexible manipulator arms. Int. J. Robot Res. **3**(3), 87–101 (1984)
6. Zhang, X., Mills, J.K., Cleghorn, W.L.: Dynamic modeling and experimental validation of a 3-PRR parallel manipulator with flexible intermediate link. J. Intell. Robot. Syst. **50**(4), 323–340 (2007)
7. Ge, S.S., Lee, T.H., Zhu, G.: A new lumping method of a flexible manipulator. In: Proceedings of the American Control Conference, Albuquerque, New Mexico, pp. 1412–1416 (June 1997)
8. Carbone, G.: Carbone: Stiffness analysis and experimental validation of robotic systems. Front. Mech. Eng. **6**, 182–196 (2011)
9. Zhang, X., et al.: Computationally efficient dynamic modelling of robot manipulators with multiple flexible-links using acceleration-based discrete time transfer matrix method. Robot. Comput-Integr. Manuf. **49**, 181–193 (2018)
10. Rui, X., He, B., Lu, Y., Lu, W., Wang, G.: Discrete time transfer matrix method for multibody system dynamics. Multibody Syst. Dyn. **14**, 317–344 (2005)

Optimized Mother Wavelet in a Combination of Wavelet Packet with Detrended Fluctuation Analysis for Controlling a Remote Vehicle with Imagery Movement: A Brain Computer Interface Study

Amin Hekmatmanesh(✉), Huapeng Wu, Ming Li, Ali Motie Nasrabadi, and Heikki Handroos

Laboratory of Intelligent Machines, Department of Mechanical Engineering, Lappeenranta University of Technology, Lappeenranta, Finland
{Amin.Hekmatmanesh,huapeng.wu,Ming.Li,heikki.handroos}@lut.fi, nasrabadi@shahed.ac.ir

Abstract. Brain computer interface (BCI) is a critical field in health care to help paralyzed or maim patients back to normal life. This study is focusing on feature extraction based on self-similarity concept in electroencephalography (EEG) signal processing. To this purpose, a combination of discrete Wavelet Packet Transform (WPT) with Detrended Fluctuation Analysis (DFA) is utilized. Also, Event Related Desynchronization (ERD) patterns are used for customizing mother wavelets in the wavelet processing. Therefore, right hand movement imagination ERDs are extracted and used as a mother wavelet in the WPT algorithm and updated automatically for individual subjects. The combination of Optimized WPT with DFA (OWPT-DFA) is utilized for feature extraction for the two classes of right hand imagination and no-imagination. The features are classified and a model is trained for online processing by Soft Margin Support Vector Machine classifier and Generalized Radial basis Function (SSVM-GRBF) kernel. The model is employed to control a remote vehicle for two state of move forward and stop. In the experiment, nine subjects are participated to record data and control the remote vehicle. Results depicted that the OWPT-DFA method's accuracy reach to 85.33% with $p < 0.001$ and 75.23% with $p < 0.05$ for offline and online processing, respectively. It is concluded that the self-similarity concept in the combination of OWPT and DFA methods with SSVM-GRBF classifier improve the results of movement imagination detection significantly.

1 Introduction

The Brain Computer Interface (BCI) application's aim in the health field is to overcome paralyzed patients limitations. This survey is controlling a remote

G. Carbone et al. (Eds.): *New Trends in Medical and Service Robotics*, MMS 65, pp. 186–195, 2019.
https://doi.org/10.1007/978-3-030-00329-6_22

vehicle with right hand imagination. To this purpose, several methods such as machine learning [1], Event Related Desynchronization (ERD) identification [2] and combination of different bio-signals [3] are implemented to control an assistant device. The concept of this study is employing self-similarity to extract imaginary movement patterns. Self-similarity in signal processing means, one part of a signal is analogous to the other parts of the signal [4]. The imaginary or real movement engender a pattern called Event Related Desynchronization (ERD) [5,6].

Wavelet is a powerful method for patterns recognition, which is based on self-similarity concept. This fashion has ability of extracting self-similarity features which is localized in time-frequency space [7]. Wavelet packet transformation (WPT) is a combination of several scales of discrete wavelet transform function [8]. The most important part of WPT computations is selecting the appropriate mother wavelet such as Daubechies families [9]. WPT is used for identifying the ERD patterns in different applications such as controlling a wheelchair [10] or a vehicle [11].

The second powerful fashion for self-similarity identification is Detrended Fluctuation Analysis (DFA) [12]. DFA is also useful in different field of studies, such as bio-signal processing [13] and economics [14]. Recently, DFA is used for identifying the real and imaginary movements [15]. Also, Combination of both WPT and DFA (WPT-DFA) generate a well-promising method base on self-similarity concept for identifying different EEG patterns. Then trustworthy of WPT-DFA is considered by comparing with chaotic features such as Lyapunov exponent. Results confirmed high reliability of this method in nonlinear systems such as EEG signals [16].

The extracted features are classified with different Fashions. One well-known fashion is Support Vector Machine (SVM) which is combined with different kernels such as polynomial, Radial Basis Function (RBF). One of the best classifiers is Soft Margin SVM (SSVM) [17] which is used by Generalized RBF (GRBF) [18–20]. In the present study, the combination of WPT-DFA with an optimized

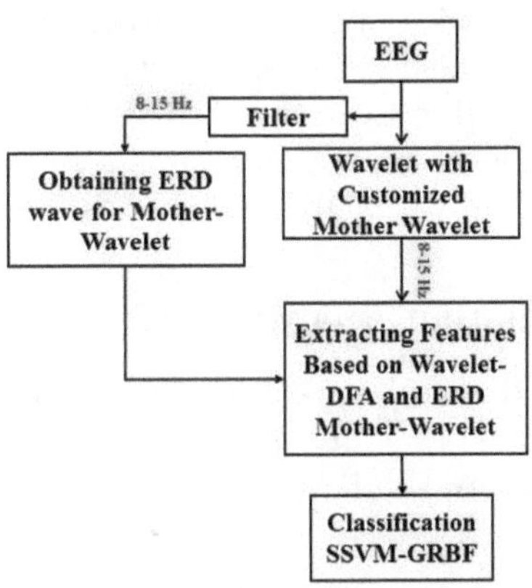

Fig. 1. Signal processing steps

mother wavelet is implemented. To this purpose, filtering techniques are applied to remove noise. Afterwards, ERD of individual subjects for imaginary right hand movements are extracted, averaged and replaced as a mother wavelet in the WPT automatically. Then, through Optimized WPT-DFA (OWPT-DFA) features based on ERD and Coif4 (exist mother wavelet in Matlab) mother wavelets, features are computed. Finally, the features are classified utilizing Soft Margin Support Vector Machine classifier and Generalized Radial basis Function (SSVM-GRBF), Fig. 1. In the second phase, the trained SSVM-GRBF is employed for real-time classification and controlling the remote vehicle.

2 Method and Material

In this part, the utilized mathematical theory for feature extraction, classification, task and montage for recording EEG are presented.

2.1 Wavelet Packet Transformation (WPT)

Wavelets have ability of self-similarity pattern identification in time-frequency domain. Wavelet constructing a time-frequency domain that gives accessibility to frequencies with correspondence time domain and vice versa [21]. Mother wavelet is the aim pattern that is playing a critical role in wavelet for identifying EEG patterns. In this study, the extracted ERD is selected as the mother wavelet which is replaced automatically for individual subjects. Wavelet computations is presented as follow [22]:

$$W_{\gamma}f(n,m) = \frac{1}{\sqrt{n}} \int_{-\infty}^{+\infty} f(t)\varphi(\frac{t-n}{m})dt, \tag{1}$$

where γ, m, n are denoted as mother wavelet, transfer function and scaling. Regarding to the wavelet concept, the input signal is decomposed by dividing the input into high and low frequency components to reach the aim frequency range, Fig. 2. Here, a WPT with optimized mother wavelet is utilized. A WPT is a combination of several linear scales of a discrete wavelet transform.

The decomposed values that are generated by WPT with the ERD mother wavelet are placed in a row to generated a time series. The time series is used for DFA to generate the OWavelet-DFA features as follow.

2.2 Detrended Fluctuation Analysis (DFA)

The DFA concept is based on self-similarity and achieved by scaling and long-term correlation approaches, that are applied on a time series (TS, $(F(i), i = 1, ..., N)$) that is generated by WPT [23]:

$$z(j) = \sum_{i=1}^{n} [F(i) - \langle F \rangle], \qquad \langle F \rangle = \frac{1}{N} \sum_{i=1}^{j} F(i), \tag{2}$$

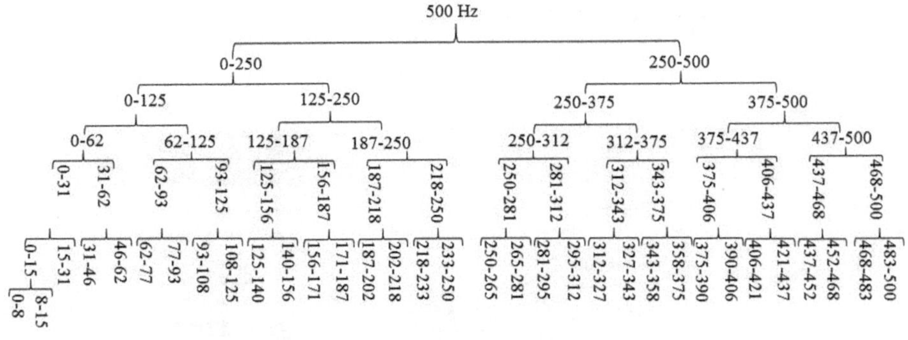

Fig. 2. The 8–15 Hz frequency range is decomposed with the wavelet technique

where, $z(j)$ is the EEG signal with j channels, which is segmented into equal segments (N_l) and length (l), $N_l = \text{int}(\frac{N}{l})$. Afterwards, for the points in each segments $(z_l(j))$ the following procedure is applied, (1) fitting an envelope with least square error (LSE) algorithm, (2) detrending the segments (Eq. (3)), and (3) Mean Square Error (MSE) function $(F(l), 4)$:

$$z_s(j) = z(j) - z_n(j), \tag{3}$$

$$K(l) = \sqrt{\frac{1}{N} \sum_{j=1}^{N} \left[z_s(j) - z_l(j)\right]^2}, \tag{4}$$

Final step of DFA implementation, regarding to the exponential law a logarithmic diagram $(K(l))$ is computed $(K(l) \propto l^{\lambda})$. Thereupon, λ is calculated by fitting a line on linear part of the obtained logarithmic diagram which is used as a feature [16]. The extracted OWPT-DFA features from two EEG channels (CP6 and FC1) are classified using SSVM-GFBF classifier as follow:

2.3 Soft Margin Support Vector Machine (SSVM)

SSVM is analogous to traditional SVM which is based on a binary classifier. In SSVM the duality problem of SVM is solved with Lagrange theory. Duality problem occurs when a plenty number of features must be classified [17]. In the SSVM two classes of $y = [0, 1]$ are defined to assign no-movement and imaginary right hand movement features $(x_i \in R^n, i = 1, ..., l)$. After assigning the labels to the features, a multi-dimension feature space is generated using the GRBF. Finally, a decision boundary is made by employing a hyper-plan $(w^T x_i + b = 0)$ in the feature space:

$$\begin{aligned} S^T Z_i + C &\geq 1 \qquad if \quad y_i = 1, \\ S^T Z_i + C &\leq -1 \qquad if \quad y_i = -1, \end{aligned} \tag{5}$$

where C, S, T symbols are bias, weights and transpose operator, respectively [17].

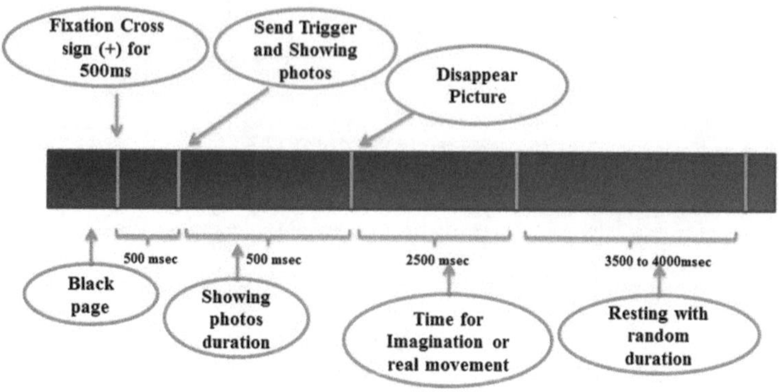

Fig. 3. The designed task for recording EEG

2.4 Generalized Radial Basis Function (GRBF)

GRBF is a well-promising method for classifying the feature space that is used as a kernel of SSVM [18–20]. The GRBF is impressively flexible by utilizing variable shape ($\tau > 0$), width ($g > 0$) and center (c) for the Gaussian distribution function:

$$H(s;c,g,\tau) = \frac{\tau}{2g\psi(^1/_\tau)}\exp(-\frac{\|s-c\|}{g^\tau}), \tag{6}$$

$$g = \sigma\sqrt{\frac{\psi(^1/_\tau)}{\psi(^3/_\tau)}},$$

$$\psi(k) = \int_0^\infty t^{k-1}e^{-k}dk, \qquad for \quad K > 0, \tag{7}$$

where the parameters g, ψ, σ and k symbols are width, factorial extension function, standard deviation and number of inputs, respectively. The τ parameter is a critical parameter which is optimized.

2.5 Data Acquisition and Experiment Setup

This experiment includes nine international participants (S1−S9, 27.3±3 years) in Lappeenranta University of Technology. The presented task for recording data is showed in detail, Fig. 3. Mainly, the task contain four stages: 1- Appearing a cross sign for 500 ms, 2- Displaying a picture of right hand for 500 ms, 3- Displaying black screen for imagination for 2500 ms, 4- Resting randomly for 3500 ms to 4000 ms. This procedure is implemented in Matlab 2016 using Cogent2000 toolbox [24]. The utilized portable amplifier for recording EEG was ENOBIO-32 with 32 gel electrodes (Fig. 4). The frequency sampling rate was selected at 500 Hz. For real-time experiment a remote vehicle is connected to the Matlab through XBEE chipset, Fig. 5.

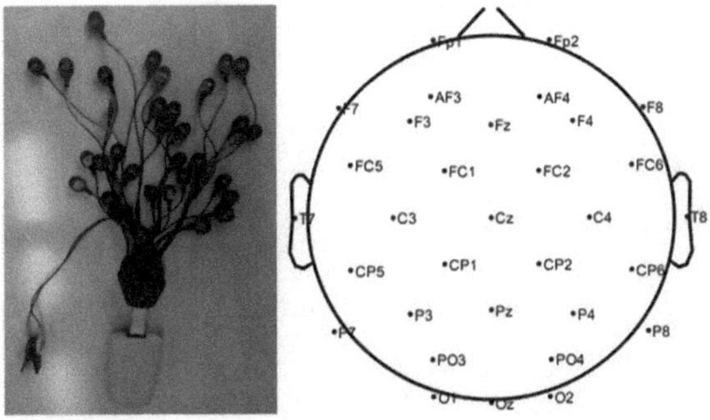

Fig. 4. The portabel Enobio-32 device and the montage for EEG recording.

Fig. 5. The utilized remote vehicle for real-time experiment.

3 Results

In this study, nine candidates are participated to record EEG. The data is used to train the SSVM-GRBF classifier for controlling a remote vehicle (Fig. 5). The scalp map in the Fig. 6a is based on the ERP's signal power to find the best channels for feature extraction. Two channels of CP6 and FC1 are selected to extract OWPT-DFA features. The classification results of the OWPT-DFA method based on the Coif4 and ERD mother wavelets are presented in the Table 1. Accuracies in Table 1 are consequence of 150 trial imagination states (75 movement imagination and 75 no-movement) for individual subjects. Alongside accuracy index, probabilistic analyzer t-test results are depicted in the Table 1.

4 Discussion

Since now, combination of the two powerful methods based on self-similarity (WPT-DFA) for BCI applications have not been considered. In addition, the combination of the WPT and DFA figured out as a reliable method for analyzing nonlinear signals such as EEG processing [16]. Therefore, it is expected that customizing this combination producing valuable features for diagnosing movement imagination. To this purpose, ERD patterns are extracted using ERPLAB

Table 1. Results of accuracies and t-test based on the SSVM-GRBF classifier for individual subjects. The OWPT-DFA with Coif4 and ERD mother wavelets were used to generate features from CP6 and FC1 channels.

Subjects	Offline accuracy% with Coif4	T-test	Real-time accuracy% with Coif4	Offline accuracy% with ERD	T-test	Real-time accuracy% with ERD
S1	64.67	$p < 0.05$	56.00	69.67	$p < 0.05$	65.00
S2	63.33	$p < 0.05$	65.00	56.00	$p < 0.05$	58.00
S3	66.67	$p < 0.05$	59.00	69.00	$p < 0.05$	63.00
S4	72.33	$p < 0.006$	65.00	73.67	$p < 0.05$	68.00
S5	59.00	$p > 0.05$	50.00	74.00	$p < 0.006$	70.00
S6	60.00	$p > 0.05$	52.00	69.00	$p < 0.05$	65.00
S7	57.00	$p > 0.05$	50.00	65.67	$p < 0.05$	61.00
S8	73.00	$p < 0.006$	64.00	84.67	$p < 0.001$	79.00
S9	73.67	$p < 0.006$	64.00	85.33	$p < 0.001$	79.00

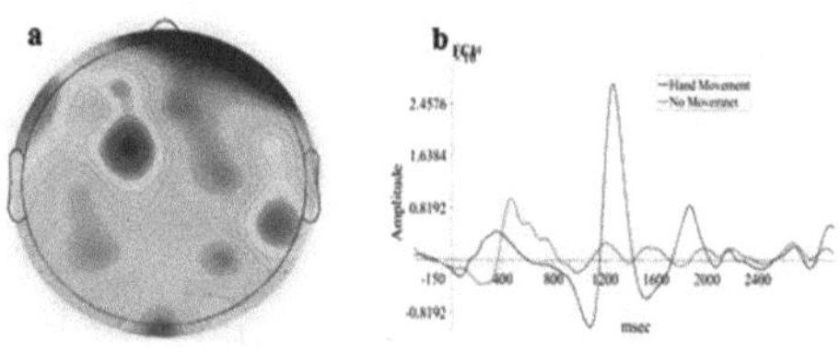

Fig. 6. (a) Average power of EEG after right hand imagination and (b) a sample of extracted ERD from FC1 channel for S8.

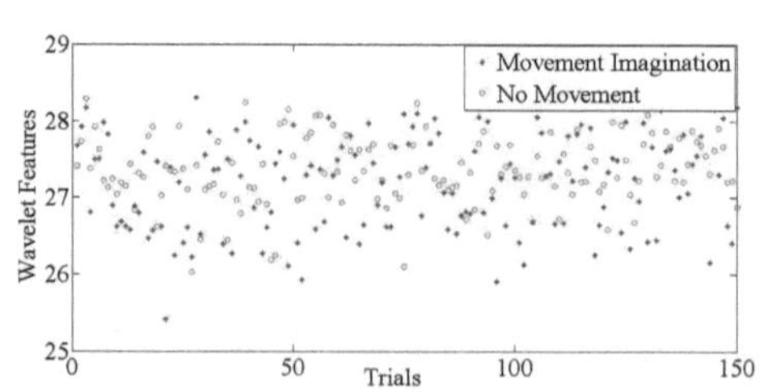

Fig. 7. Scattering of right hand movement imagination and no-movement imagination based on OWPT-DFA method with ERD and coif4 mother wavelets for S8.

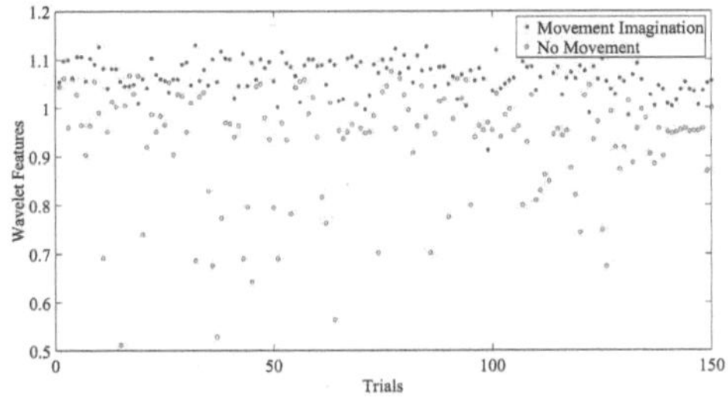

Fig. 8. Scattering plot of movement imagination and no-movement features using OWPT-DFA with ERD mother wavelet method for S8.

toolbox [25]. In the procedure, data filtered through Butterworth filter second order from 8–15 Hz. The extracted pattern is selected as a mother wavelet in the implemented WPT-DFA, Fig. 6. Amplitudes of ERD in the FC1 (Fig. 6a) and Cp6 channels in the 8–15 Hz frequency bands show informative information. Also, the depicted scalp map in the Fig. 6a approve the claim. The extracted ERDs shows 800 msec activation after stimulation. Based on the other studies, for right hand movement imagination, C4 must activated [5], that our results different location. Hence, the channels and frequency bands are fixed for signal processing.

To process the EEG signal, the input EEG data is decomposed into five levels to reach the 8–15 Hz, Fig. 2. Then, the two mother wavelets of Coif4 and ERD with similar classifier (SSVM-GRBF) are used for feature extraction. The data scattering for the two states of movement and No-movement imagination depicts efficiency of the Coif4 and ERD. Disperse scattering with less overlap between classes generates better results. Therefore, based on the Figs. 7 and 8, ERD mother wavelet has more discriminative scattering in compare to the Coif4. Also, accuracies in the Table 1 proves better capability of the ERD as a mother wavelet in compare to the Coif4 in other studies [26, 27]. The average achieved results for OWPT-DAF with Coif4 and ERD are 65.52% and 72.89%, respectively. The trained classifier is utilized for online processing that the results of the Coif4 and ERD are reached to 57.33% and 67.56%, respectively. To consider the validity of the results, T-test evaluation between the two states is applied. T-test results shows that the difference between features of in the two states was not significant for three subjects for Coif4 mother wavelet ($p > 0.05$), whilst for ERD mother wavelet the p-values were significant for the nine subjects, $p < 0.05$, Table 1. Hence, the ERD mother wavelet can obtain more meaningful results.

5 Conclusion

To recognize movement imagination a combination of WPT and DFA algorithms with the concept of self-similarity are implemented. To optimize the WPT-DFA a mother wavelet based on ERD is generated and updated automatically and individually. Two channels of FC1 and CP6 are selected as the most informative locations. Two mother wavelets of Coif4 and ERD with SSVM-GRBF classifier are considered. Results depict higher and more significant result of 72.89% for ERD mother wavelet in offline and 67.55% for online processing averagely. It is concluded that, the OWPT-DFA using ERD mother wavelet achieve significant results.

References

1. Djeha, M., et al.: A combined EEG and EOG signals based wheelchair control in virtual environment. In: 2017 5th International Conference on Electrical Engineering-Boumerdes (ICEE-B). IEEE (2017)
2. Huang, D.: Electroencephalography (EEG)-based brain-computer interface (BCI): a 2-D virtual wheelchair control based on event-related desynchronization/synchronization and state control. IEEE Trans. Neural Syst. Rehabil. Eng. **20**(3), 379–388 (2012)
3. Belkacem, A.N.: Online classification algorithm for eye-movement-based communication systems using two temporal EEG sensors. Biomed. Signal Process. Control **16**, 40–47 (2015)
4. Gifani, P.: Optimal fractal-scaling analysis of human EEG dynamic for depth of anesthesia quantification. J. Frankl. Inst. **344**(3–4), 212–229 (2007)
5. Ang, K.K., et al.: Filter bank common spatial pattern (FBCSP) in brain-computer interface. In: IEEE International Joint Conference on Neural Networks, IJCNN 2008 (IEEE World Congress on Computational Intelligence). IEEE (2008)
6. Pfurtscheller, G.: Induced oscillations in the alpha band: functional meaning. Epilepsia **44**(s12), 2–8 (2003)
7. Kumar, P., Foufoula-Georgiou, E.: Wavelet analysis for geophysical applications. Rev. Geophys. **35**(4), 385–412 (1997)
8. Burrus, C.S.: Introduction to Wavelets and Wavelet Transforms: A Primer, vol. 1. Prentice Hall, New Jersey (1998)
9. Daubechies, I.: Ten lectures on wavelets. vol. 61. SIAM (1992)
10. Kaufmann, T., Herweg, A., Kübler, A.: Toward brain-computer interface based wheelchair control utilizing tactually-evoked event-related potentials. J. Neuroeng. Rehabil. **11**(1), 7 (2014)
11. Lu, Y., et al.: The design of simulation vehicle system controlled by multichannel EEG based on imaginary movements. In: 2016 35th Chinese Control Conference (CCC). IEEE (2016)
12. Saha, D., et al.: Quantification of scaling exponent with crossover type phenomena for different types of forcing in DC glow discharge plasma. Phys. A Stat. Mech. Appl. **490**, 300–310 (2018)
13. Bunde, A.: Correlated and uncorrelated regions in heart-rate fluctuations during sleep. Phys. Rev. Lett. **85**(17), 3736 (2000)
14. Weron, R.: Estimating long-range dependence: finite sample properties and confidence intervals. Phys. A Stat. Mech. Appl. **312**(1–2), 285–299 (2002)
15. Pavlov, A.N., et al.: Detection of EEG-patterns associated with real and imaginary movements using detrended fluctuation analysis. In: Dynamics and Fluctuations in Biomedical Photonics XV, vol. 10493. International Society for Optics and Photonics (2018)
16. Vaghefi, M.: Identification of chaos-periodic transitions, band merging, and internal crisis using wavelet-DFA method. Int. J. Bifurc. Chaos **26**(04), 1650065 (2016)

17. Chang, C.-C., Lin, C.-J.: LIBSVM: a library for support vector machines. ACM Trans. Intell. Syst. Technol. (TIST) **2**(3), 27 (2011)
18. Hekmatmanesh, A., Noori, S.M.R., Mikaili, M.: Sleep spindle detection using modified extreme learning machine generalized radial basis function method. In: 2014 22nd Iranian Conference on Electrical Engineering (ICEE). IEEE (2014)
19. Noori, S.M.R., et al.: K-complex identification in sleep EEG using MELM-GRBF classifier. In: 2014 21th Iranian Conference on Biomedical Engineering (ICBME). IEEE (2014)
20. Hekmatmanesh, A.: Sleep spindle detection and prediction using a mixture of time series and chaotic features. Adv. Electr. Electron. Eng. **15**(3), 435 (2017)
21. Dongmei, H., Wen, G., Jiangqin, W.: Complexity scalable audio coding algorithm based on wavelet packet decomposition. In: 5th International Conference on Signal Processing Proceedings, WCCC-ICSP 2000, vol. 2. IEEE (2000)
22. Wickerhauser, M.V.: Adapted Wavelet Analysis: From Theory to Software. AK Peters/CRC Press, Wellesley (1996)
23. Zheng, H., Song, W., Wang, J.: Detrended fluctuation analysis of forest fires and related weather parameters. Phys. A Stat. Mech. Appl. **387**(8–9), 2091–2099 (2008)
24. Cooper, R., et al.: COGENT: an environment for the development of cognitive models. In: A Cognitive Science Approach to Reasoning, Learning and Discovery, pp. 55–82 (1998)
25. Lopez-Calderon, J., Luck, S.J.: ERPLAB: an open-source toolbox for the analysis of event-related potentials. Front. Hum. Neurosci. **8**, 213 (2014)
26. Wali, M.K., Murugappan, M., Ahmmad, B.: Wavelet packet transform based driver distraction level classification using EEG. Math. Probl. Eng. **2013**, 10 (2013)
27. Chourasia, V.S., Tiwari, A.K.: Design methodology of a new wavelet basis function for fetal phonocardiographic signals. Sci. World J. **2013**, 12 (2013)

Performance Evaluation

Optimization of Neural Network Hyperparameters for Modeling of Soft Pneumatic Actuators

Mats Wiese(✉), Gundula Runge-Borchert, and Annika Raatz

Leibniz Universität Hannover, Hannover, Germany
wiese@match.uni-hannover.de

Abstract. Especially the field of medical robotics is faced with the challenge that contact between a robotic structure is not only tolerated but at the very core of the application, for example in minimally invasive surgery. Therefore, high demands regarding safety need to be met when operating a robotic structure in such a delicate environment. Soft material robotic systems offer the potential to surpass their rigid counterparts in this area due to their material inherent compliance and adaptability. Nonetheless, soft structures can also exhibit challenging characteristics like highly nonlinear deformation. This effect is often neglected when setting up deformation models. Recent approaches tackle this challenge by applying machine learning methods to learn the nonlinear behaviour. In previous research, we gained promising results from learning the kinematics of a soft pneumatic actuator (SPA) via a feedforward artificial neural network (ANN). To overcome a trial and error approach when designing an ANN, in this article we introduce Bayesian optimization to find suitable hyperparameters. This way an ANN architecture is found that exceeds the accuracy of our previous studies by more than a factor of 20 [9].

Keywords: Soft material robotic systems · Kinematic modeling
Soft pneumatic actuators · Artificial neural networks
Hyperparameter optimization

1 Introduction

Different from classical rigid-link-robots, soft material robotic systems, as the name indicates, mainly consist of soft materials. Especially SPAs are usually made of rubber-like materials, which results in an intrinsic compliance and virtually infinite degrees of freedom. The flexibility of such structures offers high potential for increased maneuverability in cluttered environments and, at the same time, safe interaction between the robot and humans. Robotic applications in the field of medical technology demand exactly these properties, for example in minimally invasive surgery. To access internal organs through small holes, a robotic structure must be navigated through a patient's body without

G. Carbone et al. (Eds.): *New Trends in Medical and Service Robotics*, MMS 65, pp. 199–206, 2019.
https://doi.org/10.1007/978-3-030-00329-6_23

damaging delicate human tissue. The STIFF-FLOP is a stiffness variable soft pneumatic manipulator, designed for the use in minimally invasive surgery [1]. Other fields of application of soft material robotic systems are rehabilitation robotics, human-robot-collaboration and exploration robotics [2].

Despite the huge potential, there are still challenges ahead that currently obstruct widespread application of soft material robotic systems. For design studies or model based control, appropriate models need to be set up that accurately describe a soft robot's behaviour. Most approaches comprise the use of Finite Element modeling [3,4] and analytically derived models [5]. While the former are generally computationally intensive and thus not suitable for real time application, the latter goes along with comprehensive engineering work for each new robot design.

To overcome this problem, machine learning methods can be utilized to model the nonlinear relationship between actuation and deformation. [6] proposed a framework based on finite element modeling to learn an inverse kinematics model for a SPA in three dimensional space. Similarly, an inverse kinematics model is learned for a cable driven actuator in [7].

Feedforward ANNs in general are universal approximators, since with nonlinear activation functions they possess the ability to approximate any smooth function. On the other hand, little is known about the optimal design of ANNs. Hyperparameters such as the number of hidden layers, the number of neurons per hidden layer, the activation function etc. are often set via rules of thumb or trial and error [10]. However, suitable hyperparameters depend on the system considered and generally cannot be determined beforehand. Bayesian optimization has been succesfully employed by several researchers to find a suitable set of hyperparameters for various applications [11,12]. In this article we introduce Bayesian optimization into the learning cycle of the kinematic relationship to improve the ANN performance.

2 Soft Robotics Design and Modeling Framework

In recent works [8,9], we presented a methodological framework that allows soft pneumatic actuators to be designed and automatically modeled. The software framework, which is depicted in Fig. 1, consists of four building blocks. In the first step, a generic model of a soft pneumatic actuator is created, which involves defining the shape of the actuator, the arrangement of air chambers within the body and the choice of materials.

In the second step, a finite element model of the actuator is implemented and the resulting nonlinear differential equations are solved with an appropriate FE solver. The second step of the methodology corresponds to the preprocessing and processing steps in finite element analysis (FEA). This procedure is carried out to determine the robot specific relationship between the pressure inputs and the resulting deformation, which is often difficult to derive analytically. FEA preprocessing involves specifying the geometric parameters, material properties, loads, and boundary conditions of the model. To derive the nonlinear differential equations that capture the deformation of the actuator under load, suitable element types need to be assigned to the model and the geometry must be meshed.

In the third step, the forward kinematics of the actuator are computed using the piecewise constant curvature (PCC) approach proposed by Webster et al. [13]. Specifically, the PCC approach is applied to a previously fitted circular shape function that represents the so-called backbone curve of the actuator. The relationship between the curve parameters, i.e. the configuration of the actuator in three-dimensional space, and the end effector position is given by the robot independent relationship.

Finally, in the fourth step, an ANN is trained to derive a computationally efficient model representing the kinematic relationship between the pressure inputs and the configuration of the actuator. This is where the data obtained in steps two and three comes in: the finite element simulations and kinematic analysis are employed to build up a data reservoir that is subsequently used for ANN training and validation. Learning the kinematic relationships from model data allows us to amass large amounts of data within a limited time frame and pre-train the network without excessive and time-consuming data collection on a test bench.

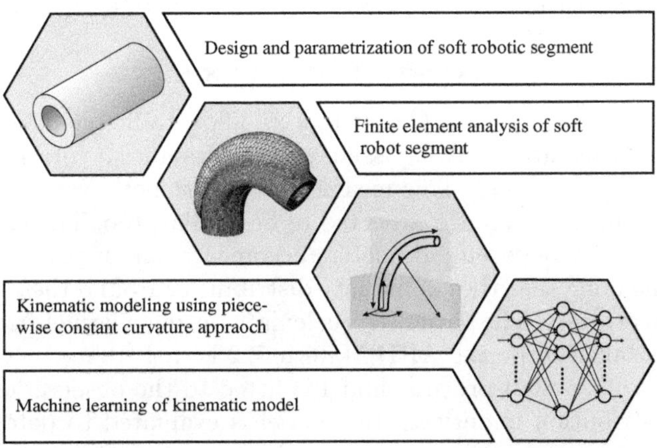

Fig. 1. Design and kinematic modeling framework for soft material robot modules [8]

3 Hyperparameter Optimization Using Bayesian Optimization

ANNs consist of an input and output layer, which in this case correspond to the pressure input and configuration output, respectively. Inbetween the input and output layer are one or more hidden layers. Each layer is composed of n artificial neurons, which process the input via activation functions $a(v)$.

To refrain from setting up ANNs via rules of thumb or by trial and error, more systematic optimization techniques can be used to derive suitable hyperparameters for a given problem. The objective is to find a set of parameters $\mathbf{x}^*$ that minimizes the mean absolute error $\bar{e}_{\text{abs}}(\mathbf{x})$ between the test target data y' and the net output y. A parameter set $\mathbf{x}$ contains hyperparameters, more specifically the number of neurons in the first and second hidden layer ($n_{\text{hid},1}$, $n_{\text{hid},2}$) and the activation functions in the first and second hidden layer ($a_1(v)$, $a_2(v)$).

This work focuses on hyperparameter optimization via Bayesian optimization [15]. This optimization technique has proven to be especially useful in cases in which generating new observations is computationally intensive.

Bayesian optimization aims to construct a probabilistic model $f(\mathbf{x})$ for an unknown relationship between parameters $\mathbf{x}$ and the corresponding response, here $\bar{e}_{\text{abs}}$. The probabilistic model is fitted with the help of Gaussian process models [14]. In Bayesian optimization, the Gaussian process model is updated with each new observation. The updated model is subsequently used to determine parameters for generating another observation. This is done by evaluating an acquisition function. In the following, a short introduction to Gaussian processes and acquisition functions is given.

A Gaussian process is a distribution over functions. Assuming that every $\mathbf{x}$ has a multivariate normal distribution, the Gaussian process can be expressed as a multivariate distribution over $f(\mathbf{x})$ by a mean function $m(\mathbf{x})$ and a variance $k(\mathbf{x}, \mathbf{x}')$.

$$f(\mathbf{x}) \sim GP(m(\mathbf{x}), k(\mathbf{x}, \mathbf{x}')) \tag{1}$$

Observations of $\bar{e}_{\text{abs},t}$, given parameters $\mathbf{x}_t$, allow to fit the prior distribution model to these observations. A Gaussian process assumes a correlation between parameter sets and response. This means, if two parameter sets are similar, the corresponding values of $\bar{e}_{\text{abs}}$ are expected to be similar, too. The variance k as a measure of similarity between parameters becomes small for parameters near an observation, meaning that the probability distribution around these observations has a lower uncertainty. The variance is calculated using a kernel-function, which in this work is taken to be the ARD Matérn 5/2 kernel [10].

Once a prior Gaussian process model is fitted to the observations, and thus a posterior distribution is derived, the model is evaluated to determine a new set of parameters $\mathbf{x}_*$ for generating a new observation. Instead of random sampling, $\mathbf{x}_*$ is chosen based on an acquisition function. In this work, an acquisition function based on *expected-improvement* is used, which chooses $\mathbf{x}_*$ to maximize the expected improvement compared to the current best observation [10].

4 ANN Hyperparameter Optimization for Soft Pneumatic Actuator

The approach is applied to two different soft pneumatic actuators. In the first case study, a feedforward ANN is tuned to learn the kinematic model for a

PneuNet actuator [16]. The second case study deals with a three degrees of freedom bending actuator (3DBA), which was studied in our previous research [8,9]. A feedforward ANN is set up for learning the relationship between pressure and actuator configuration (curvature κ and orientation θ for the PneuNet, extended PCC parameters for the three chamber actuator (κ, $\sin\phi$, $\cos\phi$, s, d) [9]. This work seeks to find the number of neurons per layer for a network consisting of two hidden layers that minimizes the deviation between target data and net output. Additionally, a suitable activation function for each layer needs to be determined. Table 1 shows the ranges for the number of neurons per layer and the possible corresponding activation functions. Setting $n_{\text{hid},2} = 0$ includes the possibility of training ANNs with a single hidden layer. A network is trained using Levenberg-Marquardt backpropagation with Bayesian regularization. The training data is divided in training data, validation data for early stopping and test data for evaluating the ANN's quality.

Table 1. ANN hyperparameters for training the direct kinematics for robot specific kinematics

	Parameter	Value
Number of neurons 1st hidden layer	$n_{\text{hid},1}$	5–30 (PneuNet)/5–50 (3DBA)
Number of neurons 2nd hidden layer	$n_{\text{hid},2}$	0–30 (PneuNet)/5–50 (3DBA)
Activation function 1st hidden layer	$a_1(v)$	Log-sigmoid, linear, tan-sigmoid
Activation function 2nd hidden layer	$a_2(v)$	Log-sigmoid, linear, tan-sigmoid

4.1 PneuNet Bending Actuator

The PneuNet actuator depicted in Fig. 2 (left) is a single-input planar bending actuator, whose configuration can be described by the curvature κ and the bending angle θ. Figure 2 (right) shows the configuration of the PneuNet for different pressures between 0 MPa and 0.0549 MPa without the influence of gravity. Bayesian optimization was carried out to determine suitable hyperparameters for an ANN that represents the relationship between pressure and the curvature and bending angle. For this purpose 994 training data pairs have been generated. The boundaries for the number of neurons in the first and second hidden layer were $n_{\text{hid},1} \in [5, 30]$ and $n_{\text{hid},2} \in [0, 30]$, respectively. Of all the configurations tested by the algorithm, the best results were obtained with $\mathbf{x}^* = \{n^*_{\text{hid},1}, n^*_{\text{hid},2}, a^*_1(v), a^*_2(v)\} = \{23, 0, \text{tan-sigmoid}, -\}$.

Table 2 lists the mean absolute error $\bar{e}_{\text{abs}}$, the standard deviation s_{abs}, and the maximum error $e_{\text{abs,max}}$ between net output and target data. The deviation of selected test data points form the output of the trained ANN for pressures between 0 MPa and 0.0546 MPa is depicted in Fig. 3.

4.2 3D Bending Actuator

The 3DBA as depicted in Fig. 1 is able to bend in three dimensional space. The configuration of the actuator is described with an extended PCC model, which

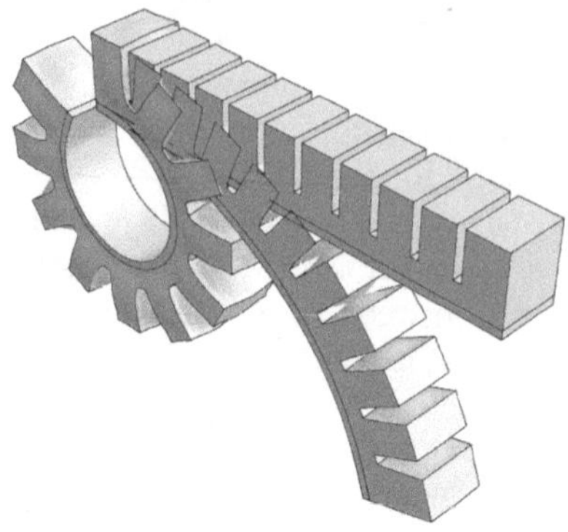

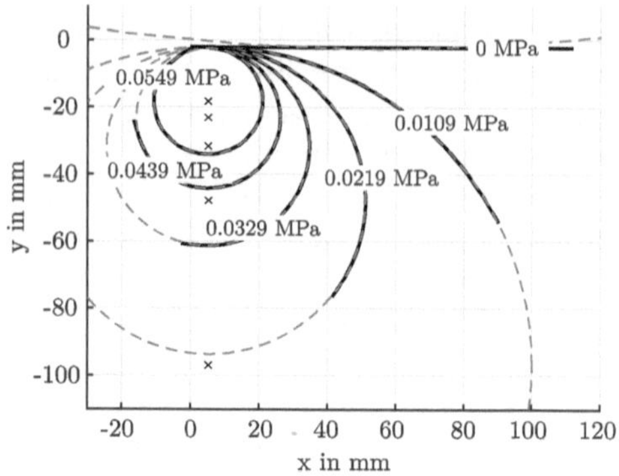

Fig. 2. PneuNet bending actuator

Table 2. Absolute error between neural network output and target configuration

	$\bar{e}_{\mathrm{abs}}$	s_{abs}	$e_{\mathrm{abs,max}}$
κ	$1.3304 \cdot 10^{-6}\,\mathrm{mm}^{-1}$	$1.7055 \cdot 10^{-6}\,\mathrm{mm}^{-1}$	$1.0181 \cdot 10^{-5}\,\mathrm{mm}^{-1}$
θ	$0.0095°$	$0.0199°$	$0.2487°$

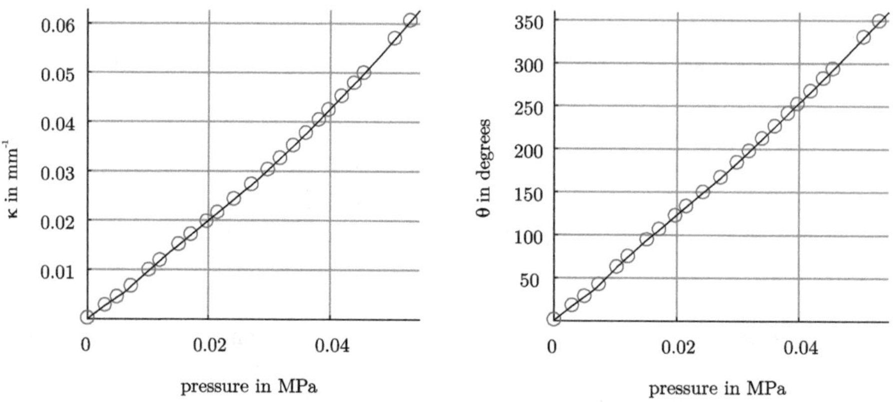

Fig. 3. Artificial neural network

considers the linear sections of the curve near the distal ends [9] (see Fig. 4 on the left). Compared to the PneuNet actuator, this results in a higher dimensional actuation and configuration space. To account for this higher complexity, the boundaries for the number of neurons as listed in Table 1 are extended. The objective for Bayesian optimization is the absolute error between the end effector position determined from FEM and the position resulting from the ANN simulation. For training an ANN 17742 training data pairs are available. Table 3 lists the mean absolute error over all test data $\bar{e}_{\mathrm{ts}}$, the standard deviation s_{ts} and the

Table 3. Absolute error between ANN simulated end effector position and target position

$\bar{e}_{ts}$ [mm]	s_{ts} [mm]	$e_{ts,max}$ [mm]
0.0206	0.0279	0.7612

maximum error $e_{ts,max}$ for the best set of hyperparameters from Bayesian optimization ($\mathbf{x}^* = \{45, 50, \text{tan-sigmoid}, \text{tan-sigmoid}\}$). To visualize the performance of the ANN, a pressure ramp from 0 MPa to 0.0228 MPa is simulated with the FE model. The same pressure ramp is used to simulate the deformation with the ANN. Figure 4 (right) shows the resulting path of the end effector. As can be seen, the path simulated with the ANN matches the path simulated with FEM.

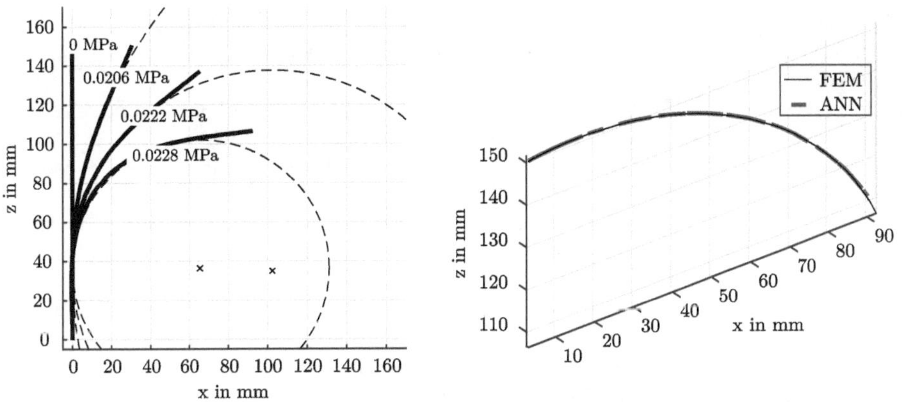

Fig. 4. Exemplary configurations of 3DBA and comparison of simulated end effector paths

5 Conclusions

In this article, the kinematics of two SPAs are modeled with ANNs. For both actuators, Bayesian optimization proved to be able to find suitable hyperparameters that minimize the error between net output and target data. In case of the 3DBA, the tuned ANN configuration outperforms the ANN from previous studies that has been set up via trial and error [9].

We showed that ANNs allow to accurately describe the kinematic relationship between pressure input and configuration. In perspective, the FEM-based framework can be used to initialize the ANN for the kinematic model of a soft pneumatic actuator. Machine learning methods promise to provide the capabilities to cope with nonlinearities and highly complex static and dynamic behaviour that is inherent to soft material robotic systems.

Moreover, machine learning approaches offer a simple solution to modeling soft robotic systems. With the help of machine learning highly specialized or individualized soft robotic system designs do not require intensive engineering work for analytical modeling. The approach helps to make system designs deployable by empowering developers to easily model and control their respective soft robotic systems for example for minimally invasive surgery or rehabilitation.

Acknowledgements. We thank our colleague Jan Peters for assisting our research and for the exchange of ideas that helped us to prepare this article.

References

1. Gerboni, G., Ranzani, T., Diodato, A., et al.: Modular soft mechatronic manipulator for minimally invasive surgery (MIS): overall architecture and development of a fully integrated soft module. Meccanica **50**(11), 2865–2878 (2015)
2. Majidi, C.: Soft robotics: a perspective - current trends and prospects for the future. Soft Robot. **1**(1), 5–11 (2013)
3. Moseley, P., Florez, J.M., Sonar, H.A.: Modeling, design, and development of soft pneumatic actuators with finite element method. Adv. Eng. Mater. **18**(6), 6–11 (2015)
4. Polygerinos, P., Wang, Z., Overvelde, J.T.B.: Modeling of soft fiber-reinforced bending actuators. IEEE Trans. Robot. **31**(3), 778–789 (2015)
5. Connolly, F., Walsh, C.J., Bertoldi, K.: Automatic design of fiberreinforced soft actuators for trajectory matching. PNAS **114**(1), 51–56 (2016)
6. Lee, K.-H., Fu, D.K.C., Leong, M.C.W.: Nonparametric online learning control for soft continuum robot: an enabling technique for effective endoscopic navigation. Soft Robot. **4**(4), 324–337 (2017)
7. Giorelli, M., Renda, F., Ferri, G., et al.: A feed-forward neural network learning the inverse kinetics of a soft cable-driven manipulator moving in three-dimensional space. In: Proceedings of the IEEE/RSJ International Conference on Intelligent Robots and Systems, pp. 5033–5039 (2013)
8. Runge, G., Raatz, A.: A framework for the automated design and modeling of soft robotic systems. CIRP Ann. Manuf. Technol. **66**(1), 9–12 (2017)
9. Runge, G., Wiese, M., Raatz, A.: FEM-based training of artificial neural networks for modular soft robots. In: IEEE International Conference on Robotics and Biomimetics (2017, in print)
10. Snoek, J., Larochelle, H., Adams, R.P.: Practical Bayesian optimization of machine learning algorithms. In: Proceedings of the 25th International Conference on Neural Information Processing Systems, vol. 2, pp. 2951–2959 (2012)
11. Shahriari, B., Swersky, K., Wang, Z.: Taking the human out of the loop: a review of Bayesian optimization. Proc. IEEE **104**(1), 148–175 (2016)
12. Eggensperger, K., Feurer, M., Hutter, F., et al.: Towards an empirical foundation for assessing Bayesian optimization of hyperparameters. In: NIPS Workshop on Bayesian Optimization in Theory and Practice (2013)
13. Webster, R.J., Jones, B.A.: Design and kinematic modeling of constant curvature continuum robots: a review. Int. J. Robot. Res. **29**, 1661–1683 (2010)
14. Rasmussen, C.E., Williams, C.K.I.: Gaussian Processes for Machine Learning. MIT Press, Cambridge (2006)
15. Mockus, J., Tiesis, V., Zilinskas, A.: The application of Bayesian methods for seeking the extremum. In: Towards Global Optimization, vol. 2 (1978)
16. Mosadegh, B., Polygerinos, P., Keplinger, C.: Pneumatic networks for soft robotics that actuate rapidly. Adv. Funct. Mater. **24**(15), 2163–2170 (2014)

A Novel Pneutronic Device for the Investigation of Compression-Induced Physiological Phenomena: Modeling and Experimental Testing

Carlo Ferraresi[1(✉)], Carlo De Benedictis[1], Daniela Maffiodo[1], Walter Franco[1], Alessandro Messere[2], Raffaele Pertusio[2], and Silvestro Roatta[2]

[1] Department of Mechanical and Aerospace Engineering, Politecnico di Torino, Turin, Italy
carlo.ferraresi@polito.it
[2] Department of Neuroscience, University of Torino, Turin, Italy

Abstract. An automatic pneumatic system designed to investigate the physiological effects induced by limb compressions, able to apply customizable pressure patterns in the physiologic range (0–200 mmHg), is here presented. A mathematical model, simulating the whole pneutronic system and its interaction with the limb is described. The model is validated by means of experimental tests. An on-going study aiming to investigate the role of different factors in the development of post compression hyperemia is presented.

Keywords: Pneutronic system · Pressure control · Hyperemia stimulation
Muscle compression · Human-machine interaction

1 Introduction

Intermittent pneumatic compression (IPC) is a technique that has been variously investigated in the biomedical field [10, 11] and that can be used for several rehabilitative applications, for example after sports injuries [13] or for the treatment of lymphoedema [6]. The IPC devices are also used in case of patients forced to immobility thanks to the effects that the IPC can exert on the cardiovascular system. When the motor capacity is compromised, the forced inactivity involves regressive phenomena affecting the cardiovascular system. In recent years, several studies tried to understand how to prevent systemic pathological conditions in these cases, and it has been demonstrated that some positive effects can be achieved with the IPC, through an active, correctly conceived biomechanical device that imitates the muscular pump function on the veins of the lower limbs by exerting coordinated pressure stimulations starting from the foot along the leg [2, 7]. This technique is therefore able to increase venous blood flow in case of immobility and to restore cardiovascular efficiency; but there is still a lack of knowledge of the parameters with which such therapy must be

G. Carbone et al. (Eds.): *New Trends in Medical and Service Robotics*, MMS 65, pp. 207–215, 2019.
https://doi.org/10.1007/978-3-030-00329-6_24

applied to achieve positive effects. Proper modeling of the pneumatic system and its interaction with humans can help the improvement of an IPC device [3, 4].

In addition, cardiovascular research is recently focusing on a previously neglected mechanism: i.e. the rapid dilatation that is evoked by a compressive stimulus [8, 9], for the relevant implications of the ensuing rapid hyperemia in IPC treatments [12]. Different underlying mechanisms have been hypothesized, including compression-induced changes in vascular filling and transmural vessel pressure and deformation of the vascular wall. The possibility to customize compressive stimuli with the addition of pre- and post- compression levels could help to discriminate the relevant factors.

Currently there is no commercially available device that can apply compressive stimuli with a desired pressure pattern, i.e. including long-lasting, low-pressure conditions as well as sharp high-pressure stimuli, with a good compromise between static performance in terms of accuracy, and dynamic performance in terms of response time.

Aim of the present study is to implement and model a new pneutronic system able to apply a customizable pressure pattern to the limb in the physiologic range (0–200 mmHg). The definition of an analytical model has been fundamental because of the demanding and particular requirements of such device, so as to define the specifications of the system and evaluate its performance. An application is then presented in which such pressure patterns are employed in healthy subjects to investigate the determinants of the compression-induced hyperemia.

2 Experimental Pneumatic System

The study has been carried out by means of an experimental pneumatic system, made up of a sleeve that must be wrapped around the area of the limb to be stimulated, and a proper control circuit. The most demanding specification was the ability to control the pressure of the sleeve with a step reference between 0 and 200 mmHg, with a good compromise between static performance in terms of accuracy (≤ 10 mmHg), and dynamic performance in terms of response time (≤ 0.3 s).

The design of the experimental system had to meet several contradictory requirements. In order to minimize the static error, it would be preferable to use pressure proportional valves with a regulation range comparable to that of the application. For our application these types of valves are normally of small size, and have therefore low maximum flow rate, which implies long sleeve inflation times, considering that the sleeve volume is about 1 dm^3. Regarding the regulation range, the pressure proportional valve SMC IT0010 would be an appropriate solution, presenting an appropriate set pressure range from 0.001 to 0.1 MPa (7.5–750 mmHg), with good static characteristics (linearity $\leq 1\%$ full scale F.S.; hysteresis $\leq 0.5\%$ F.S; repeatability $\leq 0.5\%$ F.S), and proper dynamic characteristics (response time 0.1 s). On the other hand, its maximum flow rate is only 6 dm^3/min (Standard Reference Atmosphere ANR), that means a very long sleeve inflation time of about 10 s.

Vice versa, pressure proportional valves able to regulate higher pressure ranges are of greater size, and therefore suitable in terms of flow performances, but with inadequate accuracy for the application. Moreover, they often are unable to regulate the relative pressure at low values close to zero. For example, the pressure proportional valve SMC ITV1010 would provide a significantly higher maximum flow rate (200 dm^3/min ANR), but it has an inadequate adjustment range, being unable to regulate the pressure below 0.005 MPa (37.5 mmHg). Therefore, it is not possible to achieve a good compromise between static and dynamic performance using only one commercial pressure proportional valve.

In order to overcome this issue, the pneumatic circuit of Fig. 1 has been designed. A 2/2 digital solenoid valve DV1 (direct operated solenoid normally closed valves, two ports two positions 2/2 SMC VXE2330-02F-6D01) supplies at 1.3 bar the sleeve during the transient phase, in order to minimize the inflate response time of the device. Another 2/2 digital solenoid valve (same model) DV2 is employed to discharge the sleeve during the corresponding transient phase, so as to minimize the deflate response time. A small size pressure proportional valve (SMC ITV0010) PPV is supplied at 1.8 bar for the accurate regulation of the sleeve pressure in stationary condition. A pressure transducer PT (BP-1, WPI, Sarasota USA) measures the pressure inside the sleeve, providing a feedback to the control system.

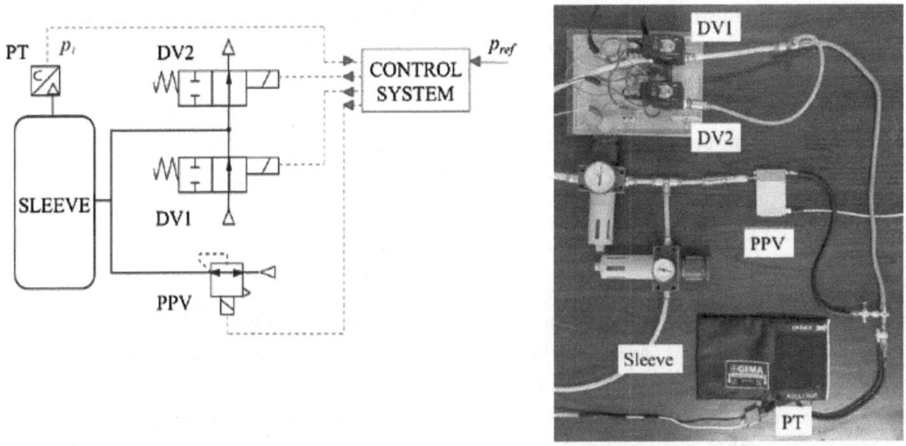

Fig. 1. The pneutronic system for hyperemia stimulation: DV1-2/2 digital solenoid valve for sleeve supply; DV2–2/2 digital solenoid valve for sleeve discharge; PPV – pressure proportional valve; PT - pressure transducer.

The control system compares the actual pressure value p_1 from the pressure transducer to a reference value p_{ref} and, by means of a PID logic, sends the corresponding command signals to the digital valves DV1 and DV2 and to the proportional valve PPV.

3 Model of the System

The core of the system consists of two 2/2 digital solenoid valves working in coordination with a pressure proportional valve. The control logic provides the appropriate commands to the valves, in order to perform the inflating and deflating of the deformable sleeve meanwhile the desired level of pressure is reached. In order to consider its irregular inflating/deflating, which occur progressively starting from the single air inlet/outlet, the sleeve has been modeled as the series of two chambers with variable volume, connected by a conveniently tuned pneumatic resistance [1] (Fig. 2).

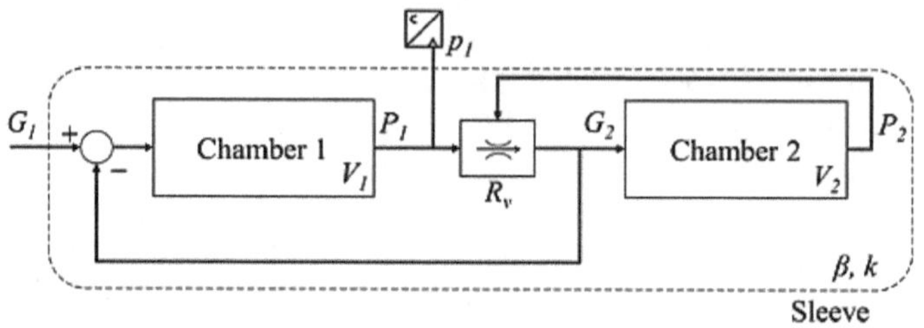

Fig. 2. Model of the sleeve.

Such discretization has been chosen to reproduce the non-uniform deformation of the sleeve volume due to the change of the internal pressure and the interaction with the human tissues, as evidenced experimentally. The variation of the inner pressure of each chamber is affected by the mechanical characteristic of the sleeve and of the limb soft tissues, and can be considered as the resultant of an elastic deformation and a viscous effect depending on the rate of volume change, as reported in Eq. (1):

$$p = k(V - V_i) + \beta \frac{dV}{dt} \tag{1}$$

where p is the variation of the relative pressure, V_i is the initial volume and V is the actual volume. β and k are the viscous constant and stiffness representing the viscoelastic behavior of both sleeve and limb tissues, defined by appropriate tuning of the model, also considering the expected volume change at maximum relative pressure.

The relationships between the mass flow rates and the pressures inside each chamber are described by the continuity Eqs. (2):

$$G_1 - G_2 = \frac{V_{i1} + \frac{p_1}{k} - \frac{\beta}{k}\frac{dV_1}{dt}}{RT}\frac{dP_1}{dt} + \frac{P_1}{RT}\frac{dV_1}{dt}, \quad G_2 = \frac{V_{i2} + \frac{p_2}{k} - \frac{\beta}{k}\frac{dV_2}{dt}}{RT}\frac{dP_2}{dt} + \frac{P_2}{RT}\frac{dV_2}{dt} \tag{2}$$

where 1 refers to the first chamber, whose internal pressure is measured to provide a feedback signal to the control system, and 2 refers to the second chamber. For each chamber, P is the absolute internal pressure and G is the mass flow rate of air. T and R represent respectively the chambers temperature and the air constant. The absolute pressures in both chambers are related by Eq. (3):

$$P_1 - P_2 = R_v G_2 \tag{3}$$

where R_v is the value of a pneumatic resistance connecting the volumes.

Each digital valve has been modeled as a pneumatic resistance, according to the ISO 6358 [1]. The mass air flow passing through the valve can be calculated in sonic or subsonic condition depending on the ratio between the downstream and upstream pressures, as presented in Eq. (4):

$$G = \rho_0 P_{us} C \text{ for } 0 < \frac{P_{ds}}{P_{us}} \leq b, \qquad G = \rho_0 P_{us} C \sqrt{1 - \left(\frac{P_{ds}/P_{us} - b}{1 - b}\right)^2} \text{ for } b < \frac{P_{ds}}{P_{us}} \leq 1 \tag{4}$$

where P_{us} is the upstream absolute pressure, P_{ds} is the downstream absolute pressure, C is the sonic conductance, b is the critical ratio and ρ_O is the air density in normal conditions. The C and b parameters have been defined respectfully of the data reported by the manufacturer (Table 1). The valve on-off status is controlled with a digital command generated by the control logic, which compares the measured sleeve pressure to selected thresholds. A first order dynamics (with time constant τ_v) has been introduced to take into account the finite response time between the command signal and the resulting variation of conductance. The upstream and downstream pressures for each valve depend on the physical connection of the device: i.e. the digital valve DV1 used to inflate the sleeve has the upstream set as supply pressure, and the P_1 pressure set as the downstream one (Fig. 1).

Table 1. Parameters of the system.

Parameter	Value	Parameter	Value	
V_{i1}	22 cm^3	τ_v	30	ms
V_{i2}	22 cm^3	τ_{ppv}	25	ms
β	1 kPa s/m^3	K_p	20	kPa/V
k	4.06 MPa/m^3	K_{ppv}	$9.22 \cdot 10^{-9}$	kg/(Pa s)
R_v	3.5 MPa s/kg	R	287.1	J/(kg K)
C	$2.3 \cdot 10^{-8}$ Nm^3/(Pa s)	T	293	K
b	0.46	ρ_O	1.188	kg/m^3

The pressure proportional valve has been modeled as a first order system [5], whose output mass flow rate G_{ppv} depends on the error between the measured pressure p_1 and the set pressure p_{set}, the latter provided by the control system. Figure 3 shows the internal pressure control loop of the device. The overall mass flow rate G_1 (Fig. 2), defined as positive when directed towards the sleeve, is given by the algebraic sum of the G_{ppv} flow rate and the ones regulated separately by the two digital valves DV1 and DV2, respectively used to inflate (G_{dv1}) and deflate (G_{dv2}) the sleeve. The K_p gain, used to convert the voltage command u_{set} in the required pressure for the proportional valve,

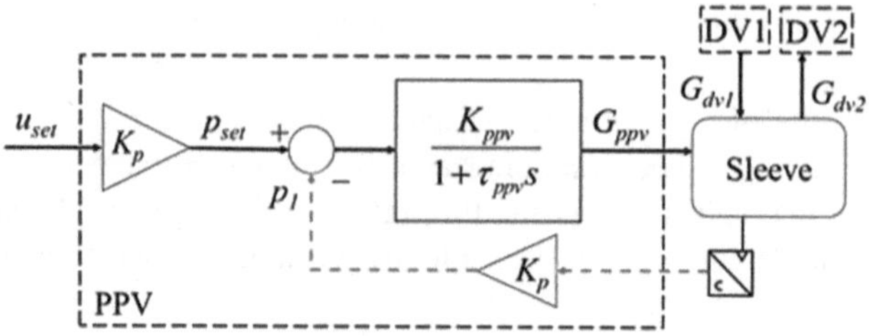

Fig. 3. Model of the pressure proportional valve.

is equal to the gain of the transducer performing the measurement of the sleeve internal pressure. The static gain of the valve K_{ppv} has been calculated since the data given by the manufacturer of the device. The main parameters of the system are presented in Table 1. The model of the system has been implemented in MATLAB-Simulink®.

4 Comparison Between Experimental Tests and Model Results

Simple patterns were created with the above-described pneumatic circuit (Fig. 1) and compared to the simulations results, in order to validate the model.

To this aim, the sleeve was applied to the calf of 3 healthy subjects, sitting upright. Figure 4 shows the model and the experimental results in case of a simple pattern obtained with the following law:

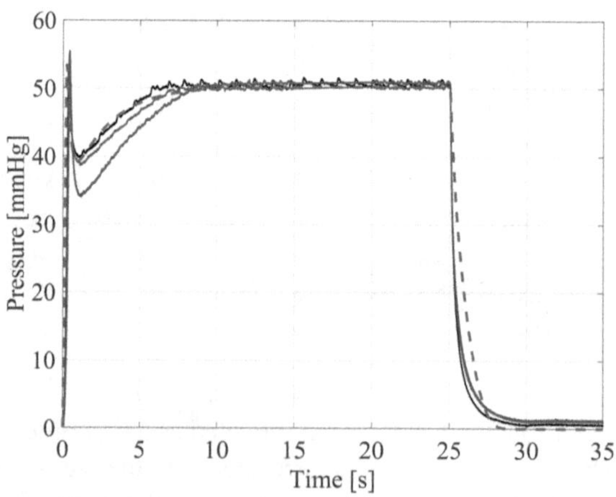

Fig. 4. Pressure inside the sleeve p_1, comparing the model (dashed line) and the experimental responses (solid lines). The experimental curves have been obtained on three different subjects.

- digital valve DV1 and pressure proportional valve PPV switched on with a set value of 50 mmHg, the control system stops this phase when set value is reached;

- DV1 switched off, set value kept as constant by the PPV up to 25 s;
- PPV switched off, DV2 switched on;
- rest phase with all valves switched off.

As shown in Fig. 4, the model response is sufficiently close to the experimental pressure signals measured by the transducer inside the sleeve. The rise time is similar in both cases, while the deflating phase of the real system appears to be faster than the model. This may be due to inaccuracies in the modeling of the sleeve, concerning the relationship between the variation of the internal pressure and the deformation of each chamber. Nevertheless, the differences between model and real system are limited and related to time intervals significantly shorter than the duration of each trial.

Both the model and the real system are able to keep a constant level of pressure set to 50 mmHg. This result is of course more relevant referring to the real system, as any losses in the pneumatic circuit need to be handled in such stationary condition by the PPV. The two-chambers modeling scheme (Fig. 2) allows to reproduce the dynamic behavior of the sleeve with sufficient accuracy, also exhibiting the drop of pressure level occurring at the end of the first phase (when DV1 is turned off), as highlighted by the experimental curves. As observed by preliminary analyses, a single chamber model could not be able to reproduce the same behavior.

5 Experimental Study on Healthy Subjects

The proposed device is currently employed in an experimental study on healthy subjects carried out at the Integrative Physiology Lab, Dept. of Neuroscience, University of Torino, to investigate the role of different factors in the development of post-compression hyperemia. For this specific study, various patterns for the lower limb compression were required: (a) single compression at 10–50 mmHg lasting 50 s, (b) a short compression at 200 mmHg, lasting 1 s, and (c) complex compression patterns consisting of a combination of the formers. During these compressions, the blood flow of the femoral vein and the femoral artery was measured by Doppler sonography (Mylab 25, ESAOTE). The sleeve is positioned around the left leg of the subject,

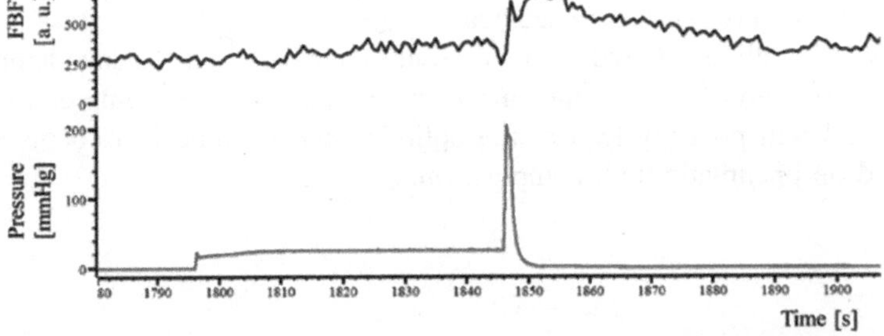

Fig. 5. Original recordings of a post-compression hyperemia from a representative subject: femoral artery blood flow (FBF) and sleeve pressure.

placed horizontally. It is particularly important that the pressure is accurately kept constant during the stationary periods and that the changes from one state to another are very rapid to improve repeatability and temporal-localization of the stimuli.

Figure 5 shows an example of a test performed with pattern (c). The bottom plot shows pressure p_1 acquired by the transducer. The upper part of Fig. 5 shows the flow pattern in the femoral artery registered by the eco-doppler probe. A significant increase of the flow is evident, after the 200 mmHg impulse.

6 Conclusions

A pneutronic system able to apply a customized pressure pattern for the compression of a limb in the range (0–200 mmHg) is here presented. The system is aimed at evoking hyperemia in the stimulated area of a limb, and has been conceived in order to deeply investigate this phenomenon and define proper and effective clinical protocols. No commercial device able to operate in this way is currently available. This application dictated several specifications particularly demanding for a normal pneumatic device, therefore it was necessary to develop a system combining high-power digital valves together with small size high-accuracy pneumo-electronic components.

The system has been studied and optimized by means of a mathematical model, which was then validated through comparison with the results of experimental tests. Particular accuracy has been devoted to modeling of the human-machine interaction: viscous-elastic parameters have been included in the definition of the sleeve structure, this latter has been modeled as a two-stage collapsible volume, to fit experimental data collected during cuff inflation.

This pneumatic system is particularly useful for investigating the physiological response of vessels to mechanical compression, because its very sharp pressure changes allow for precise temporal localization of the compressive stimulus and assessment of the latency of the vascular response.

Moreover, the precise holding of steady pressure levels for any period of time can be useful to investigate the effects of reducing vascular filling in venous compartments and of altering vascular transmural pressure levels on limb perfusion and on the hyperemic response to compression.

By combining the experimental characterization and the numerical simulations, the tuning of the device and of the control logic parameters made it possible to obtain the right compromise between these specifications.

The new possibilities offered by the present system found immediate application in research studies investigating the mechanisms behind the compression-induced hyperemia and will possibly lead to the optimization of clinical protocols and treatments based on pneumatic limb compression.

References

1. Ferraresi, C.: Modelling of pneumatic systems. Fluid Apparecchiature Idrauliche Pneumatiche **346**, 48–54 (1993)
2. Ferraresi, C., Hajimirzaalian, H., Maffiodo, D.: Identification of physical parameters in a robotized IPC device interacting with human. Appl. Mech. Mater. **490–491**, 1729–1733 (2014)
3. Ferraresi, C., Maffiodo, D., Hajimirzaalian, H.: A model-based method for the design of intermittent pneumatic compression systems acting on humans. Proc. Inst. Mech. Eng., Part H J. Eng. Med. **228**(2), 118–126 (2014)
4. Ferraresi, C., Maffiodo, D., Hajimirzaalian, H.: Simulation and control of a robotic device for cardio-circulatory rehabilitation. Adv. Intell. Syst. Comput. **371**, 357–365 (2016)
5. Ferraresi, C., Sorli, M.: Modelling and analysis of pressure control systems. Fluid Apparecchiature Idrauliche Pneumatiche **357**, 58–65 (1994)
6. Johansson, K., Lie, E., Ekdahl, C., Lindfeldt, J.: A randomized study comparing manual lymph drainage with sequential pneumatic compression for treatment of postoperative arm lymphedema. J. Lymphology **31**, 56–64 (1998)
7. Maffiodo, D., De Nisco, G., Gallo, D., Audenino, A., Morbiducci, U., Ferraresi, C.: A reduced-order model-based study on the effect of intermittent pneumatic compression of limbs on the cardiovascular system. Proc. Inst. Mech. Eng., Part H J. Eng. Med. **230**(4), 279–287 (2016)
8. Messere, A., Ceravolo, G., Franco, W., Maffiodo, D., Ferraresi, C., Roatta, S.: Increased tissue oxygenation explains the attenuation of hyperemia upon repetitive pneumatic compression of the lower leg. J. Appl. Physiol. **123**, 1451–1460 (2017)
9. Messere, A., Turturici, M., Millo, G., Roatta, S.: Repetitive muscle compression reduces vascular mechano-sensitivity and the hyperemic response to muscle contraction. J. Physiol. Pharmacol. **68**, 427–437 (2017)
10. Morris, R.J.: Intermittent pneumatic compression – systems and applications. J. Med. Eng. Technol. **32**(3), 179–188 (2008)
11. Partsch, H.: Intermittent pneumatic compression in immobile patients. Int. Wound J. **5**(3), 389–397 (2008)
12. Sheldon, R.D., Roseguini, B.T., Laughlin, M.H., Newcomer, S.C.: New insights into the physiologic basis for intermittent pneumatic limb compression as a therapeutic strategy for peripheral artery disease. J. Vasc. Surg. **58**, 1688–1696 (2013)
13. Waller, T., Caine, M., Morris, R.: Intermittent pneumatic compression technology for sports recovery. The Engineering of Sport, vol. 3 (2006)

Viscoelastic Modeling of Human Nasal Tissues with a Mobile Measurement Device

O. Işıtman[1], O. Ayit[1], E. Vardarlı[1], Ş. Hanalioğlu[2], İ. Işıkay[2], M. Berker[2], and M. İ. C. Dede[1](✉)

[1] Izmir Institute of Technology, Izmir, Turkey
{ogulcanisitman,orhanayit,erenvardarli,candede}@iyte.edu.tr
[2] Hacettepe University, Ankara, Turkey
sahinhanalioglu@gmail.com, ilkayisikay@gmail.com, mberker@hacettepe.edu.tr

Abstract. Modeling the dynamic of tool-tissue interaction for the robotic minimally invasive surgeries is one of the main issues for designing appropriate robot controllers. A mobile measurement device is produced in order to model some nasal tissues of a human. This mobile device is a hand-held one which measures the applied moments and relative angular displacements about a fixed pivot point. The ex-vivo measurements are realized by surgeons on a relatively fresh human cadaver head. The tip of the nose and the nasal concha are the two tissues that are investigated. In this study, five different viscoelastic models are considered; Elastic, Kelvin-Voight, Kelvin-Boltzmann, Maxwell and Hunt-Crossley. The results are evaluated and cross-validated on each data set. Hunt-Crossley and Kelvin-Boltzmann models provided the minimum root-mean-square (RMS) error among the other models.

Keywords: Soft tissue modeling · Ex-vivo tests · Nasal tissue
Elastic model · Kelvin-Voight model · Kelvin-Boltzmann model
Maxwell model · Hunt-Crossley model

1 Introduction

In the past years, many open surgery procedures have been replaced by minimally invasive surgeries (MIS) and in the past three decades, robots started to be included in MIS. Introduction of robotics in MIS offered various benefits such as increased dexterity, increased manipulation capability and repeatability. However, one critical consideration in such procedures is the safety.

To avoid excessive contact force to the environment, the robotic surgical instrument must behave compliantly. A pure position control is not sufficient to solve this problem alone, hence, control of the dynamic interaction is also required [4]. An accurate model of the surgical environment, which happens to a soft tissue most of the case, can result in more precise and reliable controller

G. Carbone et al. (Eds.): *New Trends in Medical and Service Robotics*, MMS 65, pp. 216–224, 2019.
https://doi.org/10.1007/978-3-030-00329-6_25

design. Previously, a force control method was proposed based on pure elastic interaction in [12]. However, this assumption is not realistic for the biological tissues [3].

In order to define the tool-tissue interaction relatively more accurately, researchers have proposed several soft tissue models. These approaches can be classified as linear elasticity or nonlinear (hyperelastic) elasticity based finite element methods (FEM), viscoelastic models, fractional models, and high order models [7]. These models can also be classified by their computational time and deformation accuracy [9]. In the case of the diagnostic and scientific analysis of the tissue, more precise FEM's are suitable in spite of their longer computational time. Due to real-time requirement, analytical models are commonly used in the robotic applications. A conventional way to represent the tissue dynamics is using combinations of ideal damper and spring components. These models are comprehensively studied and compared in literature and schematic representation of some of these models are given in Fig. 1 along with their governing equations [6,7,10]. In the equations $F(t)$ and $x(t)$ represent the applied force and the deformation of tissue, respectively. $\beta = \frac{k_1 k_2}{k_1+k_2}$, $\alpha = \frac{bk_2}{k_1+k_2}$ and $\delta = \frac{b}{k_1+k_2}$ are the linear combination of spring and damper parameters in the Kelvin-Boltzmann model.

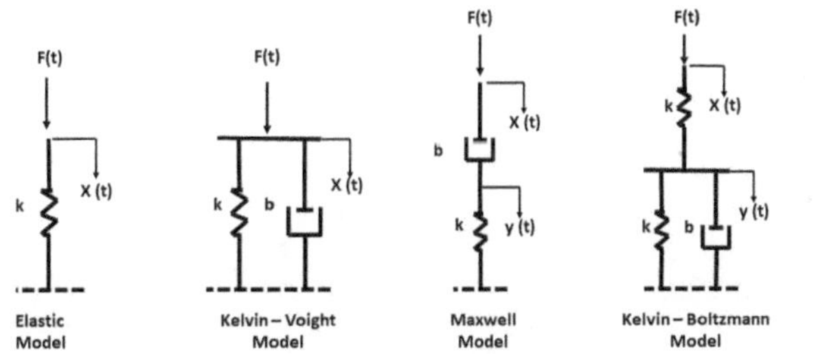

$F(t) = kx(t)$ $\quad F(t) = kx(t) + b\dot{x}(t)$ $\quad F(t) = b\dot{x}(t) - \frac{b}{k}\dot{F}(t)$ $\quad F(t) = \beta x(t) + \alpha\dot{x}(t) - \delta\dot{F}(t)$

Fig. 1. Commonly used analytical models

The simplest model is a purely elastic model with single ideal spring model. The Kelvin-Voight model is consist of a parallel connection of an ideal damper and an ideal spring. On the other hand, the Maxwell model is composed of a serial connection of an ideal spring and an ideal damper components. Serial combination of elastic and the Kelvin-Voight models describes the Kelvin-Boltzmann model. These models differ from each other by their ability to mimic tissue behaviors such as creep and relaxation. These are the two major phenomena of the biological tissues [3]. While Maxwell model provides acceptable results for relaxation, Kelvin-Voight provides a good first order approximation of creep. On the other hand, Kelvin-Boltzmann model provides a good approximation for both creep and stress relaxation [7].

In [5], Hunt and Crossley presented that linear elastic models are not consistent with physics of contact and they proposed a non-linear model which is represented mathematically in Eq. 1.

$$F(t) = kx^n(t) + \lambda x^n(t)\dot{x}(t) \tag{1}$$

In this equation, n is positive scalar it ranges from 1.1 to 1.3 for a soft tissue [2,9,10]. Several studies have experimentally showed that Kelvin-Boltzmann and Hunt-Crossley models provide relatively more realistic results with the lowest errors with respect to the aforementioned viscoelastic models [9].

In order to identify soft tissue model parameters, several types of experiments are performed in the literature. In most of these studies, a robotic system is used by driving an end-effector in a controlled manner into the in-vitro specimen [8].

The work presented in this paper is conducted to model the nasal tissues to be used in the design of an endoscopic pituitary tumor robotic surgery system. The conceptual design is presented in Fig. 2. In this robotic system, the robot is used to manipulate the endoscope inside the surgical area by rotating about a remote center of motion, which is usually called the pivot. Therefore, the endoscope handled by the robot is expected to interact with the nose tip and the nasal concha by making rotational motion.

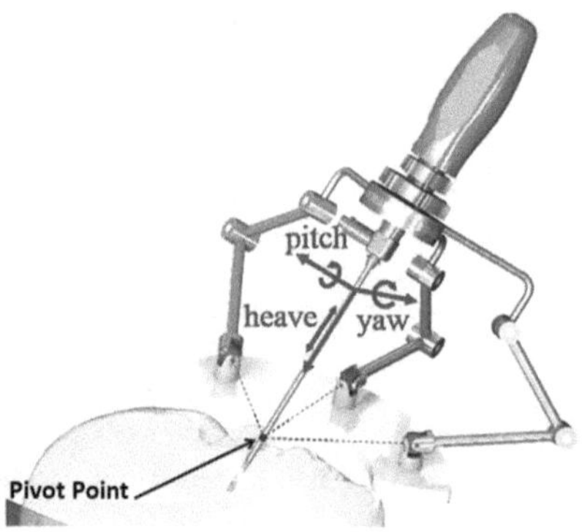

Fig. 2. Conceptual design of the robotic pituitary gland surgery system

Since preparing in-vitro specimen for this specific surgery area is not straightforward, we designed a new mobile hand-held measurement device which gives opportunity to acquire the necessary data in an ex-vivo set-up without the need of a robotic interface. However, the surgeon's hand motion cannot imitate a pure step input. For this reason, the response of soft tissue under compression is considered by moments applied on it for the evaluation instead of evaluating interaction models by their stress relaxation and creep behavior [1,11]. The soft tissue modeling is realized by using moment and angular displacement measurements due to the characteristic of the surgery which is another novelty of the study. Measurements are acquired from a relatively fresh human cadaver head in ex-vivo conditions.

In the next section, the proposed experimental set-up and test procedures are explained. In Sect. 3, obtained results are presented and we discuss the results in the final section.

2 Methodology

In order to acquire realistic data from the nasal tissue, a new mobile device is designed. An ATI MINI45 (ATI Industrial Automation) force/torque (F/T) sensor is used for the force and torque measurements and BNO055 (BOSCH Sensortech) inertial measurement unit (IMU) sensor is used to obtain angular displacements. Data acquisition is realized via MATLAB/Simulink environment with 100 Hz sampling rate. The components of the experimental set-up is presented in Fig. 3 indicating the measurement axes for angular motion and moment measurements.

The moment measurements are highly sensitive to the orientation of the device due to the weight of the F/T sensor. In order to eliminate gravitational forces, gravity compensation algorithm is developed by acquiring the absolute orientation of the device from the IMU sensor. These measurements also depend on the inertial forces and angular acceleration. To obtain more precise results, in the measurement process, the surgeons are asked to rotate the device at a constant velocity.

In order to imitate the real surgical scenario, a KARL STORZ telescope (28731 BWA Hopkins) is used and 3D printed holder is designed to provide better ergonomics for surgeons.

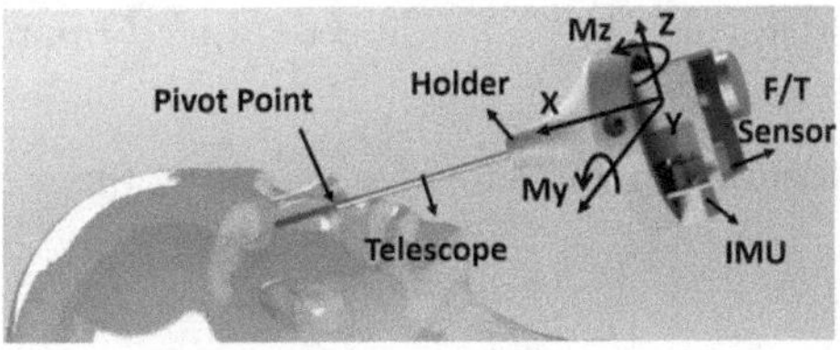

Fig. 3. CAD drawing of the proposed measurement device

Freshly frozen cadaver head is started to be thawed 48 h beforehand to prepare for the experiment. Thin section tomography images of the cadaver head are taken with the Siemens Somamtom Perspective device during the thawing process. When this procedure is completed and the tissues became closer to the live tissue strength. The cadaver head is then fixed with a three-point head clamp (Mayfield, Integra Life Sciences Co., NJ) and the measurement protocols are applied.

Two test procedures are defined to measure applied moment and angular position data. In the first test, surgeons are asked to hold the device similar to the endoscope and perform the rotational movement around the predefined pivot point and compress the tip of the nose with a constant velocity. The tip of the nose is the actual pivot point in the endoscopic pituitary tumor surgery and the endoscope is expected to be in contact with this point during the operation. In the second test, surgeons are asked to rotate the device around the tip of the nose and compress the nasal concha without interacting with pivot point (no force applied to the nose tip).

Parameter estimation process is realized as follows: (1) Each interaction model is developed in Matlab Simulink by using angular position data and the resultant moment is compared with the measured moment data to obtain the error in the model. (2) The sum of the squares of error function is minimized by using `fmincon` function in Matlab. (3) The results are evaluated by using root-mean-square errors (RMSE) and using the average of the estimated parameters' cross-validation is realized for each measurement.

3 Results

The procedure that is explained in the previous section is realized by three surgeons that have carried out endoscopic pituitary tumor surgery. A total of nine experiments carried out in order to eliminate measurement error. Four of these experiments are performed to model the tip of the nose while the others were performed for the nasal concha model.

The measured position/torque values in the experiments are used to calculate the parameters of soft tissue models. In each experiment, the maximum measured torque is approximately 0.03 Nm.

Table 1 shows the identified parameters of the nose tip models for each experiment where the unit of spring (k, k_1, k_2) and damper(b) parameters are Nm/rad

Table 1. Identified parameters for the tip of the nose models

Models		Experiments			
		#1	#2	#3	#4
Kelvin Boltzmann	k_2	0,455	1,208	0,860	1,363
	k_1	0,135	0,375	0,273	0,149
	b	16.270	28,457	24,695	25,105
Kelvin Voight	k	0,102	0,276	0,202	0,239
	b	0,209	0,115	0,448	0,593
Elastic	k	0,103	0,277	0,202	0,247
Maxwell	k	0,589	1,469	1,216	1,243
	b	101,309	125,209	127,071	127,302
Hunt Crossley	k	0,127	0,358	0,308	0,327
	λ	15,154	15,173	15,494	15,492

Table 2. RMSE between measured and calculated moments for the nose tip models

RMSE (Nm)				
Models	Experiments			
	#1	#2	#3	#4
Kelvin Boltzmann	**0,0036**	0,0056	**0,0058**	**0,0051**
Kelvin Voight	**0,0036**	0,0057	0,0059	0,0056
Elastic	**0,0036**	0,0056	0,0059	0,0058
Maxwell	0,0043	0,0069	0,0067	0,0056
Hunt Crossley	0,0046	**0,0053**	0,0060	0,0074

Table 3. RMSE between measured and calculated force with average parameters of the nose tip models

Models		Experiments				
		Avg. coeff.	RMSE (Nm)			
			#1	#2	#3	#4
Kelvin Boltzmann	k_2	0,972	**0,0216**	0,0165	**0,0062**	0,0111
	k_1	0,233				
	b	23,631				
Kelvin Voight	k	0,205	0,0235	0,0156	**0,0062**	0,0108
	b	0,341				
Elastic	k	0,207	0,0237	0,0155	**0,0062**	0,0114
Maxwell	k	1,129	0,0217	0,0135	0,0068	0,0099
	b	120,223				
Hunt Crossley	k	0,280	0,0264	**0,0132**	0,0064	**0,0096**
	λ	15,328				

and Nm.s/rad, respectively. It is observed in Table 1 that the parameters are consistent with each other.

In addition, RMSE between measured torques and calculated ones are reported in Table 2 and the minimum RMSE value received for the model of each experiment is highlighted with bold fonts. It is observed that the Kelvin-Boltzmann model resulted in the minimal errors in experiments #1, #3, #4 and Hunt-Crossley model has minimal error in experiment #2. Also, Kelvin-Voight and Elastic models resulted in the same minimal error in experiment #1.

The average values of identified parameters of the models are calculated. The soft tissue models with these new parameters are evaluated for each experiment for cross-validation and the RMSE values are calculated as shown in Table 3. According to these results, the minimum RMSE value is obtained when Kelvin-Boltzmann, Kelvin Voight, and Elastic models are applied to experiment #3.

Table 4. Identified parameters for the Nasal Concha Model

Models		Experiments				
		#1	#2	#3	#4	#5
Kelvin Boltzmann	k_2	3,829	3,700	4,198	2,993	1,795
	k_1	0,036	0,559	1,033	0,207	0,559
	b	26,357	32,038	33,878	29,029	30,979
Kelvin Voight	k	0,575	0,675	0,844	0,419	0,358
	b	0,733	0,236	1,525	1,773	0,517
Elastic	k	0,600	0,679	0,856	0,440	0,370
Maxwell	k	2,928	3,440	5,833	2,545	2,632
	b	137,166	136,950	139,842	136,699	138,644
Hunt Crossley	k	1,000	1,070	1,586	0,824	0,787
	λ	15,038	14,990	15,10	15,117	15,535

Table 5. RMSE between measured and calculated force for the Nasal Concha Models

RMSE (Nm)					
Models	Experiments				
	#1	#2	#3	#4	#5
Kelvin Boltzmann	**0,0039**	**0,0048**	**0,0034**	**0,0024**	0,0031
Kelvin Voight	0,0072	0,0068	0,0035	0,0033	0,0033
Elastic	0,0075	0,0068	0,0035	0,0036	0,0030
Maxwell	0,0068	0,0050	0,0050	0,0036	0,0059
Hunt Crossley	0,0128	0,0111	0,0049	0,0050	**0,0029**

Also, the Kevin-Boltzmann has minimum RMSE value in experiment #1. In experiment #2 and #4, Hunt-Crossley model has the best performance.

Similar studies are carried out to model the nasal concha. Table 4 presents the identified parameters and it is observed that the parameters are consistent with each other. Also, RMSE values of each model are given in Table 5 and the minimum-error ones are highlighted by bold fonts. According to Table 5, Kelvin-Boltzmann has the best performance in experiment #1, #2, #3 and #4. In experiment #5, the minimum error is obtained when Hunt-Crossley model is applied.

To obtain an approximate tissue model for each experiment, the averages of the models' parameters are calculated and these values are used for each experiment for cross-validation. The average parameters and the RMSE values for model validation are presented in Table 6.

According to these values, Kevin-Boltzmann model has the best performance in experiment #1 and #4. In experiment #2 and #5, the minimum error is observed when Hunt-Crossley model is applied. In contrast to other experiments, the elastic model has the best performance in experiment #3.

Table 6. RMSE between Measured and Obtained Force with Average Parameters of the Nasal Concha Models

Models		Experiments					
		Avg. coeff.	RMSE (Nm)				
			#1	#2	#3	#4	#5
Kelvin Boltzmann	k_2	3,303	**0,0157**	0,0085	0,0128	**0,0055**	0,0080
	k_1	0,479					
	b	30,456					
Kelvin Voight	k	0,574	0,0198	0,0114	0,0106	0,0064	0,0072
	b	0,957					
Elastic	k	0,589	0,0206	0,0110	**0,0104**	0,0066	0,0068
Maxwell	k	3,476	0,0334	0,0111	0,0107	0,0083	0,0098
	b	137,860					
Hunt Crossley	k	1,054	0,0237	**0,0050**	0,0117	0,0068	**0,0048**
	λ	15,158					

4 Discussions and Conclusions

In the endoscopic pituitary tumor surgery, the tip of nose and tissue of the nasal concha mostly interacts with the endoscope. Since the results of this modeling study will be used in the design of a remote center of motion kinematics based robotic surgery, this study investigated the behavior of tissues under rotational movement which is different from the most of the works in literature. We proposed a new mobile measurement device to acquire the torque/position data during the interactions with the focused soft tissues of a cadaver in ex-vivo tests. In the experiments, three surgeons are asked to apply torques to compress tissues with a constant angular velocity.

The soft tissues models for the nasal concha and the nose tip are obtained by suing the experimental data. The results meet the expectations since the other studies in the literature show that the Kelvin-Boltzmann and Hunt-Crossley models have better performance relative to the others.

As a result, this study provides quantitative knowledge about the expected torques during the operation and an accurate interaction model in order to design a complaint controller as a future work.

Acknowledgements. This work is supported by The Scientific and Technological Research Council of Turkey via grant numbers 115E725 and 115E726.

References

1. Bao, Y., Wu, D., Yan, Z., Du, Z.: A new hybrid viscoelastic soft tissue model based on meshless method for haptic surgical simulation. Open Biomed. Eng. J. **7**, 116 (2013)
2. Diolaiti, N., Melchiorri, C., Stramigioli, S.: Contact impedance estimation for robotic systems. IEEE Trans. Robot. **21**(5), 925–935 (2005)
3. Fung, Y., Skalak, R.: Biomechanics: mechanical properties of living tissues. J. Appl. Mech. **49**, 464 (1982)
4. Hogan, N.: Impedance control: an approach to manipulation. In: American Control Conference, pp. 304–313. IEEE (1984)
5. Hunt, K., Crossley, F.: Coefficient of restitution interpreted as damping in vibroimpact. J. Appl. Mech. **42**(2), 440–445 (1975)
6. Liu, C., Moreira, P., Poignet, P.: Viscoelastic model based force tracking control for robotic-assisted surgery. In: 4th IEEE RAS & EMBS International Conference on Biomedical Robotics and Biomechatronics (BioRob), pp. 1199–1204. IEEE (2012)
7. Misra, S., Ramesh, K., Okamura, A.M.: Modeling of tool-tissue interactions for computer-based surgical simulation: a literature review. Presence Teleoperators Virtual Environ. **17**(5), 463–491 (2008)
8. Moreira, P., Liu, C., Zemiti, N., Poignet, P.: Soft tissue force control using active observers and viscoelastic interaction model. In: IEEE International Conference on Robotics and Automation (ICRA), pp. 4660–4666. IEEE (2012)
9. Moreira, P., Zemiti, N., Liu, C., Poignet, P.: Viscoelastic model based force control for soft tissue interaction and its application in physiological motion compensation. Comput. Methods Programs Biomed. **116**(2), 52–67 (2014)
10. Pappalardo, A., Albakri, A., Liu, C., Bascetta, L., De Momi, E., Poignet, P.: Hunt-crossley model based force control for minimally invasive robotic surgery. Biomed. Signal Process. Control **29**, 31–43 (2016)
11. Takács, Á., Galambos, P., Pausits, P., Rudas, I. J., Haidegger, T.: Nonlinear soft tissue models and force control for medical cyber-physical systems. In: IEEE International Conference on Systems, Man, and Cybernetics (SMC), pp. 1520–1525. IEEE (2015)
12. Yuen, S.G., Perrin, D.P., Vasilyev, N.V., Pedro, J., Howe, R.D.: Force tracking with feed-forward motion estimation for beating heart surgery. IEEE Trans. Robot. **26**(5), 888–896 (2010)

Design of a Test Bench to Simulate Cranial Sudden Impact

J. L. Rueda Arreguín[1(✉)], C. R. Torres San Miguel[1], M. Ceccarelli[2], V. Ramírez Vela[1], and G. M. Urriolagoitia Calderón[1]

[1] Instituto Politécnico Nacional, Sección de Estudios de Posgrado e Investigación, Escuela Superior de Ingeniería Mecánica y Eléctrica, Unidad Profesional Adolfo López Mateos "Zacatenco", Edificio 5, 2do. Piso Col. Lindavista, 07320 Ciudad de México, Mexico
gemini_92_19@hotmail.com
[2] LARM: Laboratory of Robotics and Mechatronics, DiCEM-University of Cassino and South Latium Cassino, Cassino, FR, Italy

Abstract. In the history of head impacts, the focus has always been the development of helmets that reduce the damage received by the head. Few researchers have addressed the problem of the injuries that occurs at the internal organs like the brains and spinal cord. The development of a test bench to simulate cranial impacts is often costly and the analysis are only focused on frontal impacts. This paper proposes a new test bench to analyse the brain-spinal system when a sudden impact load is received at different sections of the skull. Our technique shows a clear mechanical design that can use a parallel robot 3RPS to set up various positions of a spine head system to estimate the damages on the cervical area.

1 Introduction

The brain and spinal cord are very important organs of the human body generating and transmitting impulses to perform most of the activities in a daily life. However, it is under the risk to receive impacts that can damage the integrity of the elements of the cranioencephalic area [1]. The most common ones are cervical injuries caused by vehicle crashes, these are known as the whiplash syndrome [2] and according to research conducted in Japan, 50% of car accidents trigger in this condition [3], while in North America and Western Europe the annual cumulative incidence is between 300 and 600 per 100,000 population [4]. The biomechanics of this syndrome have been determined and a sequence of events has been established during a rear impact [5], which are the most frequent and whose most important phases are shown in Fig. 1.

- Phase I: At 60 ms, the vehicle is accelerated forward, and the backrest of the seat contacts the region of the lower back, causing the seat to move back away from the upper part of the trunk. Because the head is still stationary, a cervical compression happens, this occurs before a significant acceleration forward of the trunk takes place [6].

G. Carbone et al. (Eds.): *New Trends in Medical and Service Robotics*, MMS 65, pp. 225–234, 2019.
https://doi.org/10.1007/978-3-030-00329-6_26

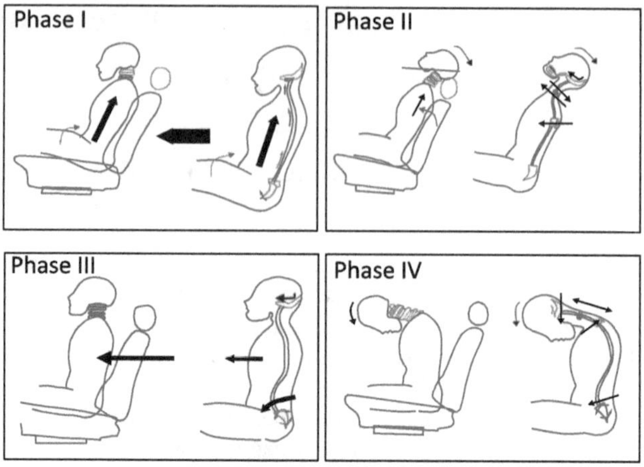

Fig. 1. Phases of whiplash.

- Phase II: At 100 ms, the body moves vertically up and forward. The upward movement compresses the cervical muscles from below, with the possibility of this movement being transmitted to the head. If the compression movement is transmitted, as a result, at 120 ms, the centre of gravity of the head begins to fall down and causes the head to rotate backwards.
- Phase III: At 150 ms, a forward movement occurs, displacing the cervical to tension and causes the trunk to act on flexion. As a consequence, the lower cervical spine is moved much faster than the upper cervical spine during impact [7]. This displacement can generate rupture forces that are transmitted from the cervicodorsal transition zone (C7-D1) to C6 and to the rest of the cervical vertebrae. Next, the soft tissues of the C7-C6 vertebrae begin to tense, transmitting the load to the rest of the cervical spine until the head is reached.
- Phase IV: At 160 ms the body pushes the base of the neck and the head forward, towards flexion. The upward displacement of the trunk reaches its greatest peak at 200 ms with an amplitude of 9 cm, while the head reaches a maximum rotation backwards of 45° at that moment.

n this event, if the speed of the vehicle goes around 60 kph, the force generated on the human body can reach 60 gforce. [X] Other events that produce a significant percentage of injuries in the human body are those produced by contact sports, such as, Football and Hockey in which, the force when two players collided helmet to helmet can goes around 80 to 100 gforce. The player is provided with a helmet to reduce the impacts received and prevent head injury. Unfortunately, this device does not totally cancel the radial accelerations of the head, so that the cervical area is still exposed to damage, but to a lesser extent when compared to a vehicle crash, but with a greater frequency [8]. There are several studies in which the brain-spinal area is analysed when receiving impacts in different areas of the skull, using, models that simulate the head and neck of a human, such as the Hybrid III 50th percentile, which is one of the dummies most used and recommended for this type of tests. In these tests, the accelerations produced are analysed, and a prediction is made of the injury that may occur depending on the area of the impact and the strength. However, these tests focus on the skull area and leave the cervical area aside [9]. It should be mentioned that the devices used in these tests are usually very large. This is because this test mainly uses the operating principle of the Charpy impact test, which depends on the weight and the radius of a pendulum. In this work the design of a test bench is proposed, simulating impacts in different regions of the head, simulating loads and forces of a propitious size to realise studies inside academic laboratories. So that, the design of a test bench was carried out using a 3-RPS platform with hydraulic cylinders, to support the model of the head to be analysed and a parallel bar element is used to simulate a free fall against an indenter to simulate a sudden impact on the desired region of the test.

2 Materials and Methods

The method used for the design of the test bench is the Methodology of Axiomatic Design, by Morris Asimow, who developed a linear structure of 7 phase chronology for the projects [10]. Before phase 1, Asimow mentions that a necessity should be established for the project. In the case of this work, the main condition of the design is the development of a device to allows a sudden impact by a mechanism that can reach several different positions and angles. Another problem to be solved is that the mechanism should be able to work with payloads to simulate an impact during a collision event that occurs usually either in sports or vehicles. Within this work, only the first 3 phases of this method are covered. These phases are feasibility design, preliminary design and detailed design [11]. For this work, we opted to implement a parallel robot, also called platform manipulators, on impact test since this mechanism can shared the external load through the actuators, so that these manipulators have large load-carrying capacity. The design of a platform 3-RPS was proposed because it is one of the most economical and practical ways to build up a test bench. The use of three hydraulic cylinders are raised in consideration of the forces needed to be applied within these studies and the number of tests that will be carried out. The design of the mechanism is presented and the equations that determine the kinematics of the test bench are considered.

For the analysis, 2 systems of Cartesian coordinates are placed in the base P (x, y, z) and in the mobile platform p (x′, y′, z′). As it is seen the points P_1, P_2, P_3 as well as p_1, p_2, p_3. are within the planes $x - y$ and $x' - y'$ respectively. The origin O of the fixed coordinate system is in the centroid of the element formed by $\Delta P_1P_2P_3$, with the x-axis in the direction of $\overline{OP}_1$. Likewise, the origin O′ of the mobile coordinate system is in the centroid of $\Delta p_1p_2p_3$, with the x′ axis in the direction of $\overline{O'p_1}$. Both elements are equilateral triangles so $\overline{OP}_i = r_1$ and $\overline{Op}_i = r_2$.

- The position vectors of the points P_i and p_i and their respective coordinate systems are obtained by

$$P_1 = [r_1, 0, 0] \quad P_2 = \left[-\frac{1}{2}r_1, \frac{\sqrt{3}}{2}r_1, 0\right] \quad P_3 = \left[-\frac{1}{2}r_1, -\frac{\sqrt{3}}{2}r_1, 0\right]$$

$$p_1 = [r_2, 0, 0] \quad P_2 = \left[-\frac{1}{2}r_2, \frac{\sqrt{3}}{2}r_2, 0\right] \quad P_3 = \left[-\frac{1}{2}r_2, -\frac{\sqrt{3}}{2}r_2, 0\right]$$

- The position vectors q_i, which are the vectors of the points p_i with respect to the fixed coordinate system, are given by

$$q_1 = \begin{bmatrix} a_x + r_2x'_x \\ a_y + r_2x'_y \\ a_z + r_2x'_z \end{bmatrix} \quad q_2 = \begin{bmatrix} a_x - \frac{1}{2}r_2x'_x + \frac{\sqrt{3}}{2}r_2y'_x \\ a_y - \frac{1}{2}r_2x'_y + \frac{\sqrt{3}}{2}r_2y'_y \\ a_z - \frac{1}{2}r_2x'_z + \frac{\sqrt{3}}{2}r_2y'_z \end{bmatrix} \quad q_3 \begin{bmatrix} a_x - \frac{1}{2}r_2x'_x - \frac{\sqrt{3}}{2}r_2y'_x \\ a_y - \frac{1}{2}r_2x'_y - \frac{\sqrt{3}}{2}r_2y'_y \\ a_z - \frac{1}{2}r_2x'_z - \frac{\sqrt{3}}{2}r_2y'_z \end{bmatrix}$$

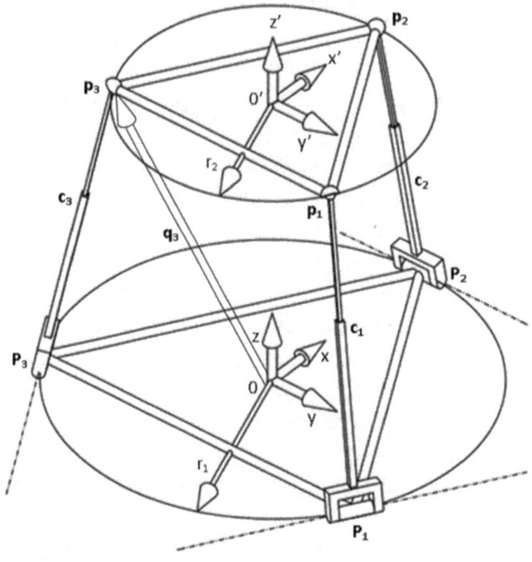

Fig. 2. Diagram of the 3-RPS platform.

- The length of cylinders c_i, shown in Fig. 2 is given by

$$c_i^2 = [q_i - P_i]^T[q_i - P_i] \text{ When } i = 1, 2, 3.$$

- By replacing the terms in the equation:

$$c_1^2 = a_x^2 + a_y^2 + a_z^2 + 2r_1\left(a_xx'_x + a_yx'_y + a_zx'_z\right) - 2r_2a_x - 2r_2r_1x'_x + r_2^2 + r_1^2$$

$$
\begin{aligned}
c_2^2 = {} & a_x^2 + a_y^2 + a_z^2 - r_1\left(a_x x_x' + a_y x_y' + a_z x_z'\right) \\
& + \sqrt{3} r_1\left(a_x y_x' + a_y y_y' + a_z y_z'\right) + r_2\left(a_x - \sqrt{3} a_y\right) \\
& - \frac{1}{2} r_2 r_1\left(x_x' - \sqrt{3} x_y'\right) + \frac{1}{2} r_2 r_1\left(\sqrt{3} y_x' - 3 y_y'\right) + r_2^2 + r_1^2
\end{aligned}
$$

$$
\begin{aligned}
c_3^2 = {} & a_x^2 + a_y^2 + a_z^2 - r_1\left(a_x x_x' + a_y x_y' + a_z x_z'\right) \\
& - \sqrt{3} r_1\left(a_x y_x' + a_y y_y' + a_z y_z'\right) + r_2\left(a_x + \sqrt{3} a_y\right) \\
& - \frac{1}{2} r_2 r_1\left(x_x' + \sqrt{3} x_y'\right) - \frac{1}{2} r_2 r_1\left(\sqrt{3} y_x' + 3 y_y'\right) + r_2^2 + r_1^2
\end{aligned}
$$

The materials used to build up the test bench are divided into two parts, the first one is concerning to the mechanism that supports the head, and the second one include all the elements associated with impact tests. The mechanism for the head support shown in Fig. 3, basically consists of a model of the platform, which is positioned by three hydraulic pistons which allow the movements of the system to perform the impact tests in different regions of the skull, simulating various events that are related to impacts on the head area.

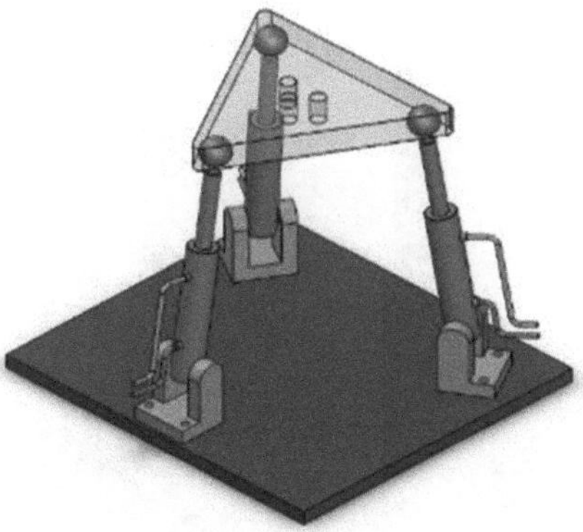

Fig. 3. Head support mechanism.

The elements used for the head support mechanism listed below:

- **Base**
  The base is designed as a plate with dimensions of 50 cm × 50 cm × 2 cm. It has 3 pairs of holes, located in a triangular configuration. The distance between the holes is 6 cm, in each of the pairs.

- **Support**
  The supports consist of metallic parts with an "L" shaped profile, in the elongated segment it has a bolt where the hydraulic piston will be placed, and in its base, it has 2 holes to fix it to the base.
- **Hydraulic cylinder**
  The hydraulic piston selected was a double-acting cylinder of the 700-series model 05 of the company CICROSA® modifying the stem to allow the union with a spherical element.
- **Spherical union**
  A metal sphere with a hole towards its centre, where the rod end of the hydraulic piston is placed.
- **Mobile Platform**
  The platform is designed with a triangular shape, and spherical holes in each of its edges, where the spherical joints are placed, likewise, it has three holes in the central part of the element, to make the fixation of the model of the head on this.

The design of the impact mechanism is a simple one, as seen in Fig. 4, it consists of 2 parallel cylindrical bars and 1 rectangular with cylinders at the ends, where the bars will be inserted to allow vertical movement.

Fig. 4. Impact mechanism.

The elements required for the impact mechanism are a follows:

- **Base**
  The base was designed as a 100 cm × 50 cm × 2 cm plate. It has 4 holes close to the edges of the plate for the placement of legs type skates, to keep its stability when placed on some surface. Likewise, it has one pair of six holes placed in a circular shape, with the centre of these along the longitudinal axis for fixing the supports.

- **Support**
  The supports for bars are circular plates with a configuration of 6 holes in a circular shape. In the central part, a bracket will act as a shirt to hold the parallel bars.
- **Parallel bars**
  Two parallel metallic steel bars are fixed as a support and are inserted inside the tubular part of the guide bar of the indenter, having a length of 4 m.
- **Indenter guide bar**
  This element consists of two parallel circular tubes, joined by a rectangular profile that has a guide where bolts will be inserted to fix the mechanism of the indenter. It has a notch locating the centre of the element.
- **Indenter support**
  It is based on a metallic C profile with 2 holes in the central part of the element and on its longitudinal axis. It has a hole in one of its projections to fix the indenter.
- **Support fixer**
  Steel plate with two circular holes that are coincident with those which support the indenter, likewise, it has a notch showing the centre of the element and rectangular slot that functions to align with the guide bar.
- **Indenter**
  It consists of a steel element in the shape of a bullet, with a hole in its flat face to make its assembly with the support of the indenter.

The last four elements are the ones that realize the impact to the head, these elements together have a mass of 30 kg. This data will help to determine the force will impact the head model.

3 Results

The use of a 3-RPS mobile platform for the execution of experimental tests allows the positioning of the cervical-cranial model under conditions that allow the configuration of sudden impact tests on the frontal and parietal areas of the skull, which are affected during impacts produced in events/accidents such as those that occurred in sports and vehicular accidents, allowing the analysis to estimate injuries [10]. Among the main study cases that could be analysed with this device, are those produced by a frontal, lateral or posterior impact.

- Case 1
  The first proposed case of study is a frontal skull impact, which is very frequent situation in sports such as Football, Hockey and Rugby, for this case the model was placed in the position shown in Fig. 5, with cylinders 2 and 3 without displacement, allowing the stroke by free fall over the frontal area of the head.
- Case 2
  The next case of study consists of the simulation of the whiplash effect, presented during a rear vehicle crash or when a very large deceleration occurs in a very short period. The whiplash is simulated by an impact on the back area of the head, for this case, the cylinder 1 is set without displacement, allowing the model to receive the hit of the indenter on the back of the head.

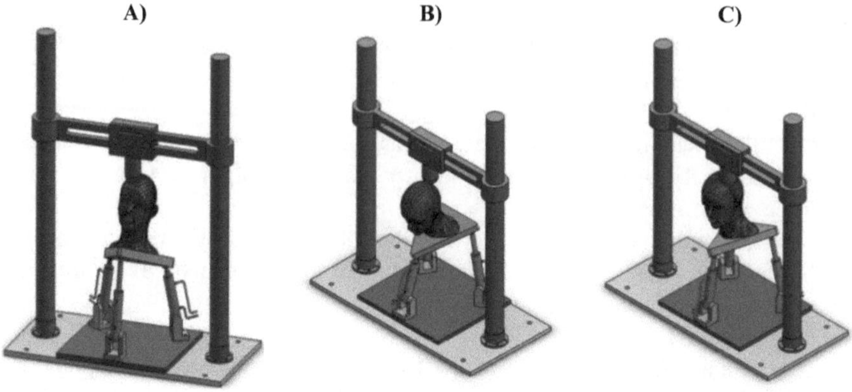

Fig. 5. Positioning of the 3 cases of impact. (A) Case 1. (B) Case 2. (C) Case 3.

- Case 3
  The last case of study is simulated by the fall of a driver of a 2 wheels vehicle, such as a motorcycle or a bicycle. For this analysis, strokes are made on the lateral zones of the head, acting out to hit on the head when the person falls over the asphalt or pavement. For this case, one of the cylinders 2 or 3 would be in its largest position, while the other is set without displacement and the last one, cylinder 1, would be at the midpoint of the stroke of its shank. [11].

In Fig. 5 we can see the mechanism in its different positions for the simulation of the 3 cases mentioned above.

The software application used to analyse the movement was SolidWorks™, which give the data about the amplitude angle of rotation made by the head, which in this case is 30°.

Finally, it can be say that the test bench is designed to allow the placement of a helmet at the head of the model to perform reliable studies for the first and third cases mentioned above (Fig. 6).

Considering the mechanical properties of the mechanism, with a mass of 30 kg, dropped from a height of 4 meters and the hypothesis of a free fall without restraints, the force received by the head reach nearly 20 g.

4 Discussion

This proposal focuses on an interesting topic opening new opportunities to conduct studies and experimental evaluations for impact tests on the head in automotive crashes or sports accidents. It is proposing innovative concepts to solve some practical issues related to the modelling of impacts that can damage the integrity of the elements of the cranioencephalic area. The use of hydraulic systems in the form of a parallel manipulator may offer several advantages for this type of tests.

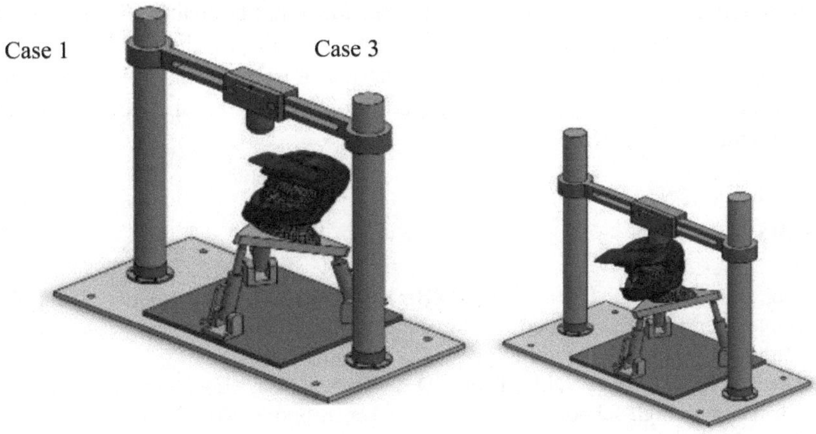

Fig. 6. Positioning of the simulator for cases 1 and 3 with a helmet.

It has been stated that hydraulic cylinders would provide a better recreation of damping that is generated within the human body providing better results than those using neumatic cylinders.

The fall of the indenter mechanism from a set height, should generate the same force in all the analyses, so the option of using a lubricant that reduces friction between the mechanism and the parallel bars should help the descent and provide a repetitiveness of the tests in a more similar situation to the real one.

5 Conclusion

The test bench shown in this paper, permit us the possibility of performing experimental analyses of impact tests on the head, allowing the mobility of the head and modifying the angle at which the impact is received. This open the possibility of conducting studies on automotive accidents to improve the passive safety of the vehicles, as well as, safety studies in sports.

This device allows the study of different cases of injury within the same mechanism, in comparison with several devices where the analyses are limited to a single position, as it is mentioned by Nightingale the gadget used for the free fall of the cranial model to the ground, where only one impact is analysed over the area of the top head [12]. In several cases of study in sports practice, the analyses are limited to the frontal area and the parietal area for specific spots. On the other hand, in the safety tests performed on helmets, the areas of analysis are more diverse, managing to stroke 4 regions of a helmet. However, these tests are limited to obtaining only helmets damage, since the model where it is mounted does not have elements to perform an analysis of head or neck damage.

The force given by the mechanism, in comparison with the ones reached during a traffic accident or an impact in sports is too low. In order to manage to realize impact test more reliable, the mass of the object in free fall must be nearly 200 kg, or in the other hand, increase the acceleration to the mechanism during the fall.

It is proposed for future work the determination of one automation method for the use of hydraulic cylinders, as well as, the addition of sensors and accelerometers to collect the data obtained from the tests.

References

1. Fiorillo, D.P., Demonti, D.H., Giuria, D.H., Aparicio, D.J., Della Vedova, D.F.: Trauma Deportivo Cervical Catastrófico en un Jugador de Rugby, Revista de la Asociación Argentina de Traumatología del Deporte, vol. 21, no 1 (2014)
2. Guskiewicz, K.M., Weaver, N.L., Padua, D.A., Garrett, W.E.: Epidemiology of concussion in collegiate and high school football players. Am. J. Sports Med. **28**(5), 643–650 (2000)
3. Ono, K., Kanno, M.: Influences of the physical parameters on the risk to neck injuries in low impact speed rear-end collisions. In: International Conference on the Biomechanics of Impacts. IRCOBI, Eindhoven (1993)
4. Johansson, M.S., Boyle, E., Hartvigsen, J., Jensen Stochkendahl, M., Carroll, L., Cassidy, J. D.: A population-based, incidence cohort study of mid-back pain after traffic collisions: factors associated with global recovery. Eur. J. Pain **19**, 1486–1495 (2015)
5. de las Peñas, C.F., Carnero, J.F., del Cerro, L.P.: Biomecanica del syndrome de latigazo cervical y su analogía osteopathic. Fisioterapia, 4ta Edicion, España, vol. 46, pp. 211–219 (2004)
6. McConnell, W.E., Howard, R.P., Guzman, H.M., Bomar, J.B., Raddin, J.H., Benedict, J.V. et al.: Analysis of human test subject kinematic responses to low-velocity rear end impacts. In: Proceedings of the 37th Stapp Car Crash Conference, San Antonio, Texas, pp. 21–30 (1993)
7. Luan, F., et al.: Qualitative analysis of neck kinematics during low-speed rear-end impact. Clin. Biomech. **15**, 649–657 (2000)
8. Guskiewicz, K.M., Weaver, N.L., Padua, D.A., Garrett, W.E.: Epidemiology of concussion in collegiate and high school football players. Am. J. Sports Med. **28**, 643–650 (2000)
9. Post, A., Hoshizaki, T.B., Gilchrist, M.D., Brien, S., Cusimano, M.D., Marshall, S.: The Influence of acceleration loading curve characteristics on traumatic brain injury. J. Biomech. **47**, 1074–1081 (2014)
10. Mills, N.J., Gilchrist, A.: Oblique impact testing of bicycle helmets. Int. J. Impact Eng. **35**, 1075–1086 (2008)
11. DeMarco, A.L., Chimich, D.D., Gardiner, J.C., Siegmund, G.P.: The impact response of traditional and BMX - style bicycle helmets at different impact severities. Accid. Anal. Prev. **92**, 175–183 (2016)
12. Nightingale, R.W., McElhaney, J.H., Richardson, W.J., Myers, B.S.: Dynamic response of the head and cervical spine to axial impact loading. J. Biomech. **29**(3), 307–318 (1996)

Nonlinear Analysis of Human Ankle Dynamics

Daniela Tarnita[1(✉)], Marius Georgescu[1], Ionut Geonea[1], Alin Petcu[1], and Danut-Nicolae Tarnita[2]

[1] University of Craiova, Craiova, Romania
tarnita.daniela@gmail.com, georgescumarius@live.com, igeonea@yahoo.com, petcu.alin.i@yahoo.com
[2] University of Medicine and Pharmacy, Craiova, Romania
dan_tarnita@yahoo.com

Abstract. In this paper finite-time Lyapunov exponents were estimated in order to quantify the local dynamic stability, based on the experimental time series of the flexion-extension and inversion-eversion angles of ankle joints, obtained from a group of five subjects with normal left ankles and right ankles suffering by repeated sprains with residual laxities walking over-ground and on plane and inclined treadmill with different speeds and inclinations.

Keywords: Human ankle · Nonlinear dynamics · Lyapunov exponents Plane and inclined treadmill

1 Introduction

In recent years, robotics researchers are increasingly interested in being inspired by biological systems and their self-stabilizing and dynamic equilibrium seen in nature in order to design robots that require less energy to maintain stability. Many papers investigate the natural behaviors of human subjects, rehabilitation devices for passive or active movements [1–7]. Numerous studies were made on robots [8–12], robotic structures and modules used for medical applications, especially rehabilitation and minimally invasive surgery [11–13] and other bio-inspired systems addressing complex problems such as modelling, simulation, kinematics, noise sensing, and robot-environment interactions leading to chaos. Chaos analysis has been used to study the dynamical behavior of systems that exhibit sensitive dependence to initial conditions such as: chaotic dynamics in bipedal locomotion [13–15], chaotic mobile robots and chaotic behavior of Robot-environment interaction [16, 17]. Human gait is one of the most widely used as biologically inspired type of locomotion for designing humanoid robots, or in medicine to design advanced prosthetic limbs. The use of nonlinear tools provides supplementary information about the nature of the step-to-step variability present in human and robotic movements [18]. The important data that contains movement destabilization are those collected especially in the frontal plane, perpendicular to the sagittal plane. In this study we apply a nonlinear analysis approach and Lyapunov exponents were estimated in order to quantify the local dynamic stability of human ankles joints to infinitesimally small perturbations during walking, based on the

G. Carbone et al. (Eds.): *New Trends in Medical and Service Robotics*, MMS 65, pp. 235–243, 2019.
https://doi.org/10.1007/978-3-030-00329-6_27

experimental time series collected for flexion-extension in sagittal plane and inversion-eversion for both ankles of each subject.

2 Experimental Protocol

The data acquisition system used is the Biometrics system, commonly used for experimental research in biomechanics, bioengineering, gait analysis, clinical rehabilitation, neurology, sports science [2, 5, 7, 18–20]. Subjects were equipped with 6 twin-axis electro-goniometer sensors: four SG 150 (for hips and knees) and two SG 110A (for ankles) (Fig. 1a), mounted in accordance with (Fig. 1b). Each of them has two separate output connectors, which simultaneously measure angles in up to two planes of movement, one measures flexion/extension in the sagittal plane, and the other measures the angle in frontal plane. In this study, the data series collected from the left and right ankles (both movements: flexion-extension and inversion-eversion) of the subjects present interest in order to quantify the local dynamic stability of this joint. The tests were performed in the Biomechanics Laboratory of the University of Craiova by a sample of five subjects with normal left ankle and with right ankle suffering by repeated sprains with residual laxities. Each subject executed ten walking tests (Table 1), on ground (OG) and on treadmill (TM) at different speeds and inclines, each test being performed during 1 min.

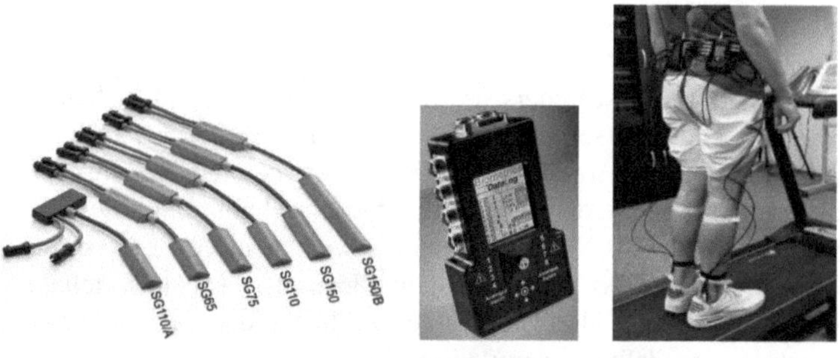

Fig. 1. (a) Electro-goniometers and DataLog device; (b) The acquisition system mounted on the subject during experimental test

The data collected by the 6 goniometers connected with two 8-channel DataLOG devices is stored in this devices and it can be transmitted in real time to a PC, processed and analyzed using Biometrics software (Fig. 2)

A total number of 600 synchronized time-series for the six joints of both legs of subjects and a number of 200 time-series of experimental data of consecutives walking cycles of ankle joints were acquired. at a frequency of 500 Hz. The subjects gave their written consent for performing the tests. In Table 2 the average values of anthropometric data of sample collected before tests are shown.

Table 1. The main characteristics of the ten walking tests.

Surface	Test name	Speed [Km/h]	Incline [deg]
TM	Test 1	2,5	0
TM	Test 2	5	0
TM	Test 3	7,5	0
TM	Test 4	2,5	7
TM	Test 5	5	7
TM	Test 6	7,5	7
TM	Test 7	2,5	11
TM	Test 8	5	11
TM	Test 9	7,5	11
OG	Test 10	2	0

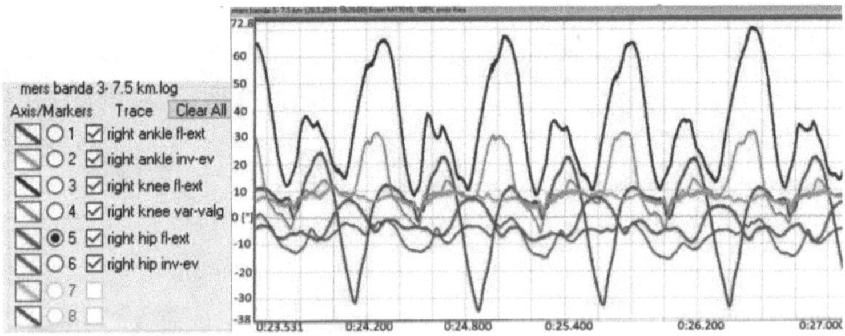

Fig. 2. The angular diagrams of right leg joints, obtained by Biometrics: flexion-extension angle for knee, hip and ankle; rotation angle in frontal plane for knee, hip and ankle

Table 2. Average values of anthropometric data of subjects sample.

Age [years]	Weight [kg]	Height [cm]	Leg length [cm]	Hip–knee length [cm]	Knee–ankle length [cm]
24.6	64.6	173.8	79.6	39.4	40.2

3 Nonlinear Dynamic Analysis

The behavior of any dynamical physical system is fully described by giving its position and momentum at each time, for every degree of freedom of the system. Standard plots in which the ankle joint angular positions are plotted function of time do not provide sufficient information about the dynamics of the system. Phase plane portraits are used to characterize the kinematics of the system when it attained this equilibrium. Using phase plane portraits one can correlate the joint rotations with the respective joint velocities. In Fig. 3, 2D Phase-plane plots for all tests performed by Subject 3 on TM for flexion-extension (a) and inversion-eversion (b) are shown. For all subjects, similar curves were obtained. The measured time-series are not sufficient to describe the dynamical system of the human ankle joints, because the flexion-extension and inversion-eversion angles are just some of the components that characterizes a real state of the system.

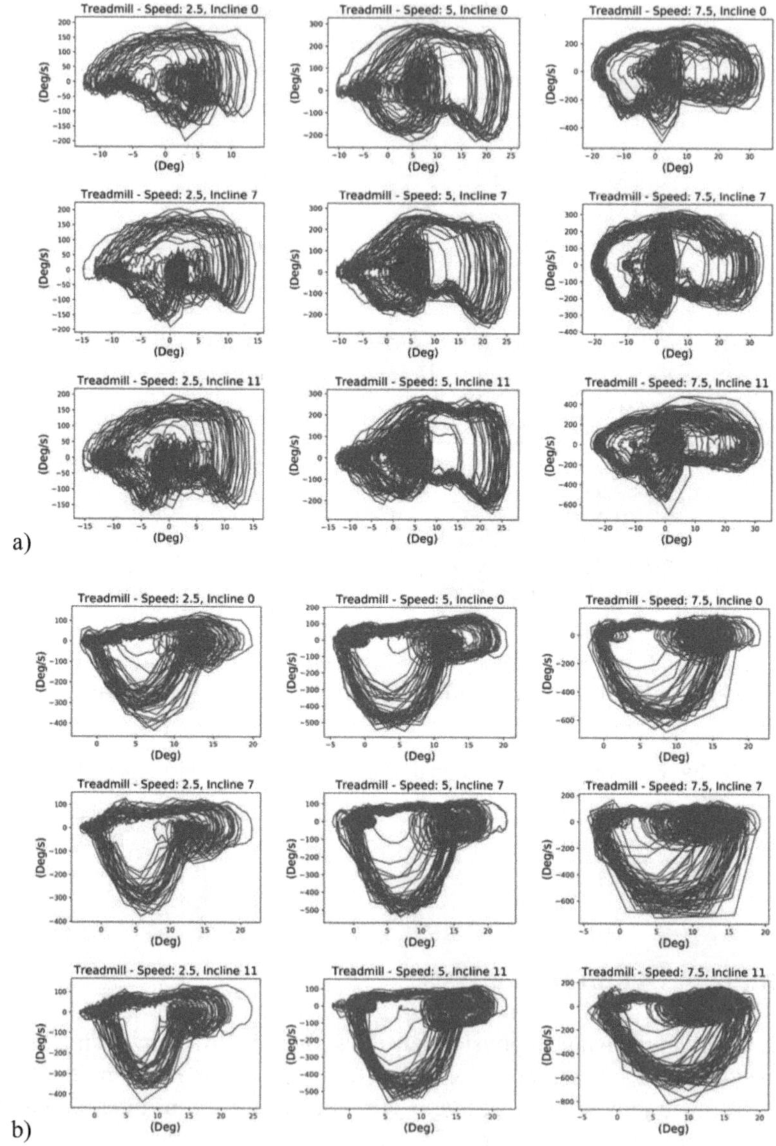

Fig. 3. 2D Phase-plane plots for all tests performed by Subject 3 on TM; (a) flexion-extension; (b) inversion-eversion

To reconstruct this space from the available data, we used the "Method of delays" [21] and TISEAN software. The appropriate state spaces for each time series were reconstructed. Vectors formed from time delayed values of a single time series are used to form a trajectory in the new space [21].

$$x_n = \{s(t_0 + nT_s),\ s(t_0 + nT_s + T), \ldots,\ s(t_0 + nT_s + (m-1)T)\} \quad (1)$$

The integer m is the embedding dimension. In (Eq. 1), s(.) is a measured scalar function and T_s is the sampling time, $n = 1, 2, \ldots, m$; and $T = dT_s$ an appropriately chosen time delay. It was demonstrated [21] that reconstructed state space is guaranteed to be topologically identical to the original state space and many of its important proprieties, such as the Lyapunov exponents are invariables [21]. The results were obtained as plots of the first minima of the average mutual information function (AMI) computed for the ankles measured time-series by using TISEAN Software. In Fig. 4 the AMI plots for Subject 3 on Test 5 are shown. This method was proposed in [22] and states that a good choice for T is the one that, given X(t), provides new information with measurement X(t + T). After choosing a value for the time-delay, the next step is to estimate the embedding dimension. This is the dimension of the reconstructed state space where the structure of the dynamical system is embedded and it represents the minimum number of variables required to form a valid state from a given time-series [22]. The method used for determining the minimal sufficient embedding dimension, (m), is False Nearest Neighbor (FNN) method [23]. The main idea of FNN to chose embedding dimension is when the fraction of FNN is close to zero. In (Fig. 5), plots of the fraction of FNN against the embedded dimension, computed for left ankle time-series of Subject 3 measured on Test 5 are presented. The m values for all time series were obtained.

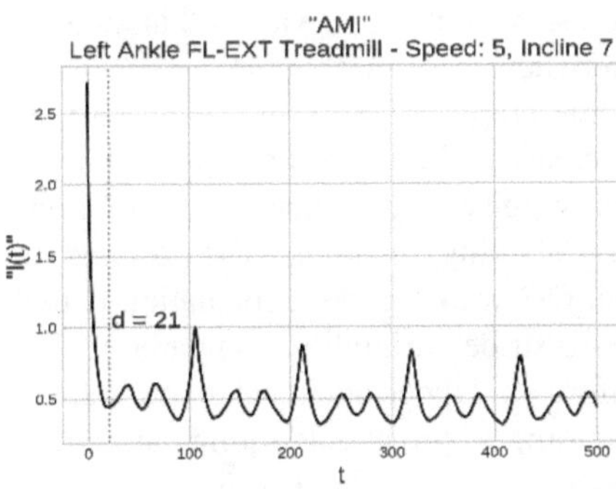

Fig. 4. Plots of AMI of left ankle Subj 3/Test 5

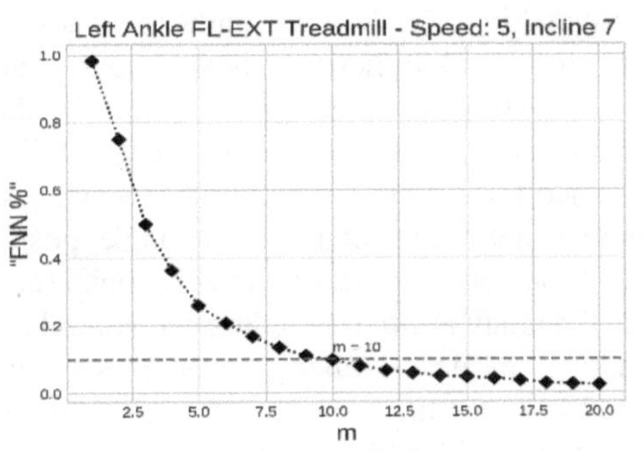

Fig. 5. Plots of FNN- left ankle of Subj 3/Test 5

The short-term Lyapunov exponent (λs), a measure of the local dynamic stability [24] is used to quantify gait stability [24]. The λs exponent quantifies the average logarithmic rate of divergence of the human ankle dynamical system after a small perturbation over a period of 0.5 or 1 stride. An increase in λs implies a decrease of local dynamic stability. The results were determined using the Rosenstein algorithm [25], 'lyap_r' routine available in TISEAN package. In Fig. 6 plots of the average logarithmic divergence for all tests corresponding to the right ankle of Subj 2 and plots corresponding to left ankle of all subjects on Test 3 are shown.

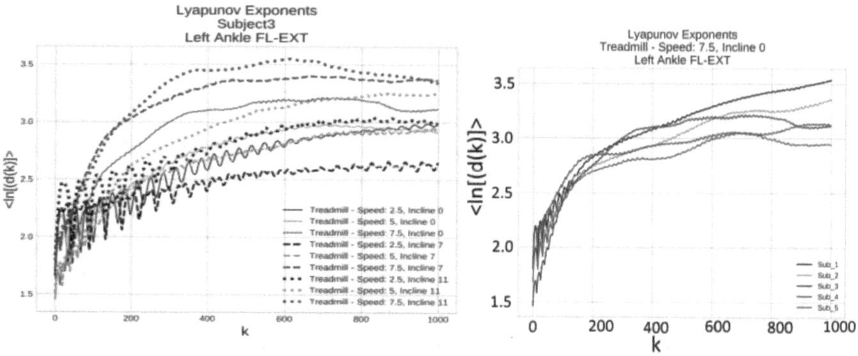

Fig. 6. Representative plots of the average logarithmic divergence < [ln[d(k)]]: (a) for left ankle of Subject 3 – all tests; (b) for left ankle of all subjects on Test 3.

4 Results

The averages of all λs exponents obtained for each test for both ankles of the entire sample on each test for both types of movements: flexion-extension and inversion-eversion are presented in (Table 3), while plots of average values of λs exponents are presented in (Fig. 7). The values of λs for each test were positive values that suggest that human ankle motions show chaotic characteristics.

Analyzing λs values, we see that for the same slope, as the speed increases, λs exponents increase so, the movement stability decreases with increasing speed. The λs obtained for each test were positive values, which suggests that human ankles motions show chaotic characteristics of angle positions. Walking on plane TM resulted in reductions of λs for most subjects compared with OG walking. An explanation is that the TMs may artificially reduce the natural stride-to-stride variability, compared to OG walking, because walking speed is strictly enforced. Under identical experimental conditions, for the same test, the values corresponding to the affected joints are larger than those of healthy joints. So, the subjects exhibit greater stability in the healthy ankles than in the affected joints, associated with pain and instabilities, with a decrease of local stability and increase of ankle movements variability, which confirms clinical investigations. These results confirm previous findings of increased stability during treadmill locomotion [24, 26].

Table 3. Average values of λs for both ankles of the entire sample for each test.

Surface	Speed	Incline	Flexion-	extension	Inversion-	eversion
			Average λs left ankle	Average λs right ankle	Average λs left ankle	Average λs right ankle
OG	2.0 km/h	0°	0.2493	0.4555	0.2573	0.2844
TM	2.5 km/h	0°	0.1559	0.2000	0.1336	0.1576
		7°	0.0467	0.1691	0.1256	0.1448
		11°	0.0479	0.2359	0.1788	0.1488
	5 km/h	0°	0.2804	0.4557	0.1638	0.1844
		7°	0.2610	0.4397	0.1229	0.2072
		11°	0.3961	0.4102	0.1543	0.2079
	7.5 km/h	0°	0.5676	0.5985	0.1681	0.2247
		7°	0.5323	0.5856	0.1425	0.2325
		11°	0.5877	0.6223	0.1640	0.1999

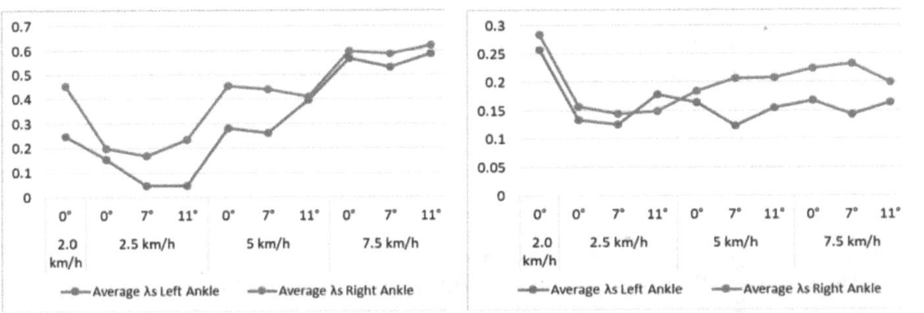

Fig. 7. Plot of average values of the Lyapunov exponents obtained for both ankles of the sample on each test: (a) flexion-extension; (b) inversion-eversion

5 Conclusions

Traditional measures of variability are insufficient to characterize the local dynamic stability properties of locomotor behavior. In this study, the exponents represent measures of the exponential divergence from trajectories in the state space domain reconstructed from ankle motions and quantified the human ankle joint movement behavior such as stability during gait. Because the values of λs quantify the sensitivity of the locomotor system to local perturbations, they are more relevant to study the neuromuscular control of balance and stability during locomotion. Larger values of λs obtained for affected ankles reflect a decrease of local stability and increase of variability, while smaller values obtained for healthy ankles reflect a local stability.

References

1. Ottaviano, E., Ceccarelli. M., et al.: An experimental evaluation of human walking. In: 3rd International Congress Design and Modelling of Mechanical Systems, CMSM (2009)
2. Tarnita, D., et al.: Experimental characterization of human walking on stairs applied to humanoid dynamics. In: Advances in Robot Design and Intelligent Control, pp. 293–301. Springer (2016)
3. Berceanu, C., et al.: About an experimental approach used to determine the kinematics of the human finger. J. Solid State Phenom. **166–167**, 45–50 (2010)
4. Ottaviano, E., et al.: An application of CaTraSys, a cable-based parallel measuring system for an experimental characterization of human walking. Robotica **28**(1), 119–133 (2010)
5. Tarnita, D., Marghitu, D.: Nonlinear dynamics of normal and osteoarthritic human knee. In: Proceedings of the Romanian Academy **18**(4), pp. 353–360 (2017)
6. Mândru, D., Rusu, C., Teuţan, E.: Robotic systems for rehabilitation. In: 11th International Research/Expert Conference "Trends in the Development of Machinery and Associated Technology", TMT 2007, Hammamet, Tunisia, pp. 1495–1498 (2007)
7. Tarnita, D., et al.: Contributions on the dynamic simulation of the virtual model of the human knee joint. Materialwissenschaft und Werkstofftechnik Mater. Sci. Eng. Technol. **40** (1–2), 73–81 (2009). Special Edition Biomaterials, Willey-Vch.
8. Osuka, K., Kirihara, K.: Analysis and experiments of passive walking robot QUARTET II. In: IEEE International Conference on Robotics and Automation (2000)
9. Geonea, I., Tarnita, D.: Design and evaluation of a new exoskeleton for gait rehabilitation. Mech. Sci. **8**(2), 307–322 (2017)
10. Carbone, G., Ceccarelli, M.: A low-cost easy-operation hexapod walking machine. Int. J. Adv. Robot. Syst. **5**(2), 21 (2008)
11. Pisla, D., Plitea, N., Vaida, C.: Kinematic modeling and workspace generation for a new parallel robot used in minimally invasive surgery. In: Advances in Robot Kinematics: Analysis and Design, pp. 459–468 (2008)
12. Vaida, C., Plitea, N., Pisla, D., Gherman, B.: Orientation module for surgical instruments-a systematical approach. Meccanica **48**, 145–158 (2013)
13. Dumitru, N., et al.: Dynamic analysis of an exoskeleton new ankle joint mechanism. In: New Trends in Mechanism and Machine Science, vol. 24, pp. 709–717. Springer (2015)
14. Goswami, A., Espiau, B., Keramane, A.: Limit cycles and their stability in a passive bipedal gait. In: Proceedings of the IEEE International Conference on Robotics and Automation (1996)
15. Goswami, A., Thuilot, B., Espiau, B.: A study of the passive gait of a compass-like biped robot: symmetry and Chaos. Int. J. Robot. Res. **7**, 1282–1301 (1998)
16. Nakamura, Y., Sekiguchi, A.: The chaotic mobile robot. IEEE Trans. Robot. Autom. **17**, 898–904 (2001)
17. Nehmzow, U.: Scientific Methods in Mobile Robotics: Quantitative Analysis of Agent Behavior. Springer (2006)
18. Tarnita, D.: Wearable sensors used for human gait analysis. Rom. J. Morphol. Embryol. **57** (2), 373–382 (2016)
19. Tarnita, D., et al.: Application of nonlinear dynamics to gait analysis on plane & inclined treadmill. In: New Trends Medical and Service Robots, vol. 39, pp. 59–73. Springer (2016)
20. Tarnită, D., et al.: Numerical simulations and experimental human gait analysis using wearable sensors. In: New Trends in Medical and Service Robots, MESROB 2016. Mechanism and Machine Science, vol. **48**, pp. 289–304. Springer, Cham (2018)

21. Takens, F.: Detecting strange attractors in fluid turbulence. In: Dynamical System Turbulence, pp. 366–381. Springer, Berlin (1981)
22. Fraser, A.M., Swinney, H.L.: Independent coordinates for strange attractors from mutual information. Phys. Rev. A **33**, 1134–1140 (1986)
23. Kennel, M.B., Brown, R., Abarbanel, H.D.I.: Determining minimum embedding dimension using a geometrical construction. Phys. Rev. A **45**, 3403–3411 (1992)
24. Dingwell, J.B., et al.: Local dynamic stability versus kinematic variability of continuous overground and treadmill walking. J. Biomech. Eng. Trans. ASME **123**, 27–32 (2001)
25. Rosenstein, M.T., Collins, I.J., Deluca, C.J.: A practical method for calculating largest Lyapunov exponents from small data sets. Phys. D **65**, 117–134 (1993)
26. England, S.A., Granata, K.P.: The influence of gait speed on local dynamic stability of walking. Gait Posture **25**(2), 172–178 (2007)

Experimental Characterizations of Operation of Exoskeleton-Assisted Wheelchair

Gao Huang[1,2,3](✉), Marco Cecarelli[1,2], Weimin Zhang[1,3](✉), and Qiang Huang[1,3]

[1] Beijing Advanced Innovation Center for Intelligent Robots and Systems, Beijing Institute of Technology, Beijing, China
huanggao@bit.edu.cn, ceccarelli@unicas.it
[2] LARM: Laboratory of Robotics and Mechatronics, University of Cassino and South Latium, Cassino, Italy
[3] Intelligent Robot Institute, School of Mechatronical Engineering, Beijing Institute of Technology, Beijing, China

Abstract. In this paper, the operation of a new exoskeleton-assisted wheelchair is characterized by experimental results. The new wheelchair works combining the action of user's legs, exoskeletons and wheelchair motors in three different operation modes. Test results are discussed to show the feasibility and efficiency of the designed operation modes.

Keywords: Exoskeleton-assisted wheelchair · Operation
Experimental characterization · EMG value

1 Introduction

The numbers of lower limb disabilities caused by stroke hemiplegia and older people are increasing gradually in today world [1]. They usually suffer from muscle weakness and gait disorders [2], which can produce a severe burden on their families and on societies. Regain the abilities of locomotion and leg exercise by themselves is the burning desire. The medical robot can be one good solution to solve the mentioned situation. However, the effectiveness of the integrated functions of locomotion and leg muscle exercise are not well validated.

She et al. proposed a three-wheel electric cart with a mounted pedal unit to enable leg exercises [3]. Tohoku University (Japan) developed Profhand, a cycling wheelchair for locomotion and leg exercise [4, 5]. They are good solutions for the locomotion and leg muscle exercise. However, the safety issues for the numb legs have not been taken consideration. Due to the lack of assisting device to keep user's legs in a right posture, the legs have the risks to be broken during cycling. In this paper, leg exoskeletons are proposed for the assisting device and experiments are carried out to show the efficiency of the designed operation modes.

G. Carbone et al. (Eds.): *New Trends in Medical and Service Robotics*, MMS 65, pp. 244–252, 2019.
https://doi.org/10.1007/978-3-030-00329-6_28

2 The Built Exoskeleton-Assisted Wheelchair in BIT

Figure 1 shows the prototype built by Intelligent Robot Institute (IRI) at Beijing Institute of Technology (BIT). The dimensions of the unit shown in Fig. 1 are suitable for home use. The operation panel has multi-functional buttons for motion modes, speed adjustments, energy display and emergency stoppage. The different functions allow the wheelchair motion to be adjusted to the specific physical needs of each user. Table 1 shows the main design parameters of the prototype. Particularly, the lengths of leg exoskeletons can be adjusted according to the length range. The prototype is designed with 14 DOFs (3 active, 11 passive). The 3 active DOFs as in Fig. 1, refer to the 2 back wheels enabling the movements of the wheelchair, and the crank running the cycling motion. The 11 passive DOFs refer to 6 lower limb revolve joints, 3 in the front wheels place, and 2 in the pedal axes. More details of the design can be found in [6–8]. The actuators of the exoskeleton are servomotors made by Huisitong Co. Ltd, with output torque 13.2 Nm and supply voltage 24 V.

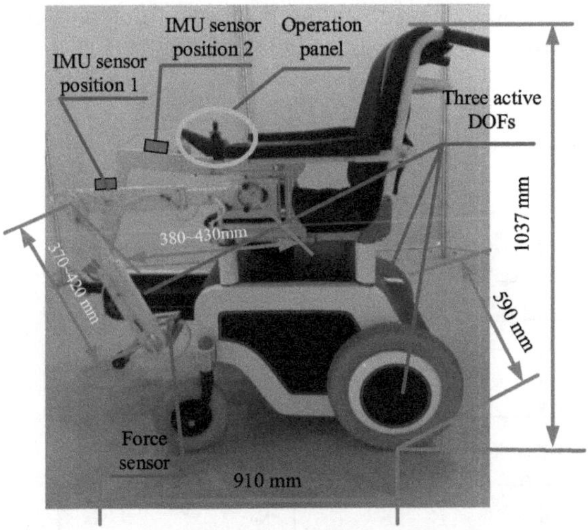

Fig. 1. Built prototype in BIT and IMU positions on the leg thighs

Table 1. Main parameters of built prototype in BIT

Design parameters	Value		
Degree of freedom (DOF)	14 (3 active, 11 passive)		
Size (mm)	910 * 590 * 1037		
Weight (kg)	35		
Motion range of leg (ROM) (degree)	0–30 (hip)	−15–20 (knee)	−30–0 (ankle)

3 Experimental Layout and Operation

To verify the effect of the integration of leg exoskeleton and wheelchair, two experiments have been carried out to characterize the system operation and to evaluate the effectiveness of leg muscles' exercise during the cycling motion. Lab experiments were conducted with six fully-abled volunteers and every experiment is repeated 3 times to get the results with statistical significance. The trajectories in the lab experiments were planned and executed to have passive and active cycling movements of the leg exoskeletons.

Figure 2 shows the experimental recording system for both sensors. The IMU sensors can record the data of angular velocities and angular accelerations during the active cycling. The data will be acquired and elaborated by the control board and sent to the PC to be stored. Before data acquisition, EMG sensors need to be calibrated to adjust the output signals. The activated situation of muscles will be judged according to the value of average voltages, $V_{averages}$ by the data recording program. If $V_{averages} > 100$ mV, the data will be stored by the PC, otherwise the data will be filtered out. The gluteus medius muscles are selected to be evaluated in the experiments. The experiments will examine both walking at 3 m/s and cycling in sitting position at different cycling speeds of 16 r/min, 32 r/min, 48 r/min, 64 r/min, 80 r/min.

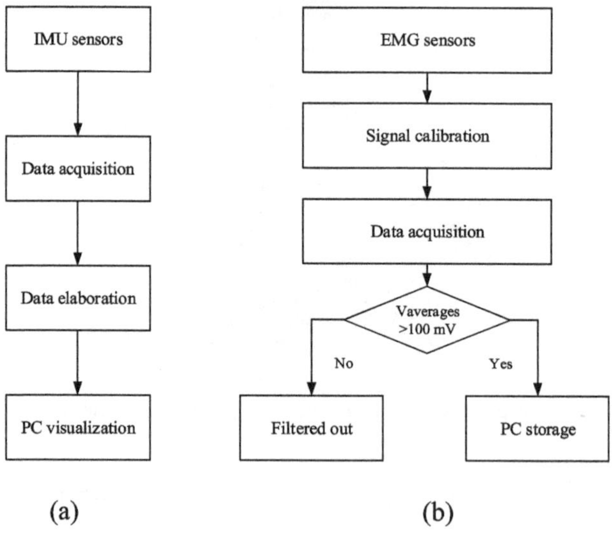

Fig. 2. Experimental recording system: (a) for IMU; (b) for EMG

4 Results

The result plots of both experiments will be shown and analyzed in this part to verify the feasibility of the proposed design.

4.1 IMU Test Results

Figure 3(a) shows acquired measurements of angles of leg exoskeletons during an active mode test in a 20 s experiment. Zoomed view from 10 s to 14 s is selected to analyze the motion, as shown in Fig. 3(b). Benefiting from the symmetrical design of the exoskeleton and crank-rocker mechanism, the angles of leg exoskeletons change

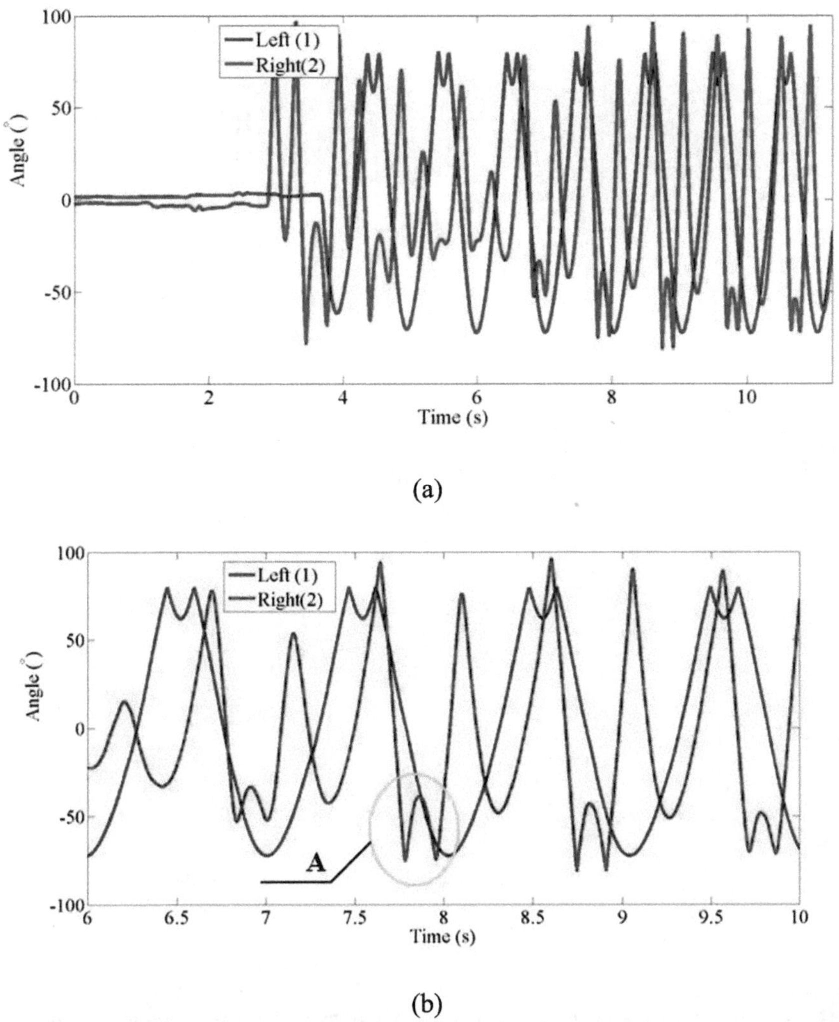

Fig. 3. Acquired measurements of angles of leg exoskeletons during an active mode test: (a) full test; (b) zoomed view

periodically, and the output angles are about 160°, with a range from about −80° to 80°. There is one phase difference between the left and right legs, indicating that the exoskeleton system runs smoothly, while the trajectories are the same. Particularly, in one cycle, a local oscillation occurs when the crank passes through the top dead center, as shown in circle A in Fig. 3(b).

Figure 4 shows acquired measurements of leg exoskeletons' angular accelerations. The acceleration range is from $-0.8\ \mathrm{m/s^2}$–$0.8\ \mathrm{m/s^2}$ and the characteristics of left leg movements are the same with right leg. The plots change periodically over time, and the interval between left and right legs is one half cycle. In summary, the experiment results can verify the correctness of exoskeleton design and cycling movements.

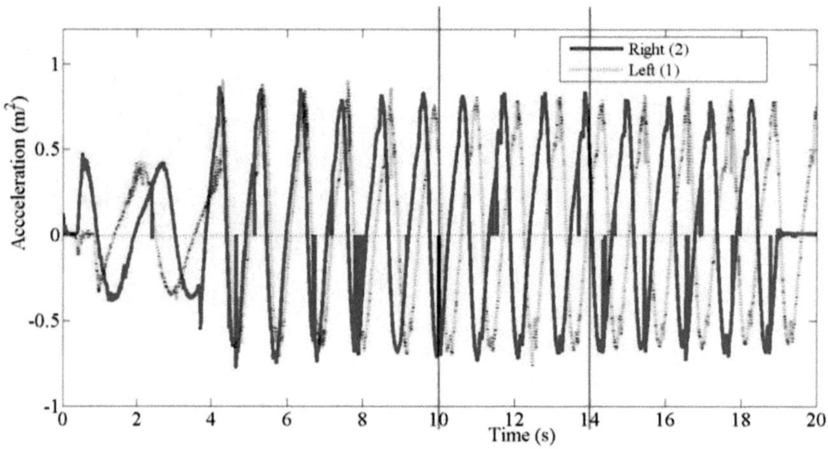

(a)

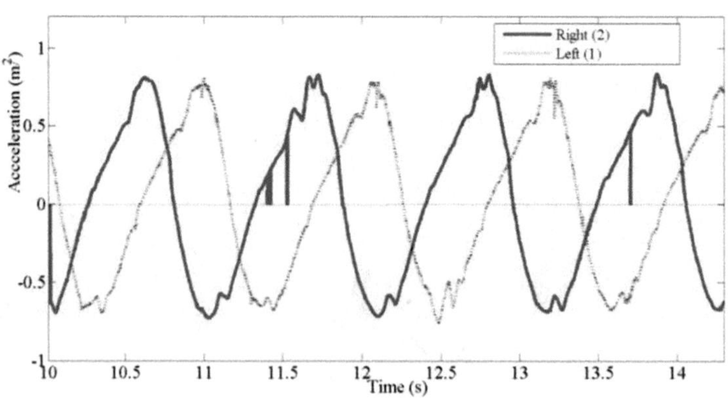

(b)

Fig. 4. Acquired measurements of angular accelerations of leg exoskeletons during an active mode test: (a) full test; (b) zoomed view

4.2 EMG Test Results

To further clarify the effectiveness of the wheelchair robot exercise, the amplitude characteristics of the EMG signals are analyzed, as shown in Fig. 5. A comparison of mean and maximum values for each muscle activation during walking and pedaling in the passive mode (at 16 r/min) shows that the degree of muscle stimulation is equal to or higher than that of the walking process for pedaling with exoskeleton assisting; however, the movements are less active than that of standard walking.

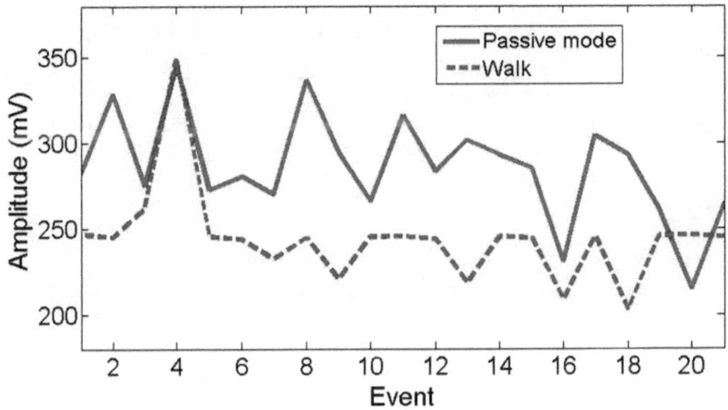

(a)

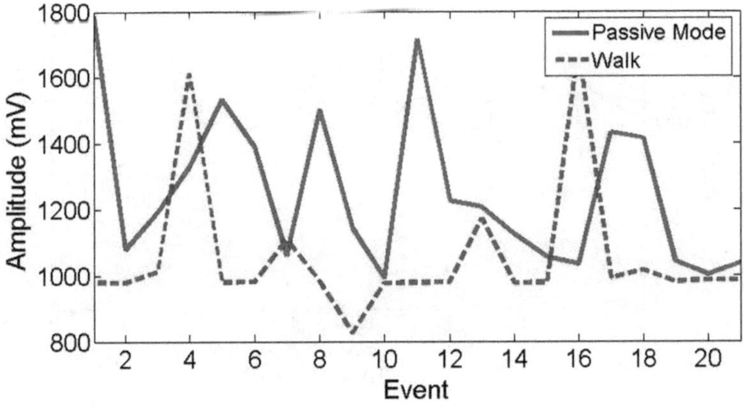

(b)

Fig. 5. Mean and maximum recorded EMG signal values at left gluteus medius muscle location for each active event during walking and pedaling in passive mode at 16 r/min: (a) Mean value at left side; (b) Maximum value at left side.

To determine the exercise effectiveness at different cycling speeds, the mean and maximum EMG signal values at the left gluteus medius muscle locations for passive pedaling at different speeds are recorded, as shown in Fig. 6. Some of the muscle activity mean and maximum values increase as the pedaling speed increases, although the rise is not significant. The frequency of the gluteus medius muscle stimulation verifies the effectiveness of the muscle exercise.

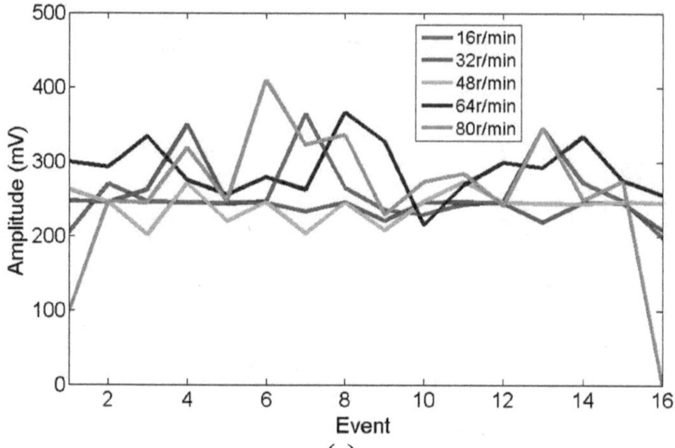

(a)

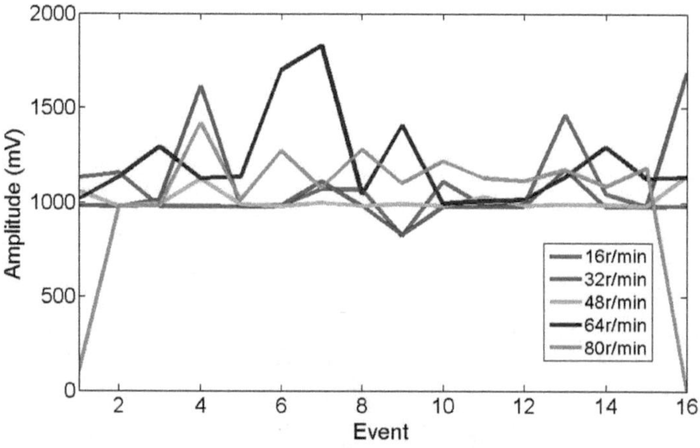

(b)

Fig. 6. Mean and maximum recorded EMG signal values at left gluteus medius muscle locations for passive pedaling at different speeds: (a) Mean value at left side; (b) Maximum value at left side.

Table 2 lists EMG signals recorded at 9 s across the different motion modes and speeds. The quantitative results are discussed and analyzed based on these values. In the walking mode, the mean value is 220 mV, while the minimum value is 240 mV for different speeds in the passive cycling mode. This indicates that passive cycling with proposed exoskeleton assisted wheelchair more effectively exercises user legs than slow walking. From the maximum values, the same result can be deduced. In addition, a comparison of the different passive cycling modes shows that faster speeds do not guarantee more effective muscle exercise. At 16 r/min, the mean value is 280 mV, while at 32 and 80 r/min, the mean values are only 240 mV. For 64 r/min, the highest mean and maximum values are obtained for the current design, it can be seemed as an optimal speed for maximum cycling exercise effectiveness.

Table 2. Recorded EMG signals at 9 s for different motion modes and speeds

Motion modes	Speed	Mean value (mV)	Max value (mV)
Walking mode (m/s)	3	220	950
Passive cycling mode (r/min)	16	280	1150
	32	240	900
	48	230	1000
	64	310	1200
	80	240	1100

5 Conclusions

Locomotion and leg rehabilitation are two key issues for persons with leg problems. In this paper, the effectiveness of the combination of two mentioned motions is verified by the proposed leg exoskeleton-assisted wheelchair that has been built in BIT. The characteristics of operation and leg muscles activation are tested to show the feasibility of the proposed design and operation modes.

Acknowledgments. The first author wished to gratefully acknowledge Beijing Institute of Technology for permitting his period of study at LARM of Cassino University in the A. Y. 2017 to 2018 within a double PhD degree program. The authors kindly acknowledge each volunteers who participate in the experiments.

References

1. Department of Economic and Social Affairs: Population fact, population division (United Nations), 4, pp. 1–4 (2014)
2. Park, B.S., et al.: The effects of aquatic trunk exercise on gait and muscle activity in stroke patients: a randomized controlled pilot study. J. Phys. Ther. Sci. **27**(11), 3549–3553 (2015)
3. She, J.H., Yokota, S., Du, E.Y.: Automatic heart-rate-based selection of pedal load and control system for electric cart. Mechatronics **23**(3), 279–288 (2013)
4. Watanabe, T., Karasawa, Y.: Comparison of muscle stimulation groups for simplified practical FES cycling control with cycling wheelchair: an experimental test with healthy subjects. IEICE Trans. Inf. Syst. **99**(5), 1345–1352 (2016)

5. Watanabe, T., Karasawa, Y., Handa, Y.: A test of controlling different muscles in FES cycling with cycling wheelchair Profhand. In: Proceeding of IEEE 19th International Functional Electrical Stimulation Society Annual Conference (IFESS), Kuala Lumpur, pp. 565–571 (2014)
6. Huang, G., et al.: Design and simulation of leg exoskeleton cycling-actuated wheelchair. Int. J. Adv. Robot. Syst. **14**(6), 1–11 (2017)
7. Huang, G., et al.: Master-slave control of an intention-actuated exoskeletal robot for locomotion and lower extremity rehabilitation. Int. J. Precis. Eng. Manuf. (2018, in print)
8. Huang, G., et al.: Design and construction of a cycling-based wheelchair prototype. In: Proceeding of IEEE International Conference on Cyborg and Bionic Systems (CBS), Beijing, China, pp. 310–313 (2017)

Special Session on Agewell Project

Design of CUBE, a Cable-Driven Device for Upper and Lower Limb Exercising

Daniele Cafolla[1(✉)], Matteo Russo[2], and Giuseppe Carbone[2]

[1] CESTER (Research Centre for Industrial Robot Simulation and Testing), Technical University of Cluj-Napoca, Cluj-Napoca, Romania
contact@danielecafolla.eu

[2] LARM (Laboratory of Robotics and Mechatronics), University of Cassino and Southern Latium, Cassino, Italy
{matteo.russo,giuseppe.carbone}@unicas.it

Abstract. This paper introduces CUBE, a cable-driven parallel robot for the assistance of patients in rehabilitation exercising for both upper and lower limb. The system is characterized by a lightweight foldable structure that is easy to set-up in different configurations. It can adapt to different exercises and to the available environment. Its cable-driven design makes it inherently safe in human/robot interactions also due to the extremely low inertia. The design is presented with its kinematic and dynamic analysis and validated through a first prototype.

Keywords: Cable-driven robots · Medical robots · Parallel robots
Rehabilitation devices · Kinematics

1 Introduction

The exercises for the upper and lower limb rehabilitation after a trauma are usually performed by a patient whose motion is guided by a therapist. However, in the last two decades, several research teams in engineering and robotics suggested the possibility to replace the therapist or clinician with a robot able to assist the patient in the exercise without requiring the constant attention and presence of a therapist, therefore allowing a single clinician to supervise a group of composed of several patients at the same time, improving the efficiency of his actions [1, 2]. Furthermore, rehabilitation robots are usually equipped with sensors that can constantly monitor the state and improvement of each patient. For this reason, several robotic rehabilitation devices are nowadays available, for example, as reported in [3–6].

The main concern when using a rehabilitation robot is safety, but no official standard clearly and fully addresses this issue, while IEC/NP 80601-2-78 is currently under development. Both medical standards and human/robot interaction standards are often adopted for robotic rehabilitation devices, as shown respectively in [7, 8]. The design of an inherently-safe structure presents obvious advantages for safety. Cable-driven parallel robots are systems based on parallel architectures with an end-effector body that is connected to the fixed frame by means of cables only, instead of using rigid links. This kind of mechanisms are characterized by extremely low moving masses,

G. Carbone et al. (Eds.): *New Trends in Medical and Service Robotics*, MMS 65, pp. 255–263, 2019.
https://doi.org/10.1007/978-3-030-00329-6_29

consisting in the mass of the end-effector only, since the mass of the cables is negligible. Thus, the risk of injuries due to errors in path planning and the collision of the robot to surrounding human being is significantly reduced. For this reason, some research groups chose cable-driven architectures for their robotic rehabilitation devices, such as SPIDAR-G [9], GENTLE/s [10], MACARAM [11], NeReBot [12] and MariBot [13].

Several cable-driven rehabilitation devices have been developed and analysed at LARM in Cassino, such as CALOWI [14, 15] and CATRASYS [16]. However, all the existing devices are characterized by a bulky frame that is not meant to be moved. This paper presents an improvement of the existing CALOWI design as CUBE (Cable-driven device for Upper and lower limB Exercising), which is characterized by a lightweight foldable and adaptable structure.

2 Limb Exercising

Physical rehabilitation can be defined as the process of helping patients in gaining control over parts of their body after a long illness or a traumatic event. In particular, upper limbs rehabilitation, often performed after a stroke [17], consists of physical exercises of the injured limb at the shoulder, elbow and wrist level. The most common exercises are based on slow, repetitive circular movements of flexion and extension of both shoulder, as in Fig. 1a, and elbow, as in Fig. 1b. The exercise for the lower limb are similar to the flexion-extension exercises performed for the arm, as shown in Fig. 1c for a knee flexion-extension exercise or in Fig. 1d for leg abduction-adduction. The workspace of a cable-driven robot fits the motion range of the exercises, while the motion of each repetition can be achieved with commercial DC motors with encoders.

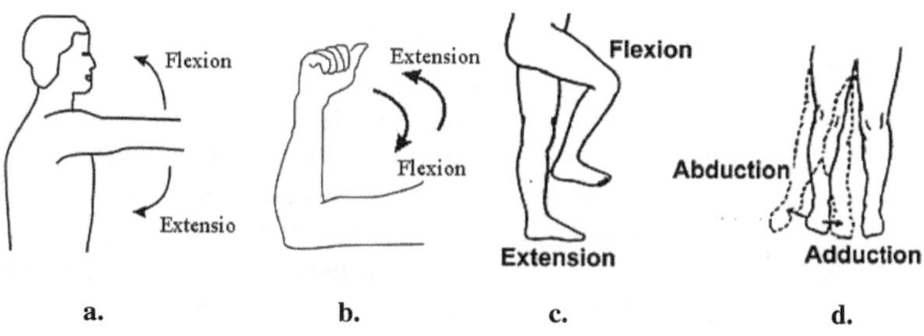

Fig. 1. Rehabilitation exercises: a. Shoulder; b. Elbow; c. Knee; d. Leg.

3 Kinematic Design of CUBE

The proposed design for CUBE is a 6-degrees-of-freedom parallel manipulator with a cable-driven architecture based on six cables. The kinematic scheme of the rehabilitation device is shown in Fig. 2 and is characterized by a fixed frame with adaptable geometry, which is defined by the variable lengths h and d. The end-effector is shaped as a ring that is worn as a wristband by the user. Its position is controlled by the six cables, giving the end-effector 3 DoFs in position, while its orientation is assumed to be constrained by a fixed support for the elbow on the Z-axis and fixed on the X-axis and Y-axis.

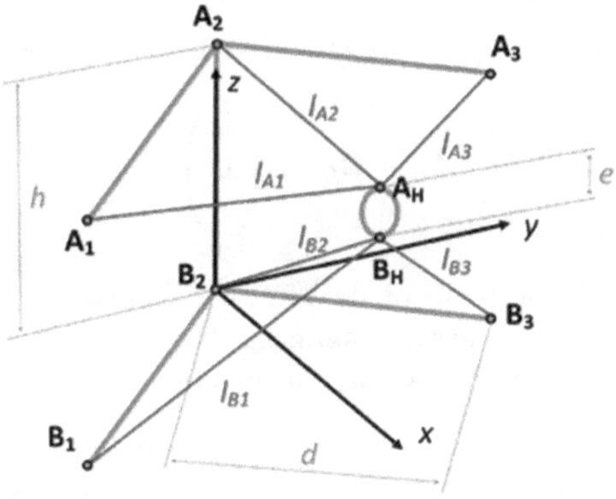

Fig. 2. Kinematic scheme of CUBE with main geometrical parameters.

Three cables are attached to the upper part of the structure in points A_1, A_2 and A_3, and they converge into point A_H of the end-effector. The remaining three cables are attached to the lower part of the structure in points B_1, B_2 and B_3, and they converge into point B_H of the end-effector. With reference to the *xyz* reference frame shown in Fig. 2, the position of the attachment points of the cables to the fixed frame can be defined as

$$\boldsymbol{A}_1 = \left(d\cos -\frac{\pi}{3}; d\sin -\frac{\pi}{3}; h\right) \tag{1}$$

$$\boldsymbol{A}_2 = (0; 0; h) \tag{2}$$

$$\boldsymbol{A}_3 = \left(d\cos \frac{\pi}{3}; d\sin \frac{\pi}{3}; h\right) \tag{3}$$

$$\boldsymbol{B}_1 = \left(d\cos -\frac{\pi}{3}; d\sin -\frac{\pi}{3}; 0\right) \tag{4}$$

$$\boldsymbol{B}_2 = (0; 0; 0) \tag{5}$$

$$\boldsymbol{B}_3 = \left(d\cos \frac{\pi}{3}; d\sin \frac{\pi}{3}; 0\right) \tag{6}$$

The end-effector is designed to work with a fixed orientation, characterized by points A_H and B_H sharing the same *x* and *y* coordinates and distant *e* from each other on the vertical axis. It is therefore possible to write the positions of the end-effector connection points A_H and B_H as

$$\boldsymbol{A_H} = \boldsymbol{A_i} - \boldsymbol{l_{Ai}} \tag{7}$$

$$\boldsymbol{B_H} = \boldsymbol{B_i} - \boldsymbol{l_{Bi}} \tag{8}$$

where $\boldsymbol{A_i}$ and $\boldsymbol{B_i}$ are the position vectors of points A_i and B_i, and $\boldsymbol{l_{Ai}}$ and $\boldsymbol{l_{Bi}}$ are the vectors of the cables, for $i = \{1, 2, 3\}$. The orientation constraint can be written as

$$\boldsymbol{A_H} - \boldsymbol{B_H} = \begin{pmatrix} 0 \\ 0 \\ e \end{pmatrix} \tag{9}$$

The inverse kinematic problem of the structure is used to compute the actuation variables, that are in this case cable lengths l_{Ai} and l_{Bi} for $i = \{1, 2, 3\}$, given the position of the end-effector points A_i and B_i. Inverse kinematics can be written from Eqs. (5) and (6) as

$$l_{Ai} = \| \boldsymbol{l_{Ai}} \| = \| \boldsymbol{A_i} - \boldsymbol{A_H} \| \quad for\ i = \{1,2,3\} \tag{10}$$

$$l_{Bi} = \| \boldsymbol{l_{Bi}} \| = \| \boldsymbol{B_i} - \boldsymbol{B_H} \| \quad for\ i = \{1,2,3\} \tag{11}$$

Once the kinematics of the system are known, its dynamics must be analysed in order to be able to formulate a proper closed-loop control. Cable-driven manipulators are characterized by a physical constraint caused by the nature of the cables, since they can transmit only a positive tension and no moment. Since the mass of the cable is much smaller than the one of the end-effector, an approximated dynamic formulation is here presented by neglecting the mass and inertia of the cables. The dynamic behaviour of the system can be studied in its general form as

$$\boldsymbol{M}(\boldsymbol{x})\ddot{\boldsymbol{x}} + \boldsymbol{c}(\boldsymbol{x},\dot{\boldsymbol{x}}) + \boldsymbol{G}(\boldsymbol{x}) = \boldsymbol{\tau} + \boldsymbol{F} \tag{12}$$

where $\boldsymbol{M}$ is the mass matrix of the system, $\boldsymbol{c}$ is the vector of centrifugal and Coriolis terms, $\boldsymbol{G}$ is the vector of gravity terms. Vector $\boldsymbol{\tau}$ is the vector of the tensions of the cables, which contains the value tension of each cable ($\tau_{A1}, \tau_{A2}, \tau_{A3}, \tau_{B1}, \tau_{B2}, \tau_{B3}$) in the given position, while $\boldsymbol{F}$ represents the external forces acting on the end-effector, which, for most rehabilitation cases, are caused by the reactions to the patient's limb weight and inertia. Since the system is designed to control the position of the limb of the patient (and, therefore, only three degrees of freedom) but actually possess 6 degrees of freedom (one for each cable), a redundancy problem has to be solved to decide how to actuate the system. The problem can be optimized on power consumption by minimizing the tensions of the cable as

$$min\ \boldsymbol{\tau} \tag{13}$$

by adding a constraint equation given by the physical characteristic of the cables of transmitting only a positive tension, which can be written as

$$\tau_{Ai} > 0; \quad \tau_{Bi} > 0; \quad for\ i = \{1, 2, 3\} \tag{14}$$

These equations allow us to control kinematically and dynamically the proposed structure, in order to achieve the required range of motion.

4 Modelling and Prototyping

A CAD model of the structure has been designed to produce a prototype, as based on the kinematic scheme shown in Fig. 2 with the dimensions in Table 1. One of its peculiarities is its telescopic structure, which makes it foldable as shown in Fig. 3. Figure 3b shows the unfolded prototype that can fit in a box of 1940 × 1180 × 990 mm, while Fig. 3a shows the folded prototype that can fit in a box of 1010 × 890 520 mm making it easier to transport. The mass of the system is 4.00 kg while the total mass of the six motors is 1.20 kg, for a total mass of 5.20 kg.

Table 1. Range of the main geometrical parameters

	Minimum value [mm]	Maximum value [mm]
d	550	1000
h	850	1000
e	80	120

To analyze the performance of the proposed mechanism before manufacturing it a Static FEM analysis has been worked out. The structure of the system has been considered as fully made in PLA plastic, Table 2 shows its characteristics. The tensions of the cables due to the weight of the end-effector and arm reaction are considered as external loads for the structural analysis. A force of 37 N directed towards the connection end-effector connection point is applied on each point where the actuator will be sited. This force has been calculated from the deadweight of an average arm in a worst-case scenario to check the stiffness of the structure for rehabilitation purposes. The structure fixed on its bottom and the system is subject to gravity. Figure 4 shows the factor of safety distribution using Maximum Normal stress criterion since PLA is a brittle material. According to the maximum principle stress failure criterion, the Factor of Safety (FOS) is

$$FOS = \frac{\sigma_{limit}}{\sigma_1} \tag{13}$$

In the simulation the minimum computed FOS is 1.942 on the frame top-left as shown in Fig. 4. Since no region of the structure reaches a FOS < 1 the material can afford the external load with no failure validating the feasibility of a prototype.

A preliminary prototype of CUBE has been rapid manufactured using FDM 3D Printing technology for further investigation, Fig. 5. The entire assembly is shown in its unfolded configuration in Fig. 5a. Figures 5b and c show the positioning of two of

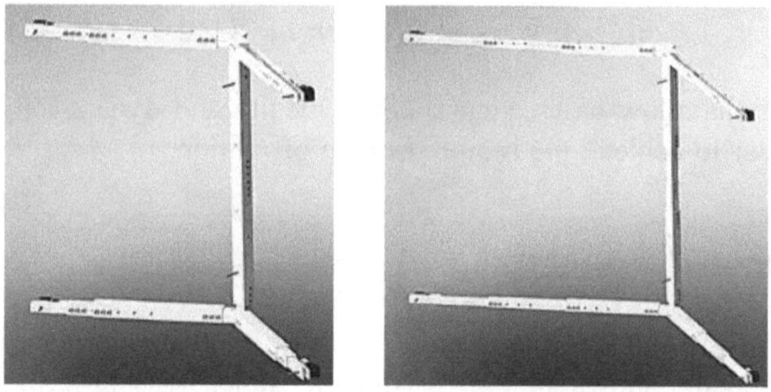

a. b.

Fig. 3. A CUBE rendering: (a) folded, (b) unfolded.

Table 2. PLA characteristics for the structure

Model type	Linear elastic isotropic
Yield strength	45.00 N/m^2
Tensile strength	3.00E+7 N/m^2
Elastic modulus	2.00E+9 N/m^2
Poisson's ratio	0.39
Mass density	1020 kg/m^3

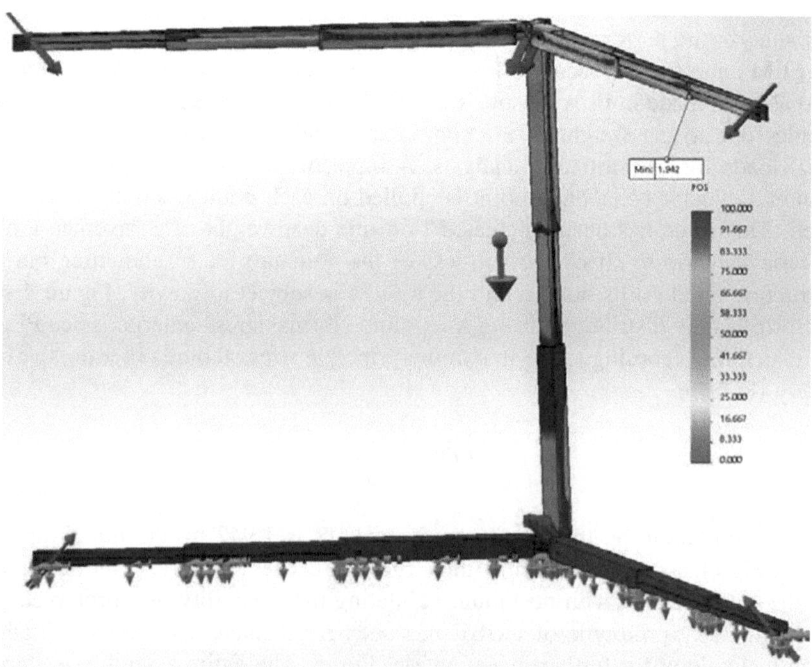

Fig. 4. FEM FOS distribution

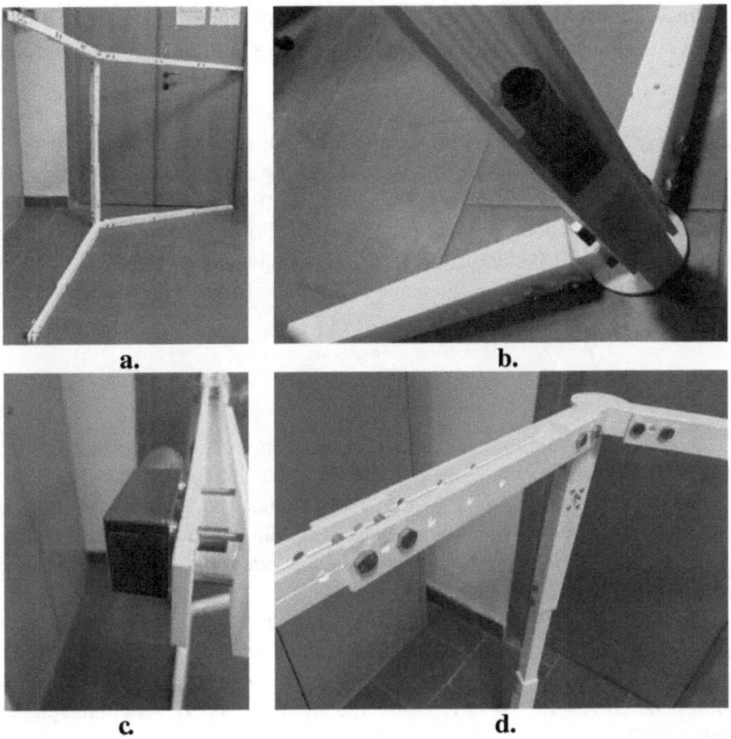

Fig. 5. CUBE preliminary prototype: a. General assembly; b. Lower motor detail; c. Upper motor detail; d. Telescopic structure.

the motors, respectively the ones actuating cable A1 and B2 in Fig. 2. Four of the motors (the external ones) are positioned in the outer part of the telescopic arms, while the remaining two are nested into the body of the frame. The telescopic arms can be fixed in place with pins or screws, such as shown in Fig. 5d.

The CUBE system in Fig. 5 is reconfigurable to achieve the desired operational workspace with the user facing the device, while minimizing its encumbrance. Moreover, the six-cable setup allows for improved position and force control when compared to the previous prototypes.

5 Conclusions

This paper introduces a cable-driven parallel robot, CUBE, for the assistance of patients in rehabilitation exercise for both upper and lower limb. CUBE structural design has been analyzed in a FEM environment to prove its feasibility. A first prototype has been built at LARM via 3D printing. CUBE presents a lightweight foldable structure that is easy to set-up in different configurations to adapt to different kind of exercises. It is portable and easy to place and assemble. Furthermore, the actuation made by cables makes it inherently safe in human/robot interactions thanks to the extremely low inertia of the system giving an extra advantage for its rehabilitation purpose.

Acknowledgments. The paper presents results from the research activities of the project ID 37_215, MySMIS code 103415 "Innovative approaches regarding the rehabilitation and assistive robotics for healthy ageing" co-financed by the European Regional Development Fund through the Competitiveness Operational Programme 20142020, Priority Axis 1, Action 1.1.4, through the financing contract 20/01.09.2016, between the Technical University of Cluj-Napoca and ANCSI as Intermediary Organism in the name and for the Ministry of European Funds.

References

1. Stein, J.: Robotics in rehabilitation: technology as destiny. Am. J. Phys. Med. Rehabil. **11**(3), S199–S203 (2012)
2. Reinkensmeyer, D.J., Boninger, M.L.: Technologies and combination therapies for enhancing movement training for people with a disability. J. Neuroeng. Rehab. **9**, 17 (2011)
3. Carrera, I., et al.: ROAD: domestic assistant and rehabilitation robot. Med. Biol. Eng. Comput. **49**(10), 1201 (2011)
4. Basteris, A., et al.: Training modalities in robot-mediated upper limb rehabilitation in stroke: a framework for classification based on a systematic review. J. Neuroeng. Rehabil. **11**(1), 111 (2014)
5. Iosa, M., et al.: Seven capital devices for the future of stroke rehabilitation. Stroke Res. Treat. **2012**, 9 (2012)
6. Sheng, B., et al.: Model based kinematic & dynamic simulation of 6-DOF upper-limb rehabilitation robot. In: Asia-Pacific Conference on Intelligent Robot Systems (ACIRS). IEEE (2015)
7. Medical Devices: Regulation (EU) 2017/745 of the European Parliament and of the Council of 5 April 2017 on medical devices. Official Journal of the European Union, L 117 (2017)
8. ISO Standards: Robots and robotic devices. Collaborative robots. PD ISO/TS 15066:2016 (2016)
9. Kim, S., et al.: Development of tension based haptic interface and possibility of its application to virtual reality. In: Proceedings of the ACM Symposium on Virtual Reality Software and Technology. ACM (2000)
10. Loureiro, R., et al.: Upper limb robot mediated stroke therapy—GENTLE/s approach. Auton. Robots **15**(1), 35–51 (2003)
11. Mayhew, D., et al.: Development of the MACARAM – a novel cable robot for upper limb neurorehabilitation. In: 9th International Conference on Rehabilitation Robotics, 2005. ICORR 2005. IEEE (2005)
12. Fanin, C., et al.: Nerebot: a wire-based robot for neurorehabilitation. In: ICORR 2003. HWRS-ERC (2003)
13. Rosati, G., et al.: Design of a new 5 DOF wire-based robot for rehabilitation. In: 9th International Conference on Rehabilitation Robotics, 2005. ICORR 2005. IEEE (2005)
14. Ceccarelli, M., et al.: Design and operation issues for parallel robotic devices in the rehabilitation of stroke patients. In: Proceedings of the RAAD 2009 18th International Workshop on Robotics in Alpe-Adria-Danube Region (2009)

15. Carbone, G., Gherman, B., Ulinici, I., Vaida, C., Pisla, D.: Design issues for an inherently safe robotic rehabilitation device. In: Ferraresi C., Quaglia G. (eds.) Advances in Service and Industrial Robotics. RAAD 2017. Mechanisms and Machine Science, vol 49. Springer, Cham (2018)
16. Hernández-Martínez, E.E., Ceccarelli, M., Carbone, G., López-Cajún, C.S., Jáuregui-Correa, J.C.: Characterization of a cable-based parallel mechanism for measurement purposes. Mech. Based Des. Struct. Mach. **38**(1), 25–49 (2010). https://doi.org/10.1080/15397730903386101
17. Major, K.A., et al.: Ranges of motion as basis for robot-assisted poststroke rehabilitation. HVM Bioflux **8**(4), 192–196 (2016)

A 3D Visual Tracking Method for Rehabilitation Path Planning

Daniele Cafolla(✉)

Frosinone, Italy
contact@danielecafolla.eu

Abstract. Rehabilitation devices are designed applying techniques that can determine the adaptability level of the patient using different type of exercises. It is important for the rehabilitation therapy applications to reproduce the exercise made by the therapist for a good recovery of the patient's motor functions. Thus, it is essential to track and use these tasks to design a path planning for controlling rehabilitation devices.

In this paper, a 3D vision tracking method is presented. This method allows to track a medical ball also in poorly-lighted and not structured environments through images captured by a camera in combination with a depth sensor, obtaining the 3D Point Cloud of the centroid as result. The Point Cloud is then fitted to a curve to give a reference trajectory to reproduce the exercise.

Keywords: Motion tracking · Object detection · Depth cloud
Rehabilitation device

1 Introduction

The goal of rehabilitation robotics is the application of robotics in therapeutic procedures for achieving a good recovery for persons with impairments due to various diseases such as neuromotor disorders or orthopaedic traumas. Assistive robotics, on the other hand, investigate to support independent living of persons who have chronic or degenerative limitations, such as the elderly and persons with severe disabilities. Nowadays advances in medical rehabilitation procedures, methodologies, and tools includes also cognitive aspects of motor control, giving an important role to flexible and programmable robotics systems that allow to set and assess several procedures [1–3].

The feature of rehabilitation devices to function in real-time, is what separates them from their less efficient counterparts, the conventional orthotics. The inclusion of this feature is intrinsic to the utilization of compact advanced actuators and smart sensors in portable and smart rehabilitation devices. These devices are computer controlled and can react in the order of milliseconds. A rehabilitation regime can be perfectly customized for each patient's individual needs [1–3]. Ideally, with closed loop control, feedback from the sensors would allow a computer to calculate the efficiency of each specific exercise and alter them in real-time accordingly to achieve optimal levels of rehabilitation. The key of a correct behaviour in such devices is a good exercise reference given by the therapist as an input for the path planning. Nowadays available tracking systems needs structured environments and needs an accurate calibration procedure, making the

G. Carbone et al. (Eds.): *New Trends in Medical and Service Robotics*, MMS 65, pp. 264–272, 2019.
https://doi.org/10.1007/978-3-030-00329-6_30

acquisition complex and far too expensive due to the necessary equipment. In this paper, a 3D vision tracking method is presented. This method allows to track a selected entity also in poorly-lighted and not structured environment without a calibration procedure obtaining a trajectory with low-cost equipment. This acquisition can be used after being post processed as a reference trajectory to reproduce the desired exercise.

2 Medical Ball Therapy

A medicine ball is a weighted sphere roughly between 30 cm and 50 cm in diameter that is designed to provide additional muscle activation during exercise movements. Contemporary, medicine balls are normally covered in rubber or leather, while the earliest examples date back to around 1000 B.C., when they were animal hides filled with sand and sewn together and used by gladiators, [4]. The term “medicine” in reference to the weighted exercise ball can be dated back to a Renaissance doctor who prescribed their use for “medicinal gymnastics.”, [4]. There are many variations of medicine balls that are used for a multitude of training and rehabilitation but the most common one used for rehabilitation purpouses is the Rubber-coated, basketball-sized balls ones that are 23 cm in diameter and weigh from 1 to 10 kg, Fig. 1. Some variations are designated “slam balls” and are designed to absorb the impact when hitting the floor or wall. The spherical shape allows for multiple hand placement for lifting, tossing and catching in play, and also makes the perfect object for strength training or optimal sports performance. Medicine balls can be found throughout history for fitness conditioning and injury rehabilitation [5].

Fig. 1. Rubber-coated Medicine ball, [4]

3 3D Tracking for Trajectory Reconstruction

In computer vision Colour object detection and tracking is an important and fundamental topic. Colour is a significant information in image analysis classification, [6].

Tracking RGB colour can be defined as the problem of estimating the trajectory of an object moving in a 3D environment [7]. The proposed method uses the RGB images generated by the Kinect v2 giving two-dimension tracking together with information coming from its depth sensor, giving the third dimension for a 3D tracking.

Moreover, Kinect v2 is composed by cameras, a depth sensor, a microphone array and an accelerometer.

The RGB camera can store colours with a resolution of 1280 × 960 pixels. Furthermore, it has an infrared (IR) emitter which emits infrared light beams and an IR depth sensor which makes it possible to read the emitted light and create a depth map from that information, [8]. The multi-array microphone contains four different microphones at different locations which provide audio recording capabilities which also can specify the location and specify the direction of the sound wave. Finally, a 3- axis accelerometer makes it possible to determine the current orientation of the Kinect v2 device, [8].

Figure 2 shows a photo of the Kinect v2 with the refence axes used in the proposed method.

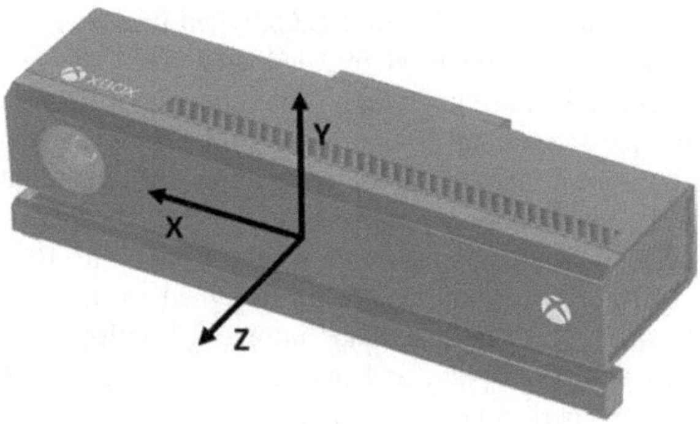

Fig. 2. Kinect v2 device with reference axes

Figure 3 shows a Flow-chart of the developed algorithm for the 3D tracking procedure.

After connecting the Kinect v2 to the computer and starting the application all the sensors in the device are initialized. The video coming from the RGB camera and the depth images coming from the depth sensor are acquired in parallel during the tracking procedure. The two sources are then merged together in one single window giving a real-time depth coloured view.

To start the tracking procedure the user can select an entity from the displayed depth coloured view using the mouse cursor. When the entity is selected by clicking on it, the algorithm stores the RGB values belonging to the selection. These values are compared to the video frames current colours using Euclidean distance. The colour is considered found if its colour distance is less than an imposed threshold that can be modified depending on how accurate the tracking is required to be. When the closest colour is found, the current location and current difference is stored. The stored current location gives the XY position of the selected entity in the acquisition area.

These XY coordinates are then matched to the depth image having back the Z coordinate respecting the reference axes shown in Fig. 2.

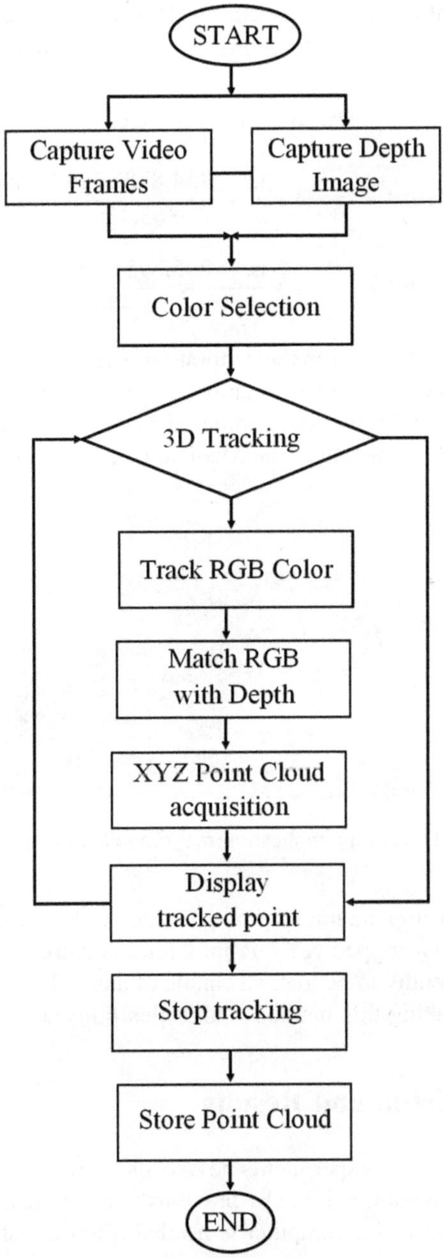

Fig. 3. Detection and tracking methods algorithms using matched colour feature and depth sensor

Thus, the XYZ Point Cloud is acquired. The Z is directly acquired in mm due to the depth sensor technology while XY positions must be converted from pixel to mm using Eqs. (1–3)

$$\boldsymbol{Point\ Z} = depth\ value \tag{1}$$

$$\boldsymbol{Point\ X} = -1 * \frac{(x - 254.878) * Point\ Z}{365.456} \tag{2}$$

$$\boldsymbol{Point\ Y} = -1 * \frac{(x - 205.395) * Point\ Z}{365.456} \tag{3}$$

where the given numbers come from the calibration parameters of the Kinect v2 camera which do not change since the camera is embedder in the device and cannot be moved.

Figure 4 shows the real-time depth coloured view where the tracked point is displayed as a black circle. In this paper the Medical ball has been chosen as the entity to be tracked.

Fig. 4. 3D tracking application real-time depth coloured view

Finally, the user can stop the tracking by pressing "e" key from the keyboard. When the tracking procedure is stopped XYZ Point Cloud is stored to be post processed.

To Summarize, the entity to be tracked can be of any colour and the procedure does not need calibration making this method a high flexibility an easiness of use.

4 Experimental Setup and Results

To test the proposed method, experiments have been worked out performing exercises used for erector spine rehabilitation. As previously mentioned, to use the proposed method only a Kinect v2 and a computer is needed. Figure 5 shows the experimental layout. An outdoor context has been chosen to analyse the performances of the proposed method in not structured environment. During the experiment a Medical ball of 1.5 kg has been used and the subject have been asked to perform two kinds of exercises: Elephant Twist and T-Twist.

Fig. 5. Experimental layout

The workouts have been chosen to show the 3D tracking capability of the proposed method since they have to be performed in two different planes that includes the X, Y and Z axes. The exercises have been carried out by a 31 years old healthy subject. After the acquisition of each test, the Point Cloud generated from the 3D tracking of the Medical ball has been post processed to obtain a reference trajectory for the given exercise.

Figure 6 shows the exercise named Elephant Twist [4]. The subject has been asked to hold the medicine ball in front of him with his arms extended to select the entity to be tracked and the exercise procedure have been initialized. The subject bends forward at the waist 90 degrees so that the ball is hanging from his arms just above his feet. The torso is slowly twisted towards left 30° to 45° or as far as the subject can comfortably go, Fig. 6a. Then, in a slow, controlled manner, the subject straightens his torso and return the ball to the central position between his legs (Fig. 6b) and repeats the previous steps moving towards right (Fig. 6c), [4].

Figure 7 shows the Elephant Twist exercise results. Figure 7a shows the acquired XYZ Point Cloud and the 3D polynomial fitting curve, while Fig. 7b shows the XY 2D polynomial fitting curve on the plane on which the exercise is performed.

a) b) c)

Fig. 6. Elephant Twist: (a) Left position, (b) Central position, (c) Right position

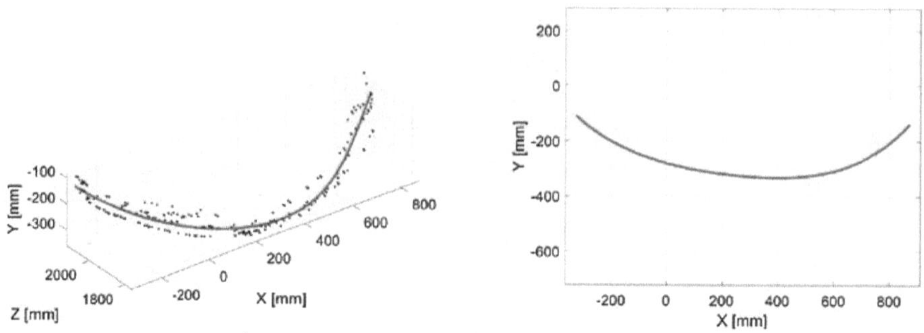

a) b)

Fig. 7. Elephant Twist test results: (a) 3D plot, (b) XY plot

The method shows a satisfactory fitting making possible to obtain a reference trajectory for the exercise. Furthermore, it is possible to analyse the subject's maximum range of motion along X that is 1188.50 mm.

Figure 8 shows the exercise named T-Twist [4]. The subject has been asked to hold the medicine ball in front of him with his arms extended to select the entity to be tracked and the exercise procedure have been initialized. Keeping the core tight and back straight, the subject raises the Medicine ball to shoulder height, directly in front of his torso with arms fully extended. With his knees slightly bent, keeps his hips facing forward while using his oblique muscles to rotate his upper body and the ball 90° or as far as he can go to the right, Fig. 8a. Then, he slowly returns to the centre position (Fig. 8b) and then repeat to the other side (Fig. 8c), [4].

a) b) c)

Fig. 8. T-Twist: (a) Left position, (b) Central position, (c) Right position

Figure 9 shows the T-Twist exercise results. Figure 9a shows the acquired XYZ Point Cloud and the 3D polynomial fitting curve, while Fig. 9b shows the XZ 2D polynomial fitting curve on the plane on which the exercise is performed.

The method shows a satisfactory fitting making possible to obtain a reference trajectory for the exercise. Furthermore, it is possible to analyse the subject's maximum range of motion along X that is 1129.60 mm.

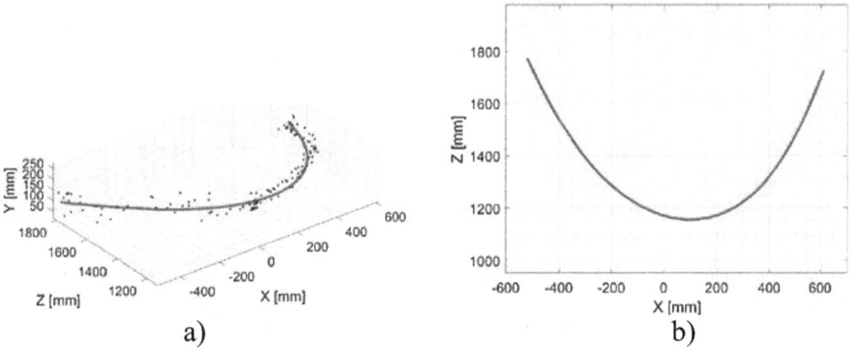

Fig. 9. T-Twist test results: (a) 3D plot, (b) XZ plot

5 Conclusions

A 3D vision tracking method has been presented. Test have been carried out with a subject to check the 3D tracking performances of the proposed solution. The results show that the introduced method is able to track a medical ball in poorly-lighted and not structured environment without a calibration procedure obtaining a reference trajectory. Furthermore, this methodology can be used to track an entity of any colour since it acquires the colour by selecting the entity with the mouse cursor making it more flexible than existing solutions allowing the use in poorly-lighted environment too.

The XYZ Point cloud after being postprocessed offered a reference trajectory for the performed exercise. This methodology has been used to know some biometrical data of the patients that can be used to monitor his recovery status in rehabilitation. Furthermore, exercises can be designed to check others needed biometrical data.

In addition, a therapist can perform the exercise and the obtained trajectory can be post processed to be used as an input for a rehabilitation device.

The exercises can be carried out at different speeds and on different plane on 2D planes or in the 3D acquisition area.

Finally, the proposed method offers a high flexibility and easiness in use since it can be used everywhere at any time with low-cost equipment.

References

1. Stein, J.: Robotics in rehabilitation: technology as destiny: S199-S203. Reinkensmeyer, D.J., Boninger, M.L.: Technologies and combination therapies for enhancing movement training for people with a disability. J. Neuroeng. Rehab. (2012)
2. Basteris, A., et al.: Training modalities in robot-mediated upper limb rehabilitation in stroke: a framework for classification based on a systematic review. J. Neuroeng. Rehabil. **11**(1), 111 (2014)
3. Medical devices. Recognized essential principles of safety and performance of medical devices. BS ISO 16142-2 (2015)

4. Stewart, B.: Medicine Ball Workouts Strengthen Major and Supporting Muscle Groups for Increased Power, Coordination, and Core Stability. Ulysses Press, United States (2013)
5. Cordasco, F.A., Wolfe, I.N., Wootten, M.E., Bigliani, L.U.: An electromyographic analysis of the shoulder during a medicine ball rehabilitation program. Am. J. Sports Med. **24**, 386–392 (1996)
6. Hemalatha, C., Muruganand, S., Maheswaran, R.: A survey on real time object detection, tracking and recognition in image processing. Int. J. Comput. Appl. **91**(16), 38–41 (2014)
7. Wang, Q, Gao, Z Study on a real-time object tracking system. In: International Symposium on Computer Science and Computational Technology, ISCSCT 2008, p. 2 (2008)
8. Microsoft. Kinect v2. https://www.xbox.com/en-US/xbox-one/accessories/kinect. Accessed 27 Mar 2018

Optimal Design of Cable Driven Robot for Rehabilitation with Prescribed Workspace

M. A. Laribi[1(✉)], G. Carbone[2], and S. Zeghloul[1]

[1] Department of GMSC, Pprime Institute, CNRS - University of Poitiers, ENSMA - UPR 3346, Poitiers, France
med.amine.laribi@univ-poitiers.fr
[2] LARM (Laboratory of Robotics and Mechantronics), University of Cassino and South Latium, via G. Di Biasio 43, 03043 Cassino, Italy
giuseppe.carbone@unicas.it

Abstract. This paper deals with an optimized design of a cable driven parallel manipulator for a prescribed workspace. The cable driven robots have characteristics that make them suitable for rehabilitation exercise purposes such as large workspace, re-configurable architecture, portability and low-cost. The proposed cable robot in this work, LAWEX, is intended for rehabilitation or exercising of upper limbs. For these purposes, a motion capture system is used to identify the range of motion of the upper limb. An optimal design method of the LAWEX robot is presented aiming to find the smallest robot sizes to achieve a prescribed workspace for upper limb exercising. The proposed objective function is based on a mathematical formulation of the power of a point with respect to a bounding surfaces.

Keywords: Robot design · Cable-driven robots · Optimization
Prescribed workspace · Power of the point

1 Introduction

Treatments in the rehabilitation clinic are costly and time consuming. Therapist has to perform stiff and repetitive efforts with the patients and thus he can only attend a limited number of patients in a day. In this context, robotics can play an important role in the process of rehabilitation by assisting the therapist and the patient. While using the robot, the patient and therapist do not get tired as motion is fulfilled by autonomous assistance of the robot. This work focuses at on the design steps for the development of cable driven robot intended for rehabilitation tasks.

Cable-driven parallel robots have unique properties which make them popular among parallel and serial robots. Cable driven robots are classified as a type of parallel robots, in which, rigid links are replaced by flexible cables. Using cables in the robot structure causes the mechanism to become very light and makes the manipulator capable of reaching high velocities and accelerations since the mass/inertia of moving parts is decreased considerably as compared with other robots having rigid links. In literature, we can enumerate significant examples of cable driven robots for rehabilitation. For instance: NeReBot (NEuro REhabilitation roBOT) is a cable-suspended

G. Carbone et al. (Eds.): *New Trends in Medical and Service Robotics*, MMS 65, pp. 273–282, 2019.
https://doi.org/10.1007/978-3-030-00329-6_31

device for upper limb rehabilitation of post-stroke patients [1]; CALOWI (Cassino Low-Cost Wire Driven), a four-cable robot for limb exercising and rehabilitation [2]; MACARM (MultiAxis Cartesian-based Arm Rehabilitation Machine), an eight-cable robot for upper limb rehabilitation [3].

Optimal design procedures have been widely addressed in literature with algorithms for optimizing for example workspace characteristics [12, 13], stiffness properties [14], global isotropy property [15]. Multi-criteria optimal designs have been proposed for example in [16, 17, 18] as applied to parallel manipulators. However, in the literature there are few studies that implement a robust multi-objective optimization of these criteria from the preliminary design phase of cable robots [10]. The contribution of this work is to fill this gap and propose an approach suited for cable driven robots. The proposed optimal design method is based on the power function of the point and the prescribed workspace definition through experimental procedure using tracking via motion capture system. The paper is organized as follows: Sect. 2 introduces the cable driven parallel robot with kinematics analysis description. The prescribed workspace obtained through analyses human shoulder joint motion for rehabilitation tasks is also defined. Section 3 formulates the optimization problem of selecting the design variables for the LAWEX robot having a prescribed workspace, defined by a set of points. The optimization algorithm as well as its formulation and numerical results are presented in Sect. 4. Section 5 provides the conclusion of the present study.

2 Characteristics and Design Requirements

The main difference between analysis of cable-driven robots and other parallel robots arises from unidirectional characteristic of cables. Since cables can only pull, it is necessary to find the workspace of a cable-driven robot where positive tensions exist in all cables. Various workspaces have been presented in literature due to this property of cables and different approaches have been introduced to find the desirable workspace. In this paper the workspace of a cable-driven robot, previously proposed by authors, is investigated. Our cable robot LAWEX, proposed in [4], uses a ring as its end-effector which makes it suitable for upper limb handling as shown in Fig. 1. This cable robot can be actuated by three active cables, which control the position of the end-effector. In this set up gravity is used to keep the cables pulling.

The general structure of a cable driven parallel robot and its main components are presented. The inverse kinematic model (IKM) consists in finding the cable lengths as function of the end-effector degrees of freedom. The length of each cable can be written in terms of difference of coordinates of its endpoints. The geometric parameters of LAWEX driven robot are defined by the coordinates of motors locations respectively at points C_1, C_2 and C_3, as shown in Fig. 1. The cable i, for $i = 1, 2, 3$, is defined by its length l_i and bounding limits $l_i^{\min}$ and $l_i^{\max}$.

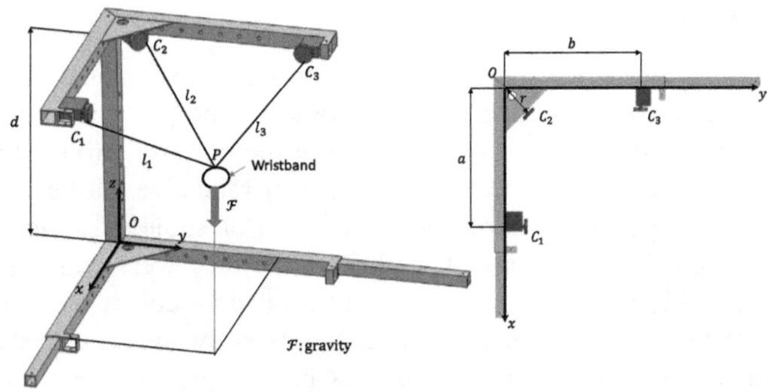

Fig. 1. The cable-driven robot and a wristband as its end-effector.

Let P be a point located on the end effector, then the geometric model can be written as

$$(X_p - x_i)^2 + (Y_p - y_i)^2 + (Z_p - d)^2 = l_i^2 \tag{1}$$

with $x_i = a_i \cos\theta_i$ and $y_i = b_i \sin\theta_i$

The workspace of the cable driven robot is defined by a volume, in the Cartesian space, reachable by the center of the platform. The geometrical model of each cable is described by the Eq. (1). It is assumed that the cables are identical, and their lengths vary between the minimal value, $l_{\min}$, and the maximum value, $l_{\max}$. The reachable points of each one of these cables are confined within a volume delimited by two concentric spheres given by (for $i = 1, .., 3$):

$$(X_p - x_i)^2 + (Y_p - y_i)^2 + (Z_p - d)^2 = l_{\max}^2 \tag{2}$$

$$(X_p - x_i)^2 + (Y_p - y_i)^2 + (Z_p - d)^2 = l_{\min}^2 \tag{3}$$

A slice of the workspace at $z = d$ is shown on Fig. 2-(a). The spatial representation of the workspace is given on Fig. 2-(b). The point P is inside the cable driven robot workspace, then the coordinates X_p, Y_p and Z_p satisfy the following relations for $i = 1, 2, 3$:

$$\mu_{\max}^i(P) = (X_p - a_i\cos\theta_i)^2 + (Y_p - b_i\sin\theta_i)^2 + (Z_p - d)^2 - l_{\max}^2 \leq 0 \tag{4}$$

$$\mu_{\min}^i(P) = (X_p - a_i\cos\theta_i)^2 + (Y_p - b_i\sin\theta_i)^2 + (Z_p - d)^2 - l_{\min}^2 \geq 0 \tag{5}$$

For every workspace to be generated by a driven cable robot, the independent design variables are a, b, d and r (see Fig. 1).

2.1 Desired Workspace

Specific tests have been carried out in collaboration with professional physiotherapists and neurologists as also reported in [5, 6] in order to properly identify the desired motions for a proper exercising of upper limbs. For the purpose, protocol based on the use of high resolution cameras detecting a number of reflective markers is performed. We have placed fourteen markers on the upper limb of subject as shown on Fig. 2. Then we have recorded the motions of markers while they were making real rehabilitation motion test. More details on the experimental protocol as well as upper limb motion specifications are given in [7]. Using Nexus software, we have been able to reconstruct each motion by regrouping markers into "segments" as shown in Fig. 3. A segment is defined as a set of markers of the same solid. Three segments have been considered: arm, forearm and torso.

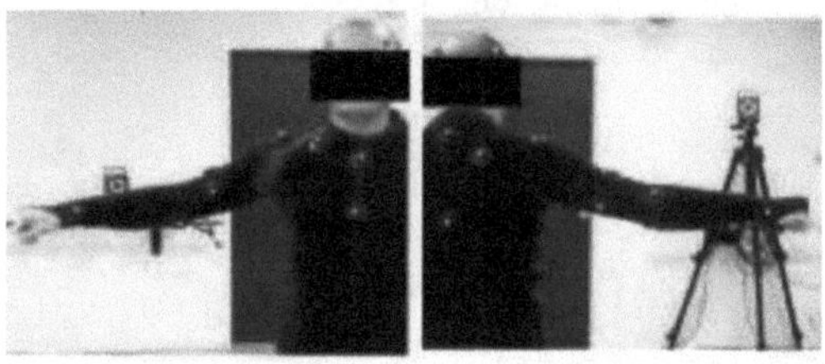

Fig. 2. Reflective markers set on the subject.

Experimental motion tracking yields after recording and processing to export the coordinates of each markers, so we can construct at any time a reference frame attached to the upper limb segment, $\mathcal{R}_s$, using Matlab software [8]. The motion of each segment can be computed using the Euler angles: precession ψ, nutation θ and rotation φ. The hemispherical range of motion for the shoulder is presented on the Fig. 4 for the case of extended arm. The subject is asked to perform with its upper limb a movement sweeping the accessible space.

The trajectory of the wrist center, point P, is computed for the case of extended arm using the experimental results obtained by motion capture. This point corresponds to the wristband location in space which the cable robot should describe and will be used as prescribed trajectory in the optimization process.

In addition to the workspace, it is important to indicate for a suitable operation of the cable driven robot that all of cables should remain in tension under any load if there is a large enough force, which can be generated by spring, gravity, dynamic force or actuator. The static force analysis, which will be addressed in future work, considers that all cables must remain in tension under any load. When a manipulator performs a given task, the end-effector exerts force and moment on the wristband in the case of cable-driven robot forces are transmitted by extending and retracting cables and by ensuring always the condition of pulling cables.

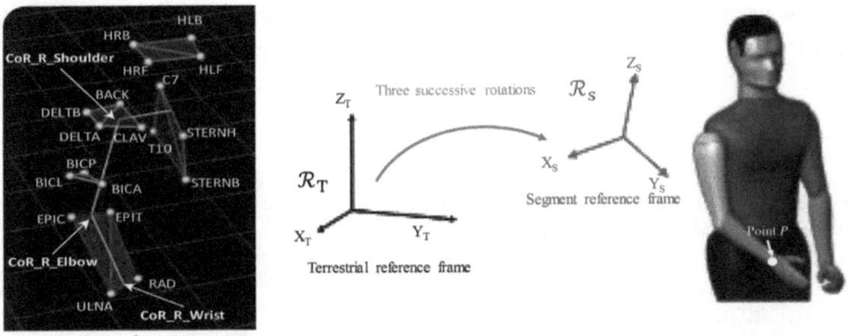

Fig. 3. Reconstruction of several detected markers into segments: arm, forearm and torso.

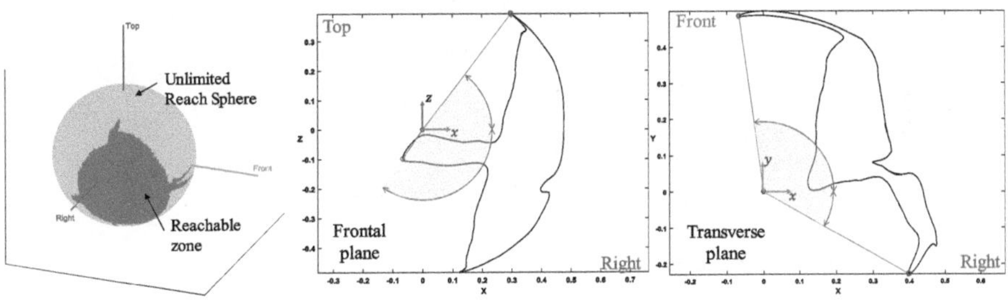

Fig. 4. Left: Range of motion for the shoulder.

3 The Optimization Problem

The aim of this section is to formulate the optimization problem of selecting the design variables for the LAWEX robot having a prescribed workspace. The prescribed workspace is given by a set of points (derived from rehabilitation movements) that can describe a trajectory $\mathbf{\Omega}$ in space as defined in Sect. 2.1.

The general optimization problem can be formulated as follows:

Given: A prescribed trajectory in space derived from rehabilitation movements.

Find: The design variables of the LAWEX robot having the smallest workspace that includes the prescribed trajectory.

The general associated optimization problem, with n parameters for a suitably chosen objective function $\mathcal{F}(\mathbf{I})$ can be stated as

$$\min \mathcal{F}(\mathbf{I}) \tag{6}$$

Subject to

$$\mu^i_{\min}(\mathbf{I}) \geq 0 \ \text{ for } \ i = 1, 2, 3$$

$$\mu^i_{\max}(\mathbf{I}) \leq 0 \ \text{ for } \ i = 1, 2, 3$$

For every trajectory to be generated by LAWEX robot, cable-based parallel robot for rehabilitation, the independent design variables are defined as follows:

$$\mathbf{I} = \left[a\, b\, r\, \theta\, d\, S_x\, S_y\, S_z\right] \tag{7}$$

In addition to the robot design variables, the location of the center of the shoulder will be considered in the optimization process. According to the Fig. 5, three distances S_x, S_y and S_z are used to locate the center of the shoulder and so then the trajectory in (x, y) plane and along z axis.

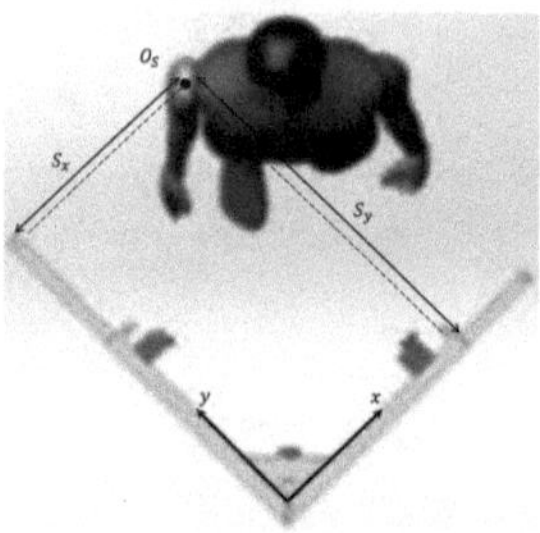

Fig. 5. Location of the center of the shoulder in (x, y) plane.

4 Optimality Criteria

The workspace is one of the most important kinematic properties of manipulators. There are several different ways of considering the workspace within an optimal design procedure. For example, one can refer to the workspace volume, shape, and/or presence of singularities. Starting from one of the above physical aspects, one can define various mathematical formulations for an objective function. In this paper, we propose an objective function $F(\mathbf{I})$ as a minimization of functions $\left|\mu^i_{\min}(\mathbf{I}, P)\right|$ and $\left|\mu^i_{\max}(\mathbf{I}, P)\right|$ which is physically equivalent to finding a surface closest to points P. We note that the constraint functions $\mu^i_{\min}$ as well as $\mu^i_{\max}$ when applied to a point P can be used as a measure of some kind of distance of this point with respect to the surface defined by $\mu^i_{\min} = 0$ as well as $\mu^i_{\max} = 0$. The objective is to find the set of parameters, given by $\mathbf{I_{opt}} = \left[a\, b\, r\, d\, S_x\, S_y\, S_z\right]$, that can yield a smallest LAWEX robot having a workspace that includes the given prescribed trajectory $\mathbf{\Omega}$ in space.

There are a number of alternative methods to solve numerically an optimization problem even in commercially available packages such as Matlab. Some algorithms such as flooding techniques, simulated annealing can be faster in finding an optimal solution with a single objective function. But, they still cannot guarantee the convergence. They cannot still guarantee that an optimal solution is a global optimum. Among the existing numerical solvers genetic algorithms have been demonstrated as particularly effective in solving highly non-linear problems [11]. Accordingly, in this paper we propose to use the Matlab GA (Genetic Algorithm) solver. A penalty function method is used to handle the constraints and to ensure that the fitness of any feasible solution is better than the infeasible ones. The objective function is accordingly constructed as

$$F = F_1 + F_2 \tag{8}$$

where F_1 is the power component of the fitness function and given by

$$F_1 = \sum_{k=1}^{K} \frac{\sum_{i=1}^{3} \left|\mu_{\min}^{i}\right| + \sum_{i=1}^{3} \left|\mu_{\max}^{i}\right|}{\sqrt{\sum_{i=1}^{3} \left(\mu_{\min}^{i}\right)^2 + \sum_{i=1}^{3} \left(\mu_{\max}^{i}\right)^2}} \tag{9}$$

and, F_2 is a penalty function defined as

$$F_2 = \sum_{k=1}^{K} \left(\sum_{i=1}^{3} \delta_{\min}^{i}(\mathbf{I}, P) + \sum_{i=1}^{3} \delta_{\max}^{i}(\mathbf{I}, P)\right) \tag{10}$$

where,

$$\delta_{\min}^{i}(\mathbf{I}, P) = \begin{cases} 0 & \mu_{\min}^{i}(\mathbf{I}, P) \geq 0 \\ \chi & \mu_{\min}^{i}(\mathbf{I}, P) < 0 \end{cases} \tag{11}$$

$$\delta_{\max}^{i}(\mathbf{I}, P) = \begin{cases} 0 & \mu_{\max}^{i}(\mathbf{I}, P) \leq 0 \\ \chi & \mu_{\max}^{i}(\mathbf{I}, P) > 0 \end{cases} \tag{12}$$

Here, χ is a large positive number. $F_2 = 0$ means that all the points, defining the prescribed trajectory $\mathbf{\Omega}$ in space, are contained within the workspace of the LAWEX robot.

4.1 A Case of Study

LAWEX robot is a cable-based parallel robot for rehabilitation, which has been designed and built at LARM in Cassino, [6]. The structure of LAWEX is a non-conventional open architecture, which allows an easy accessibility by patients under treatment. Aluminum profiles are used so that the robot is light and stiff while it can be assembled and disassembled easily for storage and transportation. Cables are connected to the end-effector which covers the arm to be trained by using a wristband, [9]. LAWEX has three servo motors and three cables, which are connected to manipulators and the end-effector through a wristband as shown in Fig. 6. Cable tensioning is guaranteed by gravity. In this example, the dimensions of the LAWEX robot are to be determined to get the smallest workspace capable of containing a prescribed trajectory $\mathbf{\Omega}$, given by a set of points P_k for $k = 1, .., K$ ($K = 2600$ for the considered trajectory given on Fig. 4). The bounding interval for each one of the design variables is presented in Table 1 (Fig. 1).

Fig. 6. A picture of LAWEX cable-driven parallel robot at LARM.

Table 1. The bounding interval for design variables.

I	a	b	r	d	S_x	S_y	S_z
max	10	10	10	10	−1500	−1500	−1500
min	1000	1000	500	1500	1500	1500	1500

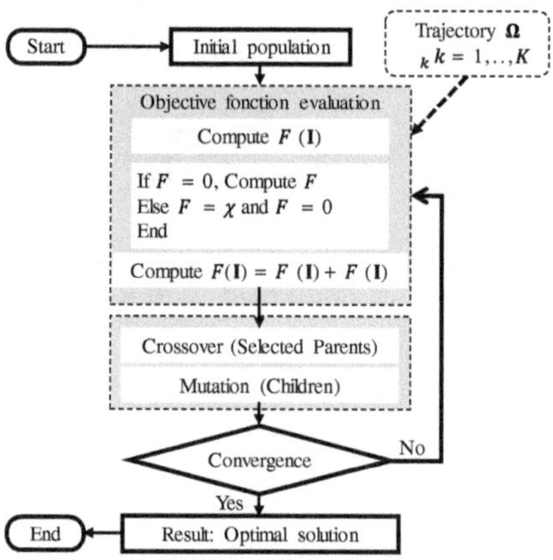

Fig. 7. Procedure of evaluating cable driven robot for prescribed trajectory.

4.2 Optimization Results

The power of points at a given height $z = d$ is represented in Fig. 8 by a mapping of the objective function F as a function of x and y for a given design vector. The minimum values of F are located at the intersections of the surfaces. The results obtained in this section are the output of the algorithm presented in previous section and developed within Matlab. The optimal solution obtained by the GA is presented in Table 2.

The calculation time, necessary for obtaining the optimum solution, is estimated at about 5 s, and the average number of analysis (function evaluation) were equal to 53000 (we use 106 generations per run, and 500 individuals in the population).

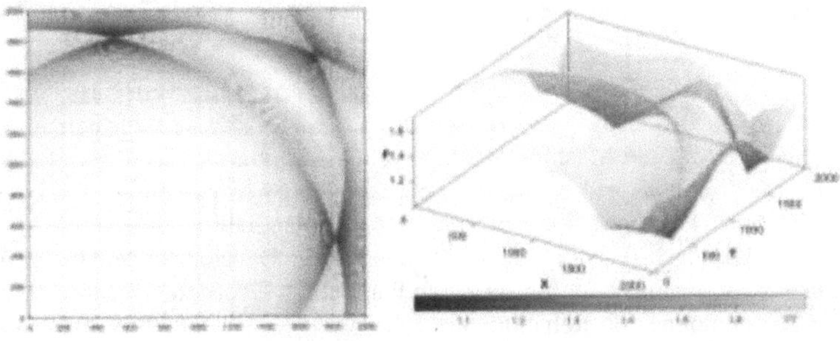

Fig. 8. Representation of the power of the point at a given high z as a function of x and y.

Table 2. The optimal solution for the cable driven robot, LAWEX.

$\mathbf{I}_{opt}$	a	b	r	d	S_x	S_y	S_z
	870	870	101	1200	820	820	550

5 Conclusions

In this work, we analyzed the workspace of the cable driven robot, LAWEX, intended for rehabilitation or exercising of upper limbs of patients. A motion capture system is used to identify the range of motion of the upper limb and specifications in terms of trajectory for a prescribed workspace. An optimal design method of the LAWEX robot is presented aiming to find the smallest workspace containing a prescribed workspace. The proposed objective function is based on the mathematical concept of the power of the point with respect to the bounding constraints. The effectiveness of the proposed method is shown through an optimal design example for a cable driven robot to reach a prescribed workspace for rehabilitation.

Acknowledgements. The paper presents results from the research activities of the project ID 37_215, MySMIS code 103415 "Innovative approaches regarding the rehabilitation and assistive robotics for healthy ageing" co-financed by the European Regional Development Fund through the Competitiveness Operational Program 2014-2020, Priority Axis 1, Action 1.1.4, through the financing contract 20/01.09.2016, between the Technical University of Cluj-Napoca and ANCSI as Intermediary Organism in the name and for the Ministry of European Funds.

The video tracking research is supported by ROBOTEX, the French national network of robotics platforms (N° ANR-10-EQPX-44-01) and by the French National Research Agency (ANR-14-CE27-0016).

References

1. Rosati, G., Masiero, S., Rossi, A.: On the use of cable-driven robots in early inpatient stroke rehabilitation. In: Boschetti, G., Gasparetto, A. (eds) Advances in Italian Mechanism Science. Mechanisms and Machine Science, vol. 47. Springer (2017)
2. Ceccarelli, M.: Problems and experiences on cable-based service robots for physiotherapy applications. New Trends Med. Serv. Robot. **16**, 27–42 (2013)
3. Mayhew, D., Bachrach, B., Rymer, W.Z., Beer, R.F.: Development of the MACARM a novel cable robot for upper limb neurorehabilitation. In: Proceedings of the 2005 IEEE 9th International Conference on Rehabilitation Robotics, Chicago, pp. 299–302 (2005)
4. Carbone, G., Aróstegui Cavero, C., Ceccarelli, M., Altuzarra, O.: A study of feasibility for a limb exercising device. In: Boschetti, G., Gasparetto, A. (eds) Advances in Italian Mechanism Science. Mechanisms and Machine Science, vol. 47. Springer (2017)
5. Carbone, G., Gherman, B., Ulinici, I., Vaida, C., Pisla, D.: Design issues for an inherently safe robotic rehabilitation device. In: 26th International conference on robotics in Alpe-Adria-Danube Region RAAD 2017, pp. 967–974. Springer, Dorchect (2017)
6. Vaida, C, Carbone, G., Major, K., Major, Z., Plitea,N., Pisla, D.: On human robot interaction modalities in the upper limb rehabilitation after stroke. Acta Teh. Napoc. Appl. Math. Mech. Eng. **60**(1), 91–102 (2017)
7. Laribi, M.A., Decatoire, A., Carbone, G., Pisla, D., Zeghloul, S.: Identification of upper limb motion specifications via visual tracking for robot assisted exercising. In: 27th International Conference on Robotics in Alpe-Adria-Danube Region RAAD, Patras, Greece. Spinger (2018)
8. Lamine, H., Laribi, M.A., Bennour, S., Romdhane, L., Zeghloul, S.: Design study of a cable-based gait training machine. J. Bionic Eng. **14**(2), 232–244 (2017)
9. Major, K.A., Major, Z.Z., Carbone, G., Pisla, A., Vaida, C., Gherman, B., Pisla, D.L.: Ranges of motion as basis for robot-assisted post-stroke rehabilitation. HVM Bioflux **8**(4), 192–196 (2016)
10. Gosselin, C.: Cable-driven parallel mechanisms: state of the art and perspectives. Mech. Eng. Rev. (2014). https://doi.org/10.1299/mer.2014dsm0004. ISSN 2187-9753
11. Essomba, T., Laribi, M.A., Zeghloul, S., Poisson, G.: Optimal synthesis of a spherical parallel mechanism for medical application. Robotica **34**(3), 671–688 (2016)
12. Merlet, J.-P.: Designing a parallel manipulator for a specific workspace. Int. J. Robot. Res. **16**(4), 545–556 (1997)
13. Schonherr, J.: Evaluation and optimum design of parallel manipulators having defined workspace. In: ASME 26th Biennial Mechanisms and Robotics Conference, Baltimore, Paper DETC2000/MECH-14092 (2000)
14. Bhattacharya, S., Hatwal, H., Ghosh, A.: On the optimum design of stewart platform type parallel manipulators. J. Robot. **13**, 133–140 (1995)
15. Takeda, Y., Funabashi, H.: Kinematic synthesis of in-parallel actuated mechanisms based on the global isotropy index. J. Robot. Mechatron. **11**(5), 404–410 (1999)
16. Gosselin, C.M.: On the design of efficient parallel mechanisms. In: Proceeding of the Advanced Study Institute of NATO: Computational Methods in Mechanics, vol 1, (Keynote Lecture), St. Konstantin and Elena, Bulgaria, pp 157–186 (1997)
17. Ottaviano, E., Carbone, G.: A procedure for the multiobjective design of parallel manipulators. Int. J. Mech. Control **04**(02), 57–62 (2003)
18. Hao, F., Merlet, J.-P.: Multi-criteria optimal design of parallel manipulators based on interval analysis. Mech. Mach. Theor. **40**, 157–171 (2005)

Design of a Mechanical Interface for a Cable-Driven Rehabilitation Device

Viorela Alexandra Lazăr[1], Daniele Cafolla[1(✉)], Doina Pisla[1], and Giuseppe Carbone[2]

[1] CESTER (Research Centre for Industrial Robot Simulation and Testing), Technical University of Cluj-Napoca, Cluj-Napoca, Romania
contact@danielecafolla.eu

[2] LARM (Laboratory of Robotics and Mechatronics), University of Cassino and South Latium, Cassino, Italy
giuseppe.carbone@unicas.it

Abstract. Movement performance in patients with neurological or orthopaedic traumas can be improved with Task-oriented repetitive movement. The application of robotics can assist, enhance, evaluate, and document neurological and orthopaedic rehabilitation of movements. A key point is the mechanical interface between the device and the user, it has to be adaptable to every patient with different anthropomorphic sizes. Furthermore, the mechanical interface must permit the needed movement for rehabilitation while avoiding dangerous ones. This paper presents design of a mechanical interface for a cable-driven rehabilitation device in terms of conceptual design and stress simulation. The interface is then tested during some rehabilitation exercises to test its usefulness.

Keywords: Cable-driven robots · Medical robots · Rehabilitation devices Mechanical interface · End-effector

1 Introduction

Robot rehabilitation devices are attracting significant research interest, since they can improve and increase effectiveness and efficiency of the rehabilitation procedure by allowing longer and more frequent therapy sessions for a patient as well as even foreseen a self-treatment with remote supervision of several patients by a single professional therapist, [1]. A disabled limb has to be moved according to its physiological limits and an assisting device should be programmed accordingly. Furthermore, the forces applied must not exceed the normal muscle strength in order to avoid any injuries, [2]. Research regarding rehabilitation shows that in most cases repeated movements of human limbs may help the patient regain functions of the limbs that were lost due to an injury or an illness. Robotic systems can be more efficient in performing these exercises than humans, and they can record information like position, trajectory, force and velocity, maximizing motor performance during active movements, [3].

In cable-driven parallel manipulators, the motion is obtained by means of cables instead of rigid bodies. The condition for a mobile platform with n degrees of freedom (DoFs) to have a fully controlled motion is having at least n+1 cables. This derives from

G. Carbone et al. (Eds.): *New Trends in Medical and Service Robotics*, MMS 65, pp. 283–292, 2019.
https://doi.org/10.1007/978-3-030-00329-6_32

the fact that due to their nature cables require positive tension. The redundant cables are needed to avoid the uncontrollable situations that can occur as a consequence of the fact that the cables can only pull, [4]. Due to this characteristic design, they present some advantages over other types of robots such as potential large workspace, higher payload to weight ratio, less inertia of the moving parts and being fairly simple, inexpensive and modular, [5]. They have received attention for several applications as reported in [4]. Specifically, for physiotherapy purposes, they present large workspace and easy reconfiguration to adapt to different exercises and patients, transportability, fairly low cost, simple maintenance, safety and potential acceptability by the patient, [5].

Fourteen randomized controlled trials comparing robot-assisted therapy that use end-effector-type devices with conventional therapies were conducted, [6]. Nine of the 14 randomized controlled trials that examined robot-assisted therapy with end-effector-type devices assessed the influence of robot-assisted training on activities of daily living (ADL) in patients with stroke. These nine reports demonstrated that robot-assisted training yielded similar or better effects on ADL in comparison with conventional therapy, [6].

The main challenge with cable-driven robots in rehabilitation is to manage to develop and integrate the engineering areas within medical frames within the development of intelligent, portable, low-cost, user-oriented systems, [7]. Moreover, from the clinical point of view, the use of compliant cables instead of rigid links makes the patient feel less constrained. Thus, a cable-driven parallel robot can easily adapt to the patient's anatomy and physiological needs, also according to rehabilitation aims. Injury during rehabilitation will be decreased if the patient collides with cables designed to break at set loads, [8].

2 The Mechanical Interface of LAWEX

The LAWEX (LARM Wire driven EXercising device) is a cable-driven rehabilitation device intended for lower and upper limb assisted exercises, Fig. 1. It has a lightweight structure so it is easy to move and store, the manufacturing and maintenance costs are low. LAWEX is composed from a rigid aluminium structure and it had a mobile platform called end-effector, shown in Fig. 2, which guides the arm during the exercises. It has holes in which the user needs to put the fingers. The end-effector is connected to the rigid structure in different points via 4 cables, each one of the cables being connected to a servomotor. Only one of LAWEX servomotors is below the platform, the remaining 3 are situated in the upper part. An initial and a final position have to be defined for the control of each motor, then the number of steps is interpolated and the angular values are calculated using MATLAB environment. This data is saved in an array for interfacing it with Arduino with 4 digits each and send the overall vector of 16 positions to the former. Afterwards, in Arduino the vector is parsed in 4 parts, each one corresponding to a servo motor. After that, each servo motor receives the assigned data from the driver. An inertial measurement unit (IMU) sensor was used for the data acquisition.

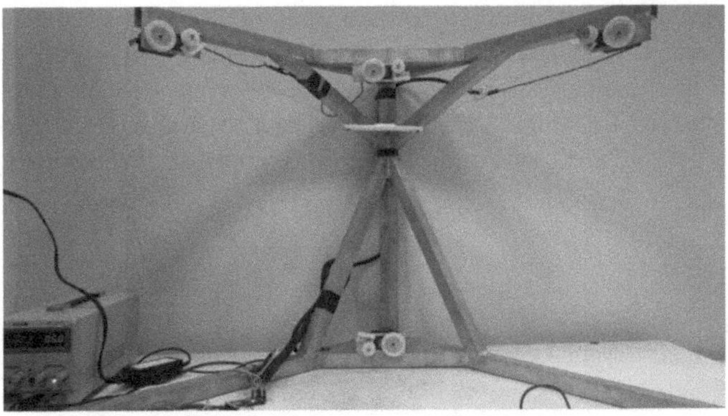

Fig. 1. A prototype of LAWEX at LARM in Cassino

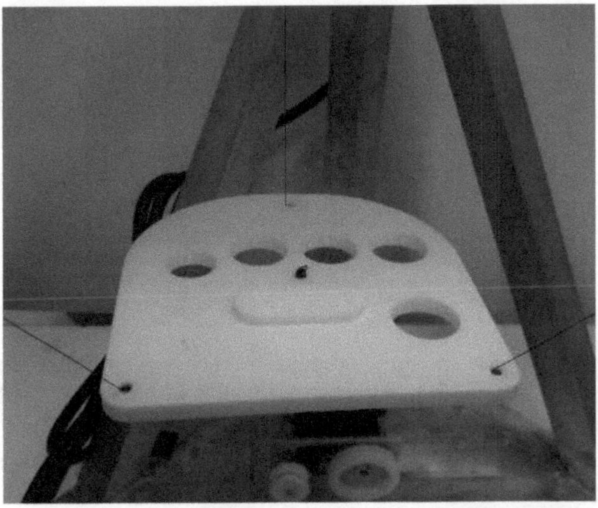

Fig. 2. First design of the mechanical interface end-effector

In order to make the system suitable for a larger range of rehabilitation, the end-effector needs to be redesigned as adjustable in dimensions between a maximum and a minimum value considered. In addition, it has to be also suitable for both arm rehabilitation as well as for leg. Another key point of the new end-effector should be its availability to fix the limb during the exercise. In this way, only the desired group of muscle will be exercised. In order to find a proper solution, several ideas have been analysed. For the dimensioning of the end-effector the average sizes of the wrist (15.9 cm in diameter) and of ankle (23.1 cm in diameter) [9] within the European population have been taken into account.

The solution shown in Fig. 3 consists in two parts: one rigid part in the shape of a ring and straps that would fix the limb in the centre. It is not considered a viable solution because of the fact that the limb would be supported only by the straps and it is both uncomfortable and it does not provide a proper fixing of the limb. Another solution of an end effector was discussed, as shown in Fig. 4a and b. This one has a rigid structure in the shape of a ring and straps attached to it in the inner inferior part. The attachment is similar as a wrist watch. The straps would converge in the middle and adjust to the dimension of the limb, fixing it to the rigid ring.

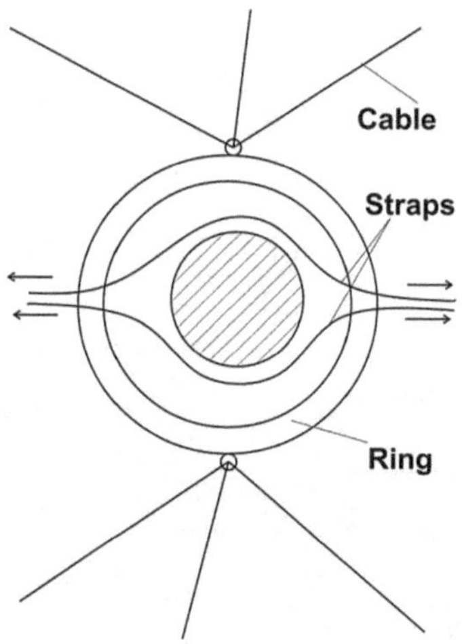

Fig. 3. First solution: rigid ring and straps

There are two options: the first could have been 3D printed rings as point of attachment and the other would have two screws that would fix the cables into the ring (the same way as the cables are fixed on the end-effector in Fig. 2). The thing that is questioned in this solution is if the straps can fix well enough the limb so that it would not rotate during the exercise.

In Fig. 5 is presented a final version of an end-effector. The ideas from Fig. 4 were combined and improved by making it in two parts for easier insertion of the limb. Some material was also added to the lower part in order to have a straight surface for comfort and stability. The fixing of the two parts will be done with two screws; one will remain in the same position having the purpose of keeping the two parts together and the other one will allow the partial detachment, inserting the limb and closing it. This version was printed and tests were carried out, both with hand and leg.

3 A FEM Analysis

To analyse the performance of the proposed mechanical interface before manufacturing it, a Static FEM analysis has been worked out using the Meshed CAD model shown in Fig. 6. The Mesh details are listed in Table 1. The structure of the system has been considered as fully made in PLA plastic, Table 2 shows its characteristics. The tensions

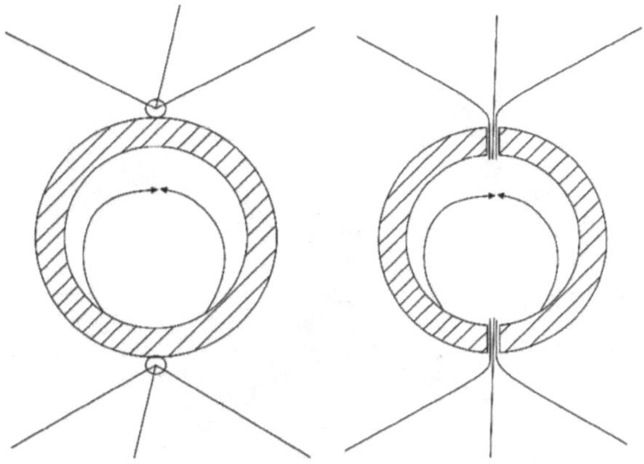

Fig. 4. Second solution: a. with printed ring connectors for cable attachment, b. with screws

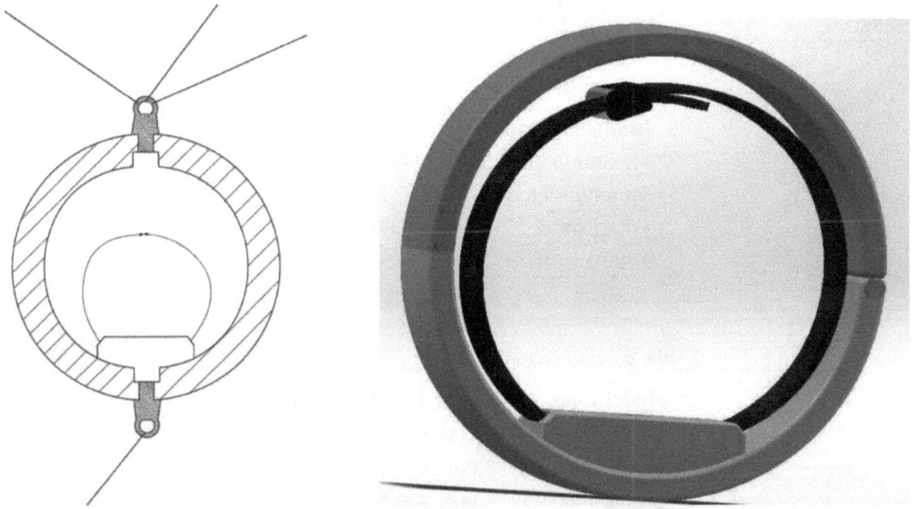

Fig. 5. Final solution: a. a scheme, b. a CAD model

of the cables due to the weight of the end-effector and arm reaction are considered as external loads for the structural analysis. A force of 50 N applied on the bottom ring of the end-effector. This force has been calculated from the deadweight of an average arm in a worst-case scenario to check the stiffness of the structure for rehabilitation purposes. The structure fixed on its bottom and the system is subject to gravity. Figure 6 shows the factor of safety distribution using Maximum Normal stress criterion since PLA is a brittle material. According to the maximum principle stress failure criterion, the Factor of Safety (FOS) is

Fig. 6. Mesh details

Table 1. Mesh details

Jacobian points	4 Points
Maximum element size	46.7324 mm
Minimum element size	2 mm
Total Nodes	284766
Total Elements	160278

Table 2. PLA characteristics for the end-effector

Model type	Linear Elastic Isotropic
Yield strength	45.00 N/m^2
Tensile strength	3.00E + 7 N/m^2
Elastic modulus	2.00E + 9 N/m^2
Poisson's ratio	0.39
Mass density	1020.00 kg/m^3

$$Fos = \frac{\sigma_{\text{limit}}}{\sigma_1} \tag{1}$$

In the simulation the minimum computed FOS is 1.178 on the frame top-left as shown in Fig. 7. Since no region of the structure reaches a FOS < 1 the material can afford the external load with no failure validating the feasibility of the end-effector.

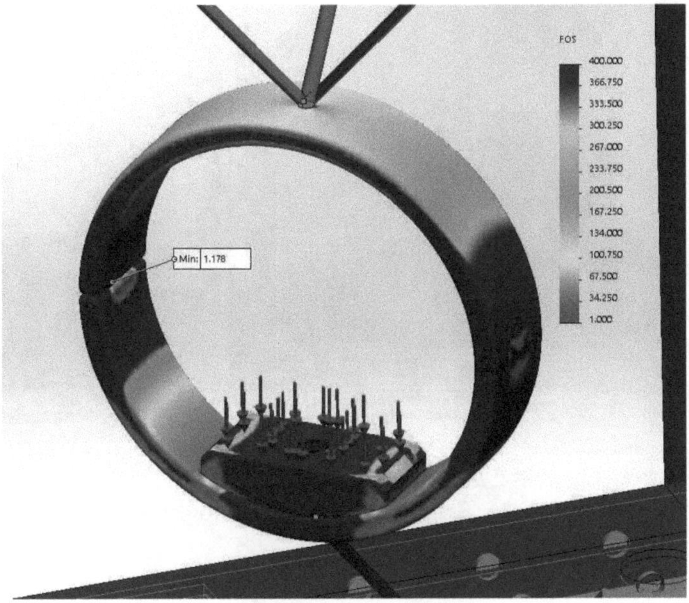

Fig. 7. FEM FOS distribution

4 Experimental Prototype and Testing Modes

In order to validate the end-effector, the natural movement of the limbs was studied and trajectories were obtained by means of image analysis using MATLAB. These trajectories were compared with the ones obtained by testing the end-effector with a healthy subject, a woman between 20 and 25 years old. The results are good; they are close to the intended trajectory. The tests show the movement within the XZ plane since it is the principal plane in which the movement is performed even if the device is able to move in XYZ, having the 3 mentioned Doffs controlled (Figs. 8 and 9).

5 Design and Operation Considerations

The end-effector was designed in a circular shape to mimic the shape of the limb; thus, it has a circular shape. The straps have the role to fix the limb in place so only the desired muscles would be stressed while exercising. It was carefully thought to be lightweight and easy to use so that the patient will not be afraid of using it. It has two ring screws in the upper and the lower part so that the cables can be easily attached and detached if needed. The distance between the attachment points needs to remain the same in order to use the same procedure for any dimension of the limb. This is the reason why the outer part of the end-effector needs to be rigid. Although, the drawback of fulfilling this necessity and wanting to make it adaptable for a large range of limb dimensions, the end-effector ended up to be quite big and influencing the range of motion for small limbs.

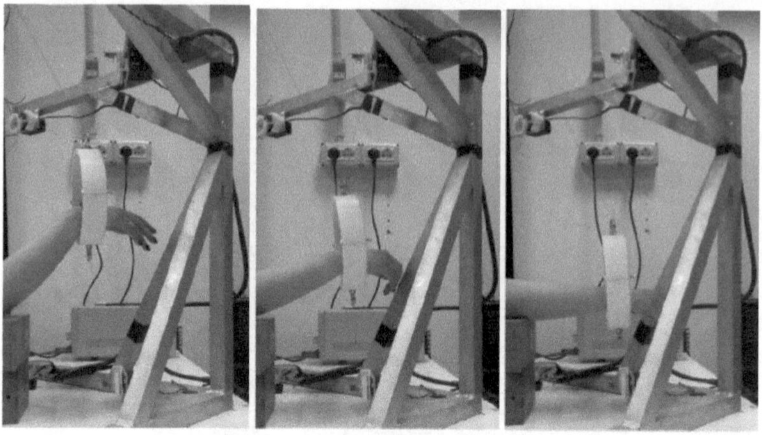

a.

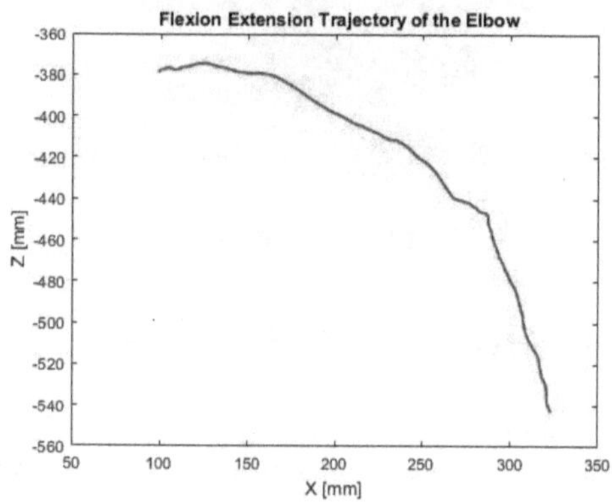

b.

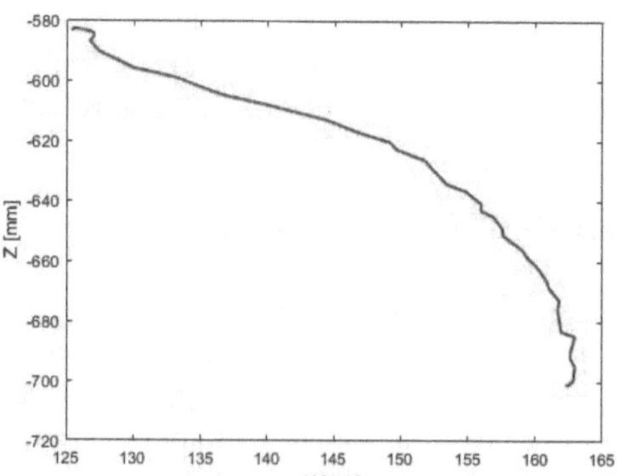

c.

Fig. 8. Experimental result with a European woman: a. arm with the prototype, b. desired trajectory, c. trajectory obtained by image analysis

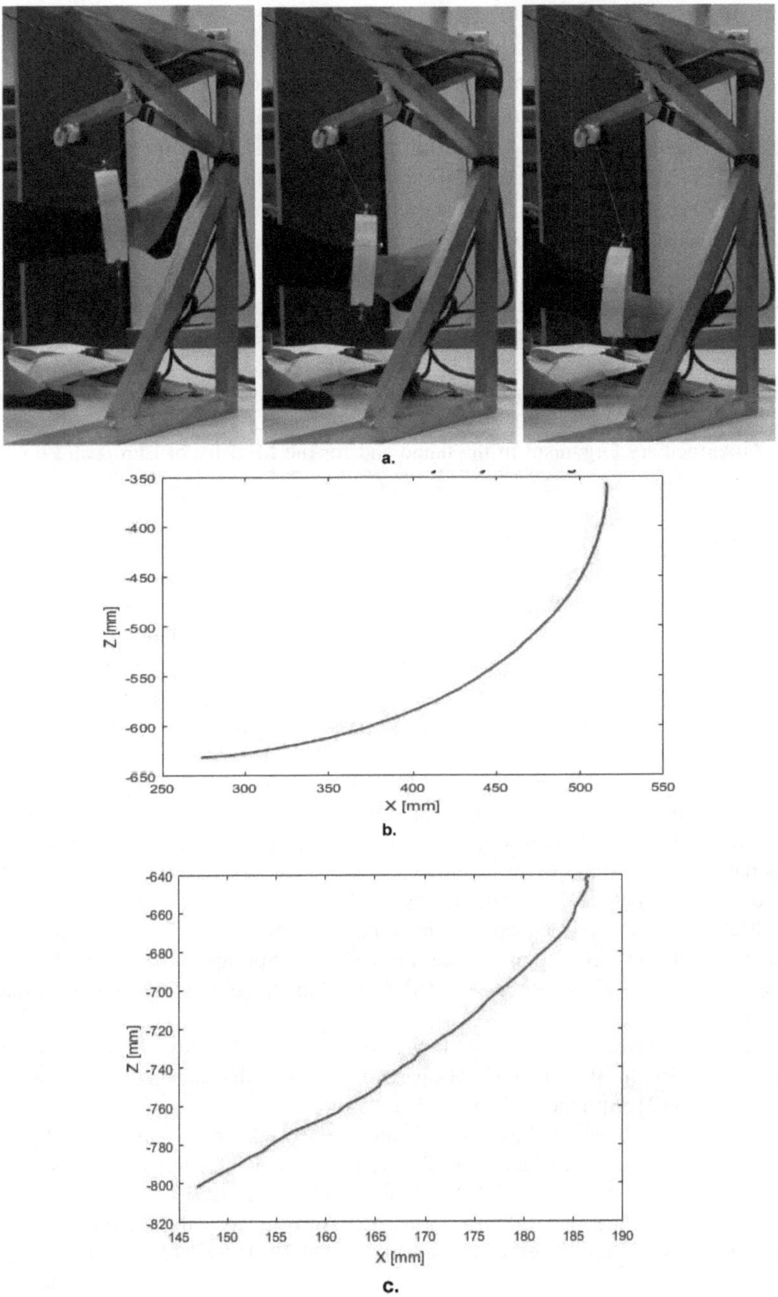

Fig. 9. Experimental result with a European woman: a. leg with the prototype, b. desired trajectory, c. trajectory obtained by image analysis

6 Conclusions

The paper presents a user-oriented mechanical interface as an end-effector for limb rehabilitation cable-driven parallel manipulator. The novel design has been developed looking for a feasible mechanical solution both for cable attachments and end-effector shape to ensure comfort, safety and efficiency in operation with user. A prototype has been properly sized considering also stress analysis. Finally, test results are discussed to show the feasibility and characterization of chosen solution.

Acknowledgements. The paper presents results from the research activities of the project ID 37_215, MySMIS code 103415 "Innovative approaches regarding the rehabilitation and assistive robotics for healthy ageing" co-financed by the European Regional Development Fund through the Competitiveness Operational Programme 2014–2020, Priority Axis 1, Action 1.1.4, through the financing contract 20/01.09.2016, between the Technical University of Cluj-Napoca and ANCSI as Intermediary Organism in the name and for the Ministry of European Funds.

References

1. Carbone G, Gherman B, Ulinici I, Vaida C, Pisla D (2017) Design issues for an inherently safe robotic rehabilitation device. In: Proceedings of the 26th international conference on robotics in alpe-adria-danube region, RAAD 2017, Dorchect, Springer, pp 967–974
2. Major, K.A., Major, Z.Z., Carbone, G., Pisla, A., Vaida, C., Gherman, B., Pisla, D.L.: Ranges of motion as basis for robot-assisted post-stroke rehabilitation. HVM Bioflux **8**(4), 192–196 (2016)
3. Gonçalves, S, Mendes Carvalho, R, JC (2012) Robot Modeling for Physical Rehabilitation, Service Robots and Robotics; Design and Applications
4. Gosselin, C., Cardou, P., Bruckmann, T., Pott, A.: Cable-driven parallel robots. Springer, Netherlands (2018)
5. Barbazza, L., Zanotto, D., Rosati, G., Agrawal, S.K.: Optimal design of a reconfigurable end-effector for cable-suspended parallel robots. In: Boschetti, G., Gasparetto, A. (eds.) Advances in Italian mechanism science, pp. 267–275. Springer, Cham (2017)
6. Chang, W.H., Kim, Y.-H.: Robot-assisted therapy in stroke rehabilitation. J. Stroke **15**(3), 174–181 (2013)
7. Carbone, G., Cavero, C.A., Ceccarelli, M., Altuzarra, O.: A study of feasibility for a limb exercising device. In: Boschetti, G., Gasparetto, A. (eds.) Mechanisms and machine science, vol. 47, pp. 11–21. Springer, Cham (2017)
8. Ceccarelli, M.: Problems and experiences on cable-based service robots for physiotherapy applications. In: Pisla, D., Bleuler, H., Rodic, A., Vaida, C., Pisla, A. (eds.) New trends in medical and service robots, pp. 27–42. Springer, Dordrecht (2013)
9. Polymeris, A., Papapetrou, P.D., Katsoulis, G.: An average body circumference can be a substitute for body mass index in women. Adv Med **2014**, 592642 (2014)

RAISE - An Innovative Parallel Robotic System for Lower Limb Rehabilitation

Calin Vaida[1](✉), Iosif Birlescu[1], Adrian Pisla[1], Giuseppe Carbone[2], Nicolae Plitea[1], Ionut Ulinici[1], Bogdan Gherman[1], Ferenc Puskas[3], Paul Tucan[1], and Doina Pisla[1]

[1] CESTER, Technical University of Cluj-Napoca, Cluj-Napoca, Romania
{calin.vaida, iosif.birlescu, nicolae.plitea, ionut.ulinici, gogdan.gherman, paul.tucan, doina.pisla}@mep.utcluj.ro, adrian.pisla@muri.utcluj.ro

[2] LARM University of Cassino, Cassino, Italy
carbone@unicas.it

[3] S.C. Electronic APRIL Ltd, Cluj-Napoca, Romania
april_private@yahoo.com

Abstract. Robotic rehabilitation is a developing field of research which could become a necessity in the next decades due to the natural shifting of the population age. The aim of the paper is to present an innovative robotic device for the rehabilitation of the lower limb of patients in the acute post stroke phase. The parallel robot has a modular construction enabling the mobility of each major joint: hip, knee and ankle. Furthermore, the robotic system kinematics is presented and the singularities are derived, to validate the proposed system for its intended medical task, considering the upper limb motion requirements for the rehabilitation exercises.

Keywords: Parallel robot · Lower limb rehabilitation · Kinematics Singularity

1 Introduction

Human advancements in the past decades, especially in the various domains of science have changed the society and lifestyle for the better, with one of the most sizeable improvements being registered in all branches of medicine, with a clear indicative of this being the decrease registered in the lethality of many diseases. One such affliction is the cardio vascular attack (CVA) or stroke. As such, the current survivability rate of stroke is 75% one year after the incident and this number is expected to increase in the future. This means that there is an increase in the number of patients suffering from the after effects of CVA's, with the most common side effect being hemiparesis i.e. a partial or complete loss of motor abilities on the side of the body corresponding to the brain hemisphere where the stroke occurred. This increase in patient numbers is also supported by the fact that the target group for strokes is comprised of people aged 65 and over, a target group that will end up representing 30% of European population by 2050 [1]. This will prove to

G. Carbone et al. (Eds.): *New Trends in Medical and Service Robotics*, MMS 65, pp. 293–302, 2019.
https://doi.org/10.1007/978-3-030-00329-6_33

be overwhelming when compared to the rate at which qualified medical personnel will become available, as such this marks the beginning of a physical therapist crisis, to which the solution is the use of advanced robotic rehabilitation devices.

Aside from solving the predicted patient number inflation, robotic rehabilitation devices would also be able to ease the rehabilitation procedure and compensate for some physically demanding aspects of the procedure, especially when it comes to the acute post stroke stage of rehabilitation. In the post stroke stage, the patient, having gone through severe mental and physical stress, is going to be mainly restricted to a hospital bed, and the ability to move any of the affected limbs will be close to zero, therefore the limbs will be entirely manipulated by the therapist. A robotic device would not encounter any sort of difficulty under those conditions, given of course that its structure will be able to be seamlessly incorporated in the bed's structure. The patient's condition is also quite sensible in this stage of post stroke rehabilitation, as such moving the patient from one point to another regardless of the distance might do more harm than good, therefore the best decision is to bring the necessary tools to the patient, and this is achievable with a rehabilitative robot.

The paper proposes a lower limb rehabilitation parallel robot for the hip, knee and ankle joints, with two modules (hip/knee module and ankle module) which are linked together to allow the rehabilitation exercises for all three joints at the same time. The robotic device should be place adjacent to a bed in order to fully sustain the limbs weight while the bed sustains the upper body weight. The kinematics of the robotic system are presented and the singularities are derived to determine the safe robot exploitation conditions, and to validate the design (for the medical exercises).

This research work is carried out within Agewell project in a team of engineers, neurologist and physiotherapist focusing at both lower and upper limb post stroke rehabilitation. Some previous works are reported in [2–4]. The paper is structured as follows: Sect. 1 continues with a state of the art of the robotics in rehabilitation (Sect. 1.1); Sect. 2 presents the RAISE modular robotic system for the lower limb rehabilitation; Sect. 3 presents the robotic system kinematics and the singularities for each rehabilitation module; Sect. 4 illustrates the conclusions and the future work.

1.1 The Present State of Robotic Rehabilitation Technologies

The implementation of robotic rehabilitation devices may provide a series of advantages to traditional therapy bringing also a paradigm change of today's rehabilitation procedures by shifting the role of the medical therapist from performing exercises to programming rehabilitation devices and the setup of personalized treatment and monitoring plans. Robotic devices allow the near seamless integration of wearable sensors in the rehabilitative process, which are essential to monitor the patient's wellbeing and progress during rehabilitation [5]. In most cases the efficiency of robotic rehabilitation and gait training, seems to be dependent on identifying the subject that would stand to benefit most from this kind of training, more specifically the compatibility between the patient's physical [6] and psychological [7] features and the available devices needs to be assessed. There are many commercial solutions or research achievements regarding the development of various devices for lower limb rehabilitation:

(1) The Lokomat device is a treadmill based 4 DOF gait rehabilitation exoskeleton robot [8]. It is composed a weight support system designed to maintain the patient in an upright position while the lower limb exoskeleton executes the necessary motions on the treadmill.
(2) The Active Ankle-Foot Orthosis or AAFO [9] developed by the Yonsei University in, Seoul, Korea is a one degree of freedom ankle-foot hinge orthosis, in this case the joint uses a hinge like structure and contact between the foot and the floor is achieved via a contact switch on the sole of the foot.
(3) The GTI or Gait Trainer [10] is a foot-plate based end effector device, developed by the University of Berlin, Germany. The device is intended to aid the patient in completing gait training, succeeding in reducing any physical effort necessary as well as saving medical resources for rehabilitation.
(4) The Rutgers ankle [11] is a platform-based end-effector robot, developed by the Rutgers University. The mechanism contained a moving platform, a fixed platform and six chains connected to the mobile platform, as such it achieved six degrees of freedom. And it includes a virtual reality interactive video game. Thus providing an additional incentive for the patient.
(5) The one degree of freedom exoskeleton like device [12], targeting the lower limb, is currently under development by the University of Craiova and presents a simple yet efficient solution for gait rehabilitation. The device uses a single rotary actuator mounted in the hip area, while the rest of the device's mechanical structure maintains the proper positioning of the device links to coincide with what is necessary during gait rehabilitation.
(6) Cable driven solutions for both upper and lower limb rehabilitation are currently being developed by the LARM laboratory in Cassino Italy. These devices showcase the advantages of adopting a parallel architecture in the development of rehabilitation technologies, as well as the simplicity encountered in the overall structure of such devices [3, 4]. LARM also proposed an exoskeleton based on cam mechanisms [13] outlining its feasibility by presenting simulations results for assisting tasks.

2 RAISE – The Parallel Robotic System

2.1 Targeted Lower Limb Rehabilitative Motions

The authors propose a parallel robotic system for the rehabilitation of the lower limb that targets patients in the acute post-stroke phase which are too weak and/or frail to use common systems like: treadmill-based exoskeleton robots, leg orthoses or foot plate-based end-effector devices [14]. The idea is to enable a patient which cannot be lifted from the bed to perform a variety of exercises for the lower limb mobilizing each single joint (hip, knee and ankle) for one or both limbs. The targeted motions are: the knee extension, the hip flexion and extension and the ankle flexion/extension and abduction/adduction (Fig. 1). The desired values for the lower limb ranges of motion are based on the maximum values [15], and are presented in Table 1. The robotic device is intended to be used on a patient in the acute stage of stroke, what this means is that the patient will have little to no control over the affected limb, moreover during this stage he/she will be constrained to a hospital bed, as such the device not only has to manipulate the leg but also sustain it completely.

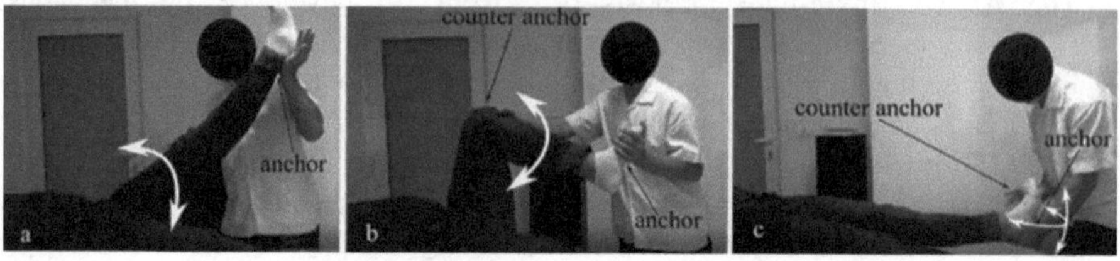

Fig. 1. Rehabilitation exercises for the lower limb

The actual motions effectuated as they should be, in the process of rehabilitation, are exemplified in by a kineto-therapist. During the rehabilitative procedure it is also necessary to respect several anchoring points which are pivotal during the manipulation of the respective limb segment, in order to maintain the proper posture and to avoid unnecessary strains in the muscles. As it can be observed considering the fact that rehabilitation in the acute stage is done with the patient laying down, the counter anchoring point is not necessary during hip motion exercise (Fig. 1), but necessary for the knee and ankle motion exercises.

2.2 The RAISE System

The lower limb rehabilitation parallel robotic system, RAISE, (designed to perform the exercises with the patient lying on its back) consists of two modules: one module for the hip and knee flexion/extension rehabilitation exercises, and the second one for the ankle flexion/extension and abduction/adduction rehabilitation exercises.

The kinematic scheme of the hip/knee rehabilitation module is illustrated in Fig. 2. A fixed coordinate frame OXYZ is chosen with its OY axis coincident with the revolute joint R_h rotation axis. For the rehabilitation exercises the angle α_1 represent the flexion/extension of the hip while the angle α_2 represents the flexion/extension of the knee. The mechanism has three kinematic chains defined as following: the *first kinematic chain* is $P\underline{P}R$ type, with a passive linear DoF (along OZ axis), an active linear DoF actuated by q_1 (along an axis parallel to OX), and a free rotation around an axis parallel to OY; the *second kinematic chain* is also $P\underline{P}R$ with a passive linear DoF (along the OZ axis), an active linear DoF along an axis parallel to OX (actuated by q_2), and a free rotation around an axis parallel with OY; the third kinematic chain is RR type with two passive rotations (one around the OY axis and the second around an axis parallel with OY). To perform the rehabilitation exercises the third kinematic chain should be placed on the outer side of the lower limb to allow anchoring the limb segments. The revolute R_1 joint of the first kinematic chain is linked with L_f (Fig. 2) at a distance l_1 from the R_h revolute joint (mounted sideways to the hip), and by actuating $q_1$the flexion/extension of the hip is achieved. Similarly, the revolute joint R_2 (from the

Table 1. Amplitude motion for the rehabilitation exercises

	Flexion [°]	Extension[°]	Abduction[°]	Adduction[°]
Hip	~*80*	~*5*	–	–
Knee	~*80*	*0*	–	–
Ankle	~*35*	~*20*	~*20*	~*20*

second kinematic chain) is linked with L_t at a distance l_2 (from R_k which is mounted sideways to the knee), and by actuating q_2 the flexion/extension of the knee is achieved.

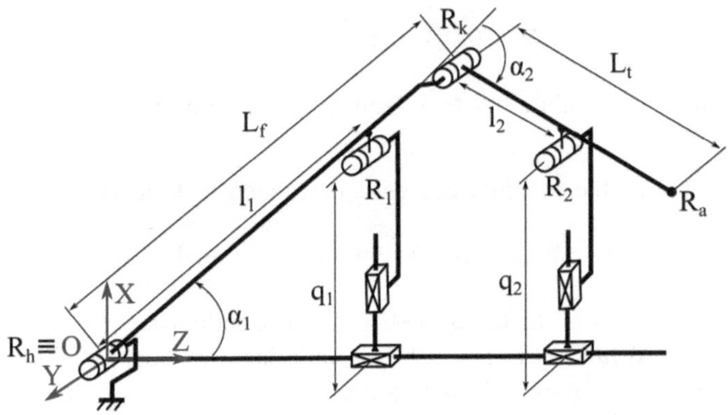

Fig. 2. Kinematic scheme of the hip-knee joint rehabilitation module

The ankle rehabilitation module kinematic scheme is presented in Fig. 3. The coordinate frame O′X′Y′Z′ is chosen with its origin located at the ankle center of rotation. The mechanism is RR type with two active rotations: around the O′Z′ axis (actuated by q_3), and around the R_{a1} axis of rotation (actuated by q_4).

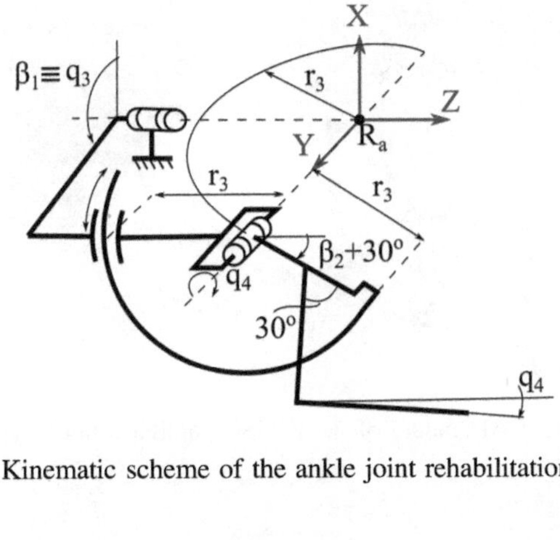

Fig. 3. Kinematic scheme of the ankle joint rehabilitation module

For the structural synthesis of the two mechanisms the next equation is used [16]:

$$M = (6 - F) \cdot N - \sum_{i=1..5} (i - F) \cdot C_i \tag{1}$$

where M is the mobility degree of the mechanism, F is the mechanism family (the number of common constraints for all mechanism elements), N is the number of mobile elements and Ci is the number of joints of i-class. The coefficients (i − F) in Eq. (1) cannot take negative values.

For the hip-knee module, the numerical values of the elements and the mobility degree are:

$$F = 3,\ N = 6,\ C_5 = 8 \Rightarrow M = 18 - 16 = 2 \tag{2}$$

Thus the hip-knee module has two degrees of freedom that correspond to the two targeted motions.

For the ankle module the mobility degree is calculated as follows:

$$F = 2,\ N = 3,\ C_5 = 4 \Rightarrow M = 6 - 4 = 2 \tag{3}$$

The preliminary design of the RAISE parallel robotic system with its main components is illustrated in Fig. 4. The ankle module has two degrees of freedom which may allow the orientation of the joint around two axes, resulting in the flexion/extension motion and the abduction/adduction one (Fig. 4.b). Since the ankle rehabilitation module is attached to the limb rehabilitation module, the RAISE robotic system may perform simultaneously the rehabilitation of all tree limb joints (hip, knee and ankle).

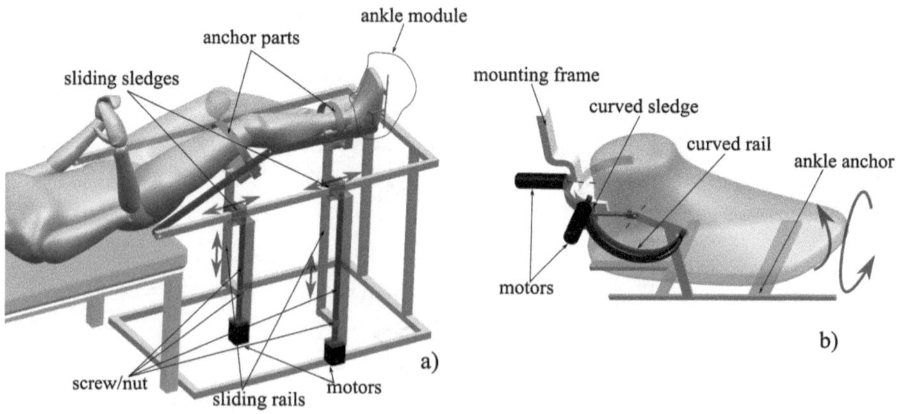

Fig. 4. CAD model of the RAISE parallel robotic system

3 Kinematic Modelling of the RAISE Robotic System

For the kinematic modelling of the RAISE robotic system the following geometric parameters are introduced:

- L_f, L_t – the distances between the rotation centers of the hip and knee and respectively knee and ankle (Fig. 2);
- l_1 – the distance between the hip joint and the first anchor point of the mechanism (Fig. 2);
- l_2 – the distance between the knee joint and the second anchor point of the mechanism (Fig. 2);
- q_1, q_2 – active prismatic joints for the hip-knee joints rehabilitation module;
- q_3, q_4 – active revolute joints for the hip-knee joints rehabilitation module.

All the geometrical elements have adjustable values to adapt to the different anthropometric characteristics and will be used without numerical values in the kinematic model.

For the inverse kinematic model, the values of the four angular values, α_1, α_2, β_1, β_2 are given as input data. It results that:

$$q_1 = l_1 \cdot \sin(\alpha_1) \tag{4}$$

The knee joint has the following coordinates:

$$\begin{cases} X_{Rk} = l_f \cdot \sin(\alpha_1) \\ Y_{Rk} - 0 \\ Z_{Rk} = l_f \cdot \cos(\alpha_1) \end{cases} \tag{5}$$

The second anchor point, R2 has the following coordinates:

$$\begin{cases} X_{R2} = X_{Rk} + l_2 \cdot \sin(\alpha_1 - \alpha_2) \\ Y_{R2} = 0 \\ Z_{R2} = Z_{Rk} + l_2 \cdot \cos(\alpha_1 - \alpha_2) \end{cases} \tag{6}$$

And finally, the Ra coordinates, that defines the center of rotation for the ankle module that is attached to the hip-knee module, has the following coordinates:

$$\begin{cases} X_{Ra} = X_{Rk} + L_t \cdot \sin(\alpha_1 - \alpha_2) \\ Y_{Ra} = 0 \\ Z_{Ra} = Z_{Rk} + L_t \cdot \cos(\alpha_1 - \alpha_2) \end{cases} \tag{7}$$

The coordinates of the active joints are:

$$\begin{aligned} q_2 &= Z_{R2} \\ q_3 &= \beta_1 \\ q_4 &= \beta_2 \end{aligned} \tag{8}$$

Based on the determined equations, (1) ÷ (8), the closure equations are the following:

$$\begin{cases} f_1 : q_1 - l_1 \cdot \sin(\alpha_1) = 0 \\ f_2 : q_2 - L_f \cdot \cos(\alpha_1) - l_2 \cdot \cos(\alpha_1 - \alpha_2) = 0 \\ f_3 : q_3 - \beta_1 = 0 \\ f_4 : q_4 - \beta_2 = 0 \end{cases} \tag{9}$$

Using Eq. (9), the Jacobi matrices A and B can be easily computed, resulting in the following expressions:

$$A = \begin{bmatrix} -l_1 \cdot \cos(\alpha_1) & 0 & 0 & 0 \\ \mathrm{L_f}\sin(\alpha_1) + \mathrm{l_2}\sin(\alpha_1 - \alpha_2) & -\mathrm{l_2}\sin(\alpha_1 - \alpha_2) & 0 & 0 \\ 0 & 0 & -1 & 0 \\ 0 & 0 & 0 & -1 \end{bmatrix} B = \begin{bmatrix} 1 & 0 & 0 & 0 \\ 0 & 1 & 0 & 0 \\ 0 & 0 & 1 & 0 \\ 0 & 0 & 0 & 1 \end{bmatrix} \tag{10}$$

Singularities Analysis

Using Eq. (10) the determinants of the Jacobi matrices can be computed in order to perform the singularities analysis. Thus, it results:

$$\begin{aligned} \det(\mathrm{A}) &= l_1 \cdot l_2 \cdot \sin(\alpha_1 - \alpha_2) \cdot \cos(\alpha_1) \\ \det(\mathrm{B}) &= 1 \end{aligned} \tag{11}$$

Based on (11) there are two cases of possible singular configurations for the RAISE parallel robotic system:

1. $\sin(\alpha_1 - \alpha_2) = 0 \Leftrightarrow \alpha_1 - \alpha_2 = 0 \Leftrightarrow \alpha_1 = \alpha_2$ - this position describes, from the geometrical point of view, the case when all the rotational joints are collinear, that would represent the boundary line between the two working modes of the robot. However, as the angle α_2 can take only negative values the singularity can be met only when both angles are equal to 0, position that can be avoided from the initial configuration of the robot;
2. $\cos(\alpha_1) = 0 \Leftrightarrow \alpha_1 = \frac{\pi}{2} \equiv \alpha_1 = 90^\circ$ - this describes, from the geometrical point of view the situation when the thigh is lifted vertically. However, the system does now allow the thigh to reach an angle of more than 80° (for safety reasons to avoid any pressure upon the joints).

It can be seen (from the above singularity analysis) that there are no singular configurations inside the workspace of the robot, validating it as safe solution for medical rehabilitation.

4 Conclusions

A parallel robotic system, RAISE, designed for the rehabilitation of the lower limb for patients in the acute post-stroke phase that require bed rehabilitation has been designed and presented in the paper. The robotic system has a modular construction having a module with 2 DoF for the rehabilitation of the hip and knee and a second one with 2 DoF for the rehabilitation of the ankle. The robotic modules are linked together to allow the rehabilitation exercises of all three joints to be performed at the same time (when needed). To validate the proposed robotic system for the medical exercises, the inverse kinematics and singularities analysis have been presented illustrating that the robot has a safe behavior.

Future work is intended to develop the first constructive design, and to simulate multiple medical relevant rehabilitation exercises (including complex motion such as gait rehabilitation).

Acknowledgments. The paper presents results from the research activities of the project ID 37_215, MySMIS code 103415 "Innovative approaches regarding the rehabilitation and assistive robotics for healthy ageing" cofinanced by the European Regional Development Fund through the Competitivness Operational Programme 2014-2020, Priority Axis 1, Action 1.1.4, through the financing contract 20/01.09.2016, between the Technical University of Cluj-Napoca and ANCSI as Intermediary Organism in the name and for the Ministry of European Funds.

References

1. Robotics 2020, Multi-Annual Roadmap for Robotics in Europe, Release B 02/12/2016
2. Vaida, C., et al.: On human robot interaction modalities in the upper limb rehabilitation after stroke, ACTA TECHNICA NAPOCENSIS, Applied Mathematics, Mechanics, and Engineering, Vol. 60, Issue I, March, 2017
3. Carbone, G., et. al.: A study of feasibility for a limb exercising device. In: Boschetti, G., Gasparetto, A. (eds.) Advances in Italian Mechanism Science, pp. 11–21. Springer International Publishing AG (2017). https://doi.org/10.1007/978-3-319-48375-7_2
4. Carbone, G., et. al.: Design Issues for an inherently safe robotic rehabilitation device. In: Proceedings of the 26th International Conference on Robotics in Alpe-Adria-Danube Region RAAD 2017, Dorchect, pp. 967–974. Spinger (2017)
5. Tarnita, D.: Wearable sensors used for human gait analysis. Rom. J. Morphol. Embryol. **57**(2), 373–382 (2016)
6. Morone, G., et al.: Who may have durable benefit from robotic gait training? a 2-year follow-up randomized controlled trial in patients with subacute stroke. Stroke **43**(4), 1140–1142 (2012)
7. Straudi, S., et al.: tDCS and robotics on upper limb stroke rehabilitation: effect modification by stroke duration and type of stroke. Biomed. Res. Int. **2016**, ID. 5068127 (2017)
8. Barroso F., et al., Influence of the robotic exoskeleton Lokomat on the control of human gait: an electromyographic and kinematic analysis. Portuguese Meeting in Bioengineering, Braga, Portugal, pp. 1–6 (2013)
9. Blaya, J.A., Herr, H.: Adaptive control of a variable-impedance ankle-foot orthosis to assist drop-foot gait. IEEE Trans. Neural Syst. Rehabil. Eng. **12**(1), 24–31 (2004). https://doi.org/10.1109/TNSRE.2003.8232660
10. Iosa, M., et al.: Driving electromechanically assisted gait trainer for people with stroke. J. Rehabil. Res. Dev. **48**(2), 135–145 (2011). https://doi.org/10.1682/JRRD.2010.04.0069
11. Yoon, J., et al.: A 6-DOF gait rehabilitation robot with upper and lower limb connections that allows walking velocity updates on various terrains. IEEE/ASME Trans. Mechatron. **15**(2), 201–215 (2010)

12. Geonea, I., Tarnita, D.: Design and evaluation of a new exoskeleton for gait rehabilitation. Mech. Sci. **8**(2), 307–322 (2017)
13. Copilusi, C., et al.: Design and numerical characterization of a new leg exoskeleton for motion assistance. Robotica **33**(05), 1147–1162 (2014)
14. Zang, X., Yue, Z., Wang, J., Robotics in lower-limb rehabilitation after stroke. Behavioural Neurology, 13 p. (2017), https://doi.org/10.1155/2017/3731802
15. Norkin, C., White, I.: Measurement of Joint Motion: A Guide to Goniometry, 4th edn. F.A. Davis, Philadelphia (2009)
16. Plitea, N., et al.: Structural analysis and synthesis of parallel robots for brachytherapy. New Trends Med. Serv. Robot. **16**, 191–204 (2014). Springer, MMS

Design of Dual-Arm Exoskeleton for Mirrored Upper Limb Rehabilitation

Iuliu Nadas[1,3(✉)], Doina Pisla[1], Marco Ceccarelli[2], Calin Vaida[1], Bogdan Gherman[1], Paul Tucan[1], and Giuseppe Carbone[2]

[1] CESTER-Research Center for Industrial Robots Simulation and Testing, Technical University of Cluj-Napoca, Cluj-Napoca, Romania
iuliu.nadas@itim-cj.ro, {doina.pisla, calin.vaida, bogdan.gherman, paul.tucan}@mep.utcluj.ro
[2] University of Cassino and South Latium, Cassino, Italy
{marco.ceccarelli, carbone}@unicas.it
[3] National Institute for Research and Development of Isotopic and Molecular Technologies, Cluj-Napoca, Romania

Abstract. The paper addresses the upper limb rehabilitation for post-stroke patients. A dual-arm exoskeleton for the elbow and wrist flexion-extension is proposed. The added benefit of this robotic system is given by its portability and its ability to perform mirrored exercises using the motions provided by the unimpaired limb. The dual-arm exoskeleton addresses patients in the chronic phase after the stroke and even the patients in the post-acute phase allowing them to do rehabilitation exercises themselves.

Keywords: Design · Exoskeletons · Rehabilitation

1 Introduction

One of the most frequent causes of limb disabilities worldwide is generated by stroke. Due to the aging of the population in the next 20 years is estimated that stroke rate will increase [1], therefore the number of physiotherapists that will do rehabilitation physiotherapy will be exceeded by the number of cases of vascular accidents. In order to prevent this imminent problem, a solution that could replace these shortcomings would be robotic rehabilitation. The advantage of robotic systems used in rehabilitation training compared with conventional therapies consists of increasing training intensity and frequency and the possibility to train independently. From kinematic aspect this paper proposes an exoskeleton structure as a robotic rehabilitation solution.

An exoskeleton is a wearable mechanical structure that has its joints and links corresponded with the human limbs joints [2]. The advantage of an exoskeleton is the reduced necessary workspace compared with robotic systems using an end-effector for performing the same therapeutic exercises. The proposed robotic rehabilitation system intends to combine the simplicity and effectiveness of exoskeletons that use two simple motions, with passive-mirror rehab therapy method, resulting a solution that helps patients to perform their own rehabilitation training even at home. This solution consists in using the less affected arm equipped with the master exoskeleton as a driver and

G. Carbone et al. (Eds.): *New Trends in Medical and Service Robotics*, MMS 65, pp. 303–311, 2019.
https://doi.org/10.1007/978-3-030-00329-6_34

the other impaired limb is equipped with the slave exoskeleton, the second exoskeleton being a mirrored mechanism of the first exoskeleton.

This bilateral upper training method is achieved using dual-arms robotic devices and it has been previously used an example being ARAMIS [3], a robotic system with 6 DOF. In the case of ARAMIS the position of the rotation center of the shoulders may change during exercises and this leads to a complex control algorithm.

An interesting kind of bilateral upper limb training method after stroke is presented in MIME [4] concept that is a rehabilitation system that uses two industrial robots Puma 560 having 6 DOFs and being able to perform rehabilitation exercises for shoulder and elbow.

Another example of an exoskeleton robotic device is the ReExRob [5] designed for the following motions: elbow flexion/extension, pronation/supination, wrist flexion/extension and adduction/abduction. It has been designed for single arm use for patients in the post-acute phase of stroke.

Table 1 shows a few examples of bilateral therapy devices compared with proposed device highlighting the motion types, type of rehabilitation training and number of degree of freedom.

Table 1. Comparison table between bilateral rehabilitation systems

Name	Motion types	Rehabilitation training type	DOF's
The ARAMIS [3]	Shoulder flexion/extension and adduction/abduction Elbow flexion/extension Wrist adduction/abduction	Synchronous exercises Asynchronous exercises Training in immersive virtual reality settings	6
BATRAC [6]	Shoulder flexion/protraction Elbow flexion/extension	Exercises with both upper limb simultaneously (in-phase) Alternatingly (anti-phase)	2
Reha-Slide Duo [7]	Shoulder adduction/abduction Elbow flexion/extension Wrist flexion/extension	Synchronous exercises	2
APBT (the Rocker) [8]	Wrist flexion/extension	Mirror symmetrical exercises Near-symmetrical exercises (60° phase lag)	1
DualEx [REF]	Elbow flexion/extension Wrist flexion/extension	Passive-mirrored therapy Unilateral training therapy	2

The robotic system presented in this paper addresses patients in the chronic or even post-acute phase. It has a simple 2 DOF architecture and focuses on the rehabilitation of two basic motions of the upper limb: flexion/extension of elbow and flexion/extension of the wrist, combining the advantage of simplicity with portability.

2 Design Requirements

The main goal for patients after stroke is to regain the postural and motoric control. In the case of the upper limb the main reason for rehabilitation is to provide the ability to reach different positions in order perform daily activities.

The role of therapists is to perform rehabilitation exercises aiming to correct the passive posture and the limitation of active movements of the upper limb. The predicted lack of therapists in next decades will lead to increased costs of therapy after stroke. Thus, a change of paradigm needs to be adopted meaning that rehabilitation therapy after stroke has to move forward in being performed mostly with the help of robotic systems so the announced shortage of human specialists in motoric rehabilitation after a stroke will be covered by robotic systems.

To measure the motions range of human upper limb joints and the forces involved in this motion, the physiotherapists use two instruments: goniometer and dynamometer. The acquired data are necessary in designing the proper therapeutic interventions.

Major et al. [9] made a set of measurements on a group of 21 patients' men and women having common sings of comorbidity. The specific measurements are made using a goniometer to measure the range of motion around a joint in the upper limb. The study concluded by indicating an influence of age over the range of motion, older patients having stiffer joints that reduce the motion amplitudes, also a higher range of motion presented at women patients, these data represent starting point in the design of the robot. For safety reason all patients might use data taken from men studies or the therapy might be individualized by sex.

In Fig. 1 are shown the amplitudes of motion necessary in upper limb rehabilitation and the maximum mean ranges necessary for each target motion of the arm.

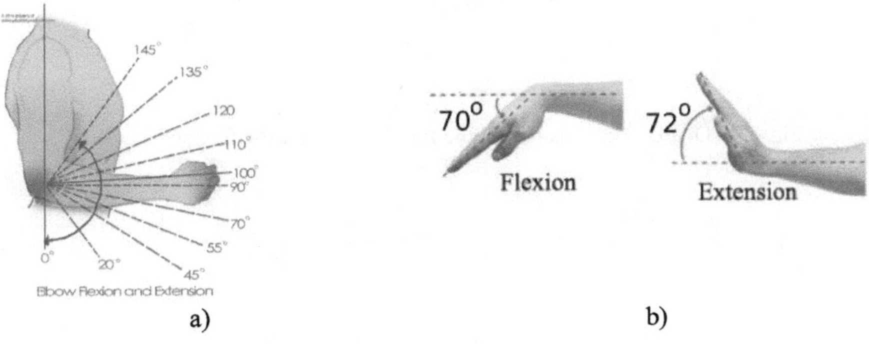

Fig. 1. Upper limb motion amplitudes (a) elbow flexion/extension; (b) wrist flexion/extension [5]

3 A Novel Design Architecture

Based on the rehabilitation motion of the upper limb a 2 DOF serial robotic structure is designed. The kinematic scheme of this structure is presented in Fig. 2.

The forward kinematic model for this robot system uses as input data the values of the active joints q1, q2 and length of links l1 and l2, having to determine the coordinates of the end-effector (XE, YE, ZE).

The fixed system of coordinates OXYZ was placed in the middle of the first rotation. Relation (1) is used to obtain coordinates of E, the results being presented in (5). Relations (2) and (3) show the transformation matrices with rotations around Y and Z, relation (4) is transformation matrix with translation along X.

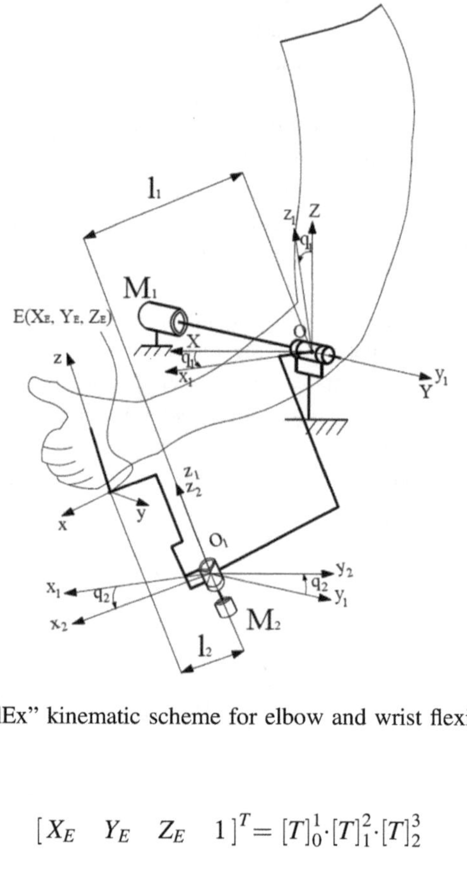

Fig. 2. "DualEx" kinematic scheme for elbow and wrist flexion/extension

$$[X_E \quad Y_E \quad Z_E \quad 1]^T = [T]_0^1 \cdot [T]_1^2 \cdot [T]_2^3 \tag{1}$$

$$[T]_0^1 = \begin{bmatrix} \cos(q_1) & 0 & \sin(q_1) & 0 \\ 0 & 1 & 0 & 0 \\ -\sin(q_1) & 0 & \cos(q_1) & 0 \\ 0 & 0 & 0 & 1 \end{bmatrix} \tag{2}$$

rotation around OY axis with q_1

$$[T]_1^2 = \begin{bmatrix} \cos(q2) & -\sin(q2) & 0 & l1 \\ \sin(q2) & \cos(q2) & 0 & 0 \\ 0 & 0 & 1 & 0 \\ 0 & 0 & 0 & 1 \end{bmatrix} \tag{3}$$

Geometric parametric translation along O_1X_1 with l_1 followed by Rotation OZ axis with q_2

$$[T]_2^3 = \begin{bmatrix} l2 \\ 0 \\ 0 \\ 1 \end{bmatrix} \tag{4}$$

Geometric parametric translation along O_2X_2 with l_2

$$\begin{aligned} XE &= l2\cos^2(q_1) + l1\cos(q_1) \\ YE &= l2\sin(q2) \\ ZE &= -l_2\sin(q_1)\cos(q_1) - l_1\sin(q_1) \end{aligned} \tag{5}$$

X_E, Y_E, Z_E represents the position coordinates of end-effector.

The passive-mirror physiotherapy is the starting point of this concept, is dedicated to patients who are in the post-acute and chronic phases after stroke. The idea of the device started from the premises that patients who suffered a stroke in most cases are paralyzed on one side of their body, the other one remains functional. The robotic system can be worn using a harness attached through an anchor long enough to be used as a support point for both arms to perform flexion/extension motions of elbows.

The operating mode is based on passive-mirrored therapy. The healthy arm performs strictly the physiotherapy motions that are recorded by the exoskeleton of the healthy upper limb and based on these motions, the motors of exoskeleton that serves the affected arm will execute the same motions in mirror, thus providing therapy for the impaired arm.

During the therapy the motors of exoskeleton attached to the healthy arm will be in the passive mode allowing free motion of arm within the limits imposed by the mechanical configuration, namely flexion/extension of the elbow and flexion and extension of the wrist. Moreover, when the patient registers progress he can switch from the passive-mirrored therapy to active-assistive therapy, corrective therapy, active therapy, path-guidance therapy or resistive therapy by using only one side of the exoskeleton, the other part being detachable (Fig. 3).

The double arm exoskeleton is a portable rehabilitation system, it is fixed in a harness that give a better stiffness through the back anchor.

The elbow anchors of both arms are fixed in the back plate at one point, having a rotation degree in OXZ plane allowing a free adduction/abduction of the shoulder. The motor that serves the elbow joint motion is collinear with the elbow joint axis and it performs elbow flexion/extension. The second motor is placed under the forearm and it

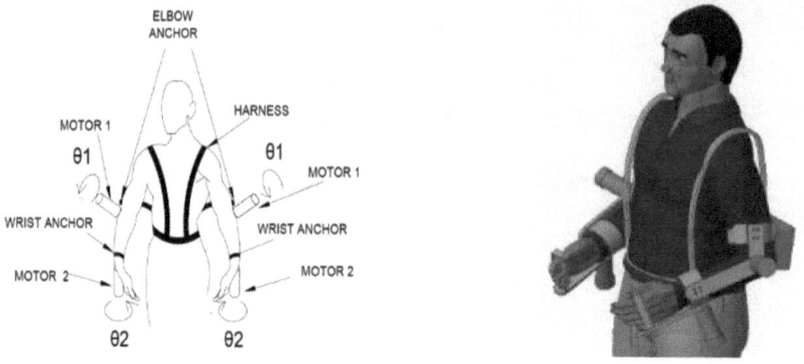

a) b)

Fig. 3. A scheme of dual exoskeleton robot (a) conceptual design; (b) a CAD solution

makes wrist flexion/extension motions moving the last link where the hand is placed. The healthy arm will grab the handle and will perform wrist flexion/extension motions in horizontal plan.

The DualEx (dual exoskeleton) concept is a simple device that is able to do two basic motions of upper limb; the advantage comes with the possibility of bilateral rehabilitation training which can be done by the patient in chronic phase, of course with the speed and torques adjusted by specialists.

This concept has come as a simple and efficient solution, the easy training of elbow and wrist motions using bilateral training can be performed at patient's home. The novelty brought by this device consists in its portability. The energy source is provided by Li-ion rechargeable battery that is the best choice of power source for portable devices. The exoskeleton links are adjustable; thus, it can be fitted on a different length of human upper limb segments. The contact between the robotic device and the human arm will be carried out at the anchor points of the arm and the method of clamping the arm in exoskeleton will be secure and ergonomic.

The Fig. 4(b) shows a section view through the motors so that it can be seen how the transmission is made from the motor to the rotational segment.

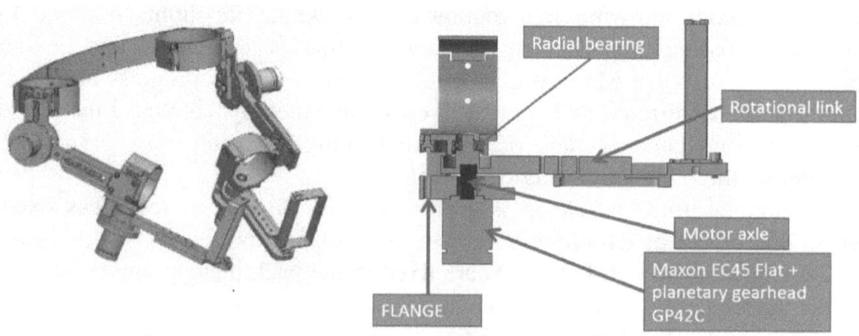

a) b)

Fig. 4. DualEx 3D CAD model (a) 3D CAD design; (b) section view through motor

The DualEx is highlighted also by the following advantages: the device is built using lightweight materials so it can be easily worn; the physiotherapy exercises can be done both in the sitting position and in standing position; beside the passive-mirrored therapy, the patients can perform active-assistive therapy, corrective therapy, active therapy, path-guidance therapy or resistive therapy by using only one side of the exoskeleton, the other part being detachable.

4 Performance Characteristics

The plots presented in Figs. 5, 6 and 7 were obtained in Solid Works Motion Analysis based on the 3D model presented in Fig. 4(a), using rotary motors for two rotational joins, taking also in consideration weight of exoskeleton arm plus upper limb and reaction force of muscles. The torque curve of elbow motor has started at non-zero value for the reason that the total weight of upper limb and exoskeleton are taken in consideration when the elbow motions has started, also the elbow motion started with the upper limb flexed about 90° where a torque is needed to keep the arm in this position. The resulted torque value of elbow joint is around 6Nm that reach the required elbow torque in rehabilitation after stroke based of Ramlee et al. [10], the wrist joint torque of exoskeleton is lower because the joint doesn't need to sustain any weight of arm or exoskeleton performing only wrist flexion/extension motion. In the Fig. 7(a) and (b) are presented the power consumption of elbow and wrist rotational joins, observing a noticeable increase in consumption in favor of the elbow joint. The plot results where power consumption is presented are given in watts in a certain time. In the Figs. 5(a) and 6(a) are given the plots of the angular velocities measured in degree/second in a certain time.

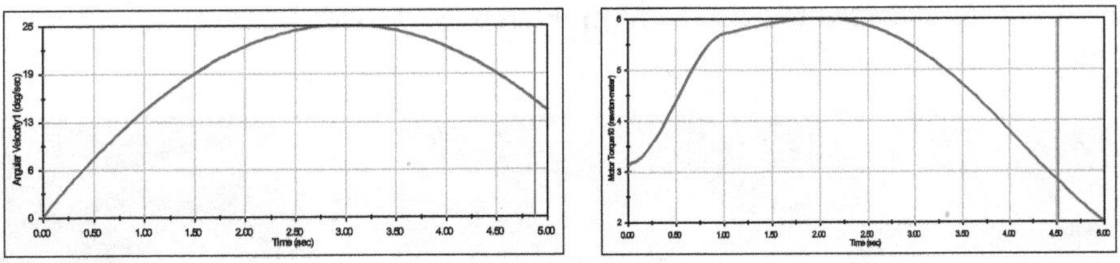

a) b)

Fig. 5. Components results for the elbow motor (a) angular velocity; (b) torque

As it can be seen the torque plot for elbow joint is around 6 Nm. To reach this values mentioned above, the suitable motor is Maxon EC45 Flat, but to reach the proper torque for joint elbow the motor will come with planetary gearhead GP 42 C that develop a moment of force between 3–15Nm. This type of motor is ideal from technical characteristics point of view but has a big cost. The design presented in Fig. 4(a) is made using

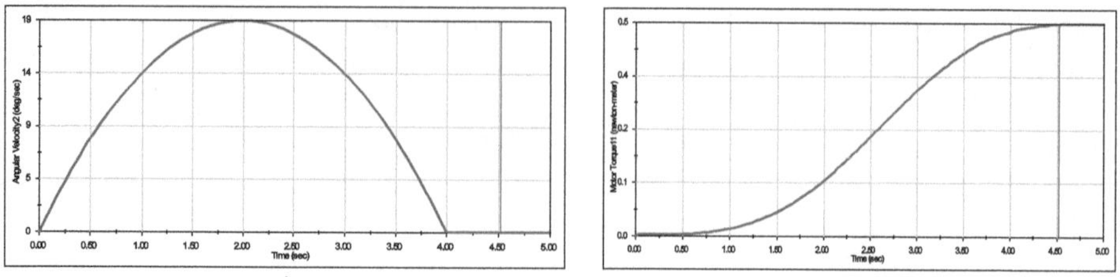

a) b)

Fig. 6. Components results for the wrist motor (a) angular velocity; (b) torque

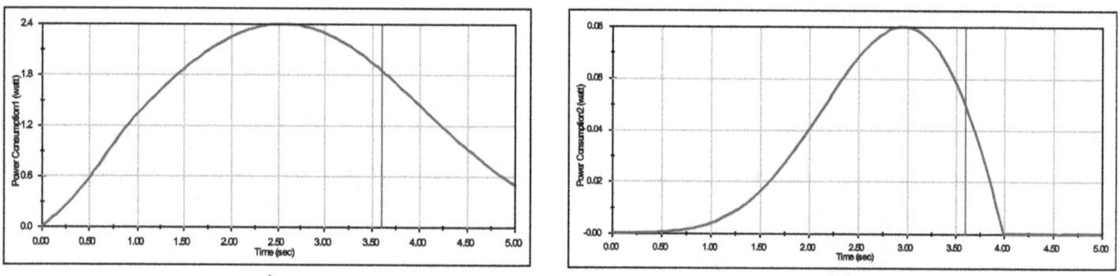

a) b)

Fig. 7. Power consumption (a) for elbow; (b) for wrist

a dummy Maxon EC45 Flat motors with GP 42 C gearhead 3D printed. Anyway, the motors will be chosen to have a good ratio between performances meaning small size and torque/force necessary for elbow and wrist joints and cost efficiency.

In Fig. 8 the prototype of DualEx robotic system is presented. This prototype was built using a 3D printing for each component of the assembly and afterwards all the components were joined together to form the robotic device.

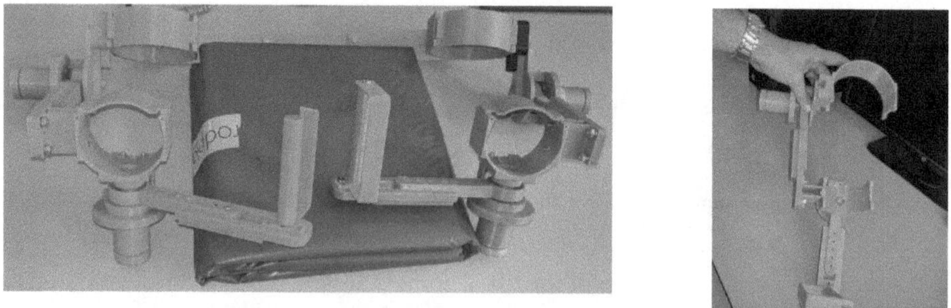

Fig. 8. The prototype of DualEx

5 Conclusions

This paper presents the actual stage of a dual exoskeleton for mirror therapy after stroke. The "Dual-Ex" robotic system is a portable device built predominately from ABS material; its parts are printed using 3D printing technology. The motions performed by this robotic system are not complex but very easy to execute due to their simplicity. Its advantage consists in the possibility of being worn due of its lightweight structure, also to actuate the elbow and wrist joints are used motors able to confer the proper torque and force with a reduced weight of motors. This project will continue with the developing of exoskeleton presented in this paper, the aim is to make this concept functional and ready to use the advantages of mirrored therapy.

Acknowledgements. The paper presents results from the research activities of the project ID 37_215, MySMIS code 103415 "Innovative approaches regarding the rehabilitation and assistive robotics for healthy ageing" co-financed by the European Regional Development Fund through the Competitiveness Operational Programme 2014-2020, Priority Axis 1, Action 1.1.4, through the financing contract 20/01.09.2016, between the Technical University of Cluj-Napoca and ANCSI as Intermediary Organism in the name and for the Ministry of European Funds

References

1. Donnan, G.A., Fisher, M., Macleod, M., Davis, S.M.: Stroke. Lancet **371**, 1612–1623 (2008)
2. Panich, S.: Kinematic analysis of exoskeleton suit for human arm. J. Comput. Sci. **6**, 1272–1275 (2010)
3. Collizi, L., Lidonnici, A., Pignolo, L.: The ARAMIS project: a concept robot and technical design. J. Rehabil. Med. **41**(12), 1011–1015 (2009)
4. Burgar, C.G., Lum, P.S., et al.: Development of robots for rehabilitation therapy: the Palo Alto VA/Stanford experience. J. Rehabil. Res. Dev. **37**(6), 663–673 (2000)
5. Plitea, N., Gherman, B., Carbone, G., Ceccarelli, M., Vaida, C., Banica, A., Pisla, D., Pisla, A.: Kinematic analysis of an exoskeleton based robot for elbow, and wrist rehabilitation. In: Carvalho, J.C.M., et al. (eds.) Multibody Mechatronic Systems, p. 424. Springer (2018). https://doi.org/10.1007/978-3-319-67567-1_40
6. Whitall, J., Waller, S.M., Silver, K.H.C., Macko, R.F.: Repetitive bilateral arm training with rhythmic auditory cueing improves motor function in chronic hemiparetic stroke. Stroke **31**(10), 2390–2395 (2000)
7. Buschfort, R., Brocke, J., Heß, A., Werner, C., Waldner, A., Hesse, S.: Arm studio to intensify upper limb rehabilitation after stroke: concept, acceptance, utilization and preliminary clinical results. J. Rehabil. Med. **42**(4), 310–314 (2010)
8. Stinear, J.W., Byblow, W.D.: Rhythmic bilateral movement training modulates corticomotor excitability and enhances upper limb motricity poststroke: a pilot study. J. Clin. Neurophysiol. **21**(2), 124–131 (2004)
9. Major, K., Major, Z., Carbone, G., Pisla, A., Vaida, C., Gherman, B., Pisla, D.: Ranges of motions for robot-assisted post-stroke rehabilitation. HVM Bioflux **8**, 192–196 (2016)
10. Ramlee, M.H., Gan, K.B.: Function and biomechanics of upper limb in post-stroke patients - a systematic review. J. Mech. Med. Biol. **17**, 1–19 (2017)

Author Index

G. Carbone et al. (Eds.): *New Trends in Medical and Service Robotics*, MMS 65, pp. 313–315, 2019.
https://doi.org/10.1007/978-3-030-00329-6

FSC
www.fsc.org
MIX
Papier | Fördert
gute Waldnutzung
FSC® C083411

Kolibri 360 GmbH
Ferdinand-Jühlke-Straße 7
99095 Erfurt, Deutschland
produktsicherheit@kolibri360.de